Recent Advances in Chemical Engineering

I. Regupathi · Vidya Shetty K.
Murugesan Thanabalan
Editors

Recent Advances in Chemical Engineering

Select Proceedings of ICACE 2015

Springer

Editors
I. Regupathi
Department of Chemical Engineering
National Institute of Technology Karnataka (NITK)
Surathkal, Mangalore, Karnataka
India

Vidya Shetty K.
Department of Chemical Engineering
National Institute of Technology Karnataka (NITK)
Surathkal, Mangalore, Karnataka
India

Murugesan Thanabalan
Department of Chemical Engineering
Universiti Teknologi PETRONAS
Bandar Seri Iskandar
Malaysia

ISBN 978-981-10-1632-5 ISBN 978-981-10-1633-2 (eBook)
DOI 10.1007/978-981-10-1633-2

Library of Congress Control Number: 2016947205

Printed on acid-free paper

This Springer imprint is published by Springer Nature
The registered company is Springer Science+Business Media Singapore Pte Ltd.

Preface

During the last few years, chemical engineering is changing its facet towards increasingly a multidisciplinary approach due to the demand for safer, cleaner technologies and novel materials. The industrial practice and academic research are oriented towards the development of better strategies on management of energy, raw materials and the environment. Research and practice in the field of chemical engineering applies the knowledge of mathematics, physics, chemistry, biology and engineering to solve any technical challenges safely and economically. The advances in chemical engineering are towards developing novel materials and processes. Realizing the need for spread of knowledge on advances in the area of chemical engineering, the **Department of Chemical Engineering at National Institute of Technology Karnataka (NITK), Surathkal** has taken initiatives in organizing a global event to commemorate its Golden Jubilee year in 2015. The **International Conference on Advances in Chemical Engineering (ICACE - 2015)** was held during 20–22 December 2015 at NITK and more than 250 delegates attended and presented their papers. The full-length papers were reviewed by experts from pioneering academic and research organizations from India and abroad and the selected papers under the broad theme of Recent Advances in Chemical Engineering are included in this book.

The book introduces the outcomes of latest research in the field of chemical engineering. Also it illustrates the application of chemical engineering principles to provide innovative and state-of-the-art solutions to problems associated with chemical industries. It covers a wide spectrum of topics in chemical engineering such as transfer operations, novel separation processes, adsorption, photooxidation, process control, modelling and simulation. This book provides timely contribution towards implementation of recent approaches and methods in chemical engineering research. It presents chapters focussed on several chemical engineering principles and methodologies of wide multidisciplinary applicability.

The editors, on behalf of faculty, staff, students and alumni of the Department of Chemical Engineering, NITK Surathkal, and the Organizing Committee members of ICACE-2015, wish to thank the contributors, speakers, researchers, academicians for their valuable technical contributions and co-ordination towards publishing this book. The editors would also like to thank NITK administration and Technical Education Quality Improvement Programme –II (TEQIP-II) for the support.

Surathkal, India K. Vidya Shetty
Surathkal, India I. Regupathi
Bandar Seri Iskandar, Malaysia Thanabalan Murugesan

Contents

About the Editors

Dr. I. Regupathi is Assistant Professor at the Department of Chemical Engineering, National Institute of Technology Karnataka. He has produced more than 60 research publications and is actively engaged in the research work in the areas of transfer operations, extraction engineering including aqueous two-phase extraction, reverse micellar extraction and downstream processing including separation and purification of proteins and bioactive compounds. He also executed several research projects funded by the ministries of Government of India.

Dr. Vidya Shetty K. is Associate Professor at the Department of Chemical Engineering, National Institute of Technology Karnataka. She has authored more than 40 research publications and her area of research includes application of chemical engineering in the field of wastewater treatment, bioremediation, bioleaching, environmental biotechnology, bioreactors, chemical and biological synthesis of nanostructures and nano photocatalysis. Her research interests also include theoretical and artificial neural network based modelling and simulation of chemical engineering systems.

Dr. Murugesan Thanabalan is Professor at the Department of Chemical Engineering, Universiti Teknologi PETRONAS, Malaysia. He has published more than 175 articles in international journals apart from many conference proceedings. His research expertise includes ionic liquids/deep eutectic solvents for chemical engineering applications; new solvents/functionalized clay-based adsorbents/mixed matrix membranes for CO_2 capture; green approaches to the utilization of biomass; biodegradation/application of CNTs/advanced oxidation processes for water treatment; aqueous two-phase extraction of biomolecules, microwave drying of non-conductive materials; hydrodynamics and mass transfer operations involving two-phase (Gas–Liquid, Liquid–Solid) and three-phase (Gas–Solid–Liquid) systems in mechanically agitated contactors and design and development of equipment for the liquid–liquid extraction of value-added products.

Part I
Transfer Operations

Numerical Study of Heat Transfer Characteristics of Nano-fluids in Channel Containing Different Shapes of Submerged Tube

I. Mohammad, A. Khapre and A. Keshav

List of symbols

C_p	Specific heat (J/kg K)
D	Diameter of in-line tube (cm)
H	Diameter of channel (cm)
L	Length of single module (cm)
Nu	Nusselt number
Re	Reynolds number
$\dot{S}_{gen}$	Total entropy generation per unit volume $\left(\mathrm{J}/(\mathrm{s\,K})\ \mathrm{m}^3\right)$
$\dot{S}_{HT}$	Entropy generation due to heat transfer per unit volume $\left(\mathrm{J}/(\mathrm{s\,K})\ \mathrm{m}^3\right)$
$\dot{S}_{VD}$	Entropy generation due to viscous dissipation per unit volume $\left(\mathrm{J}/(\mathrm{s\,K})\ \mathrm{m}^3\right)$
V	Velocity (m/s)
ρ	Density $\left(\mathrm{kg/m^3}\right)$
μ	Viscosity (kg/m s)
Φ	Concentration
λ	Thermal conductivity (W/m K)
ΔP	Pressure Drop $\left(\mathrm{N/m^2}\right)$

I. Mohammad · A. Khapre (✉) · A. Keshav
Department of Chemical Engineering, National Institute of Technology, Raipur, India
e-mail: akhileshkhapre@gmail.com

I. Mohammad
e-mail: Mohammad.isti@gmail.com

A. Keshav
e-mail: akeshav.che@nitrr.ac.in

I. Regupathi et al. (eds.), *Recent Advances in Chemical Engineering*,
DOI 10.1007/978-981-10-1633-2_1

1 Introduction

The nanofluid is a liquid containing a suspension of nano size solid particles (nanoparticles) (Choi 1995). The nanofluids have characteristic properties which are used for the enhancement of thermal conductivity of liquid (Masuda et al. 1993). Due to this superior property, the nanofluids are generally used in comparison to conventional heat transfer liquids, and liquids containing micro-sized metallic particles (Manca et al. 2011). Manufacturing of nanofluids is affected by a number of factors like the disagreement between results, poor categorization of suspensions, and the lack of theoretical understanding of the mechanism (Wang and Majumdar 2007). The most important parameter accountable to the enhanced heat transfer is thermal conductivity. Alumina (Al_2O_3) and copper oxide are the most common and inexpensive nanoparticles and have been used in many experimental works (Chun et al. 2008). In convective flows, Choi (1995) has found that heat transfer coefficients increased with fluid flow rates or with increasing thermal conductivities of the fluid. The results in the study was observed by keeping the other properties of the nanofluid, such as heat capacity, density, and viscosity, same as the base fluid. Li and Xuan (2002) experimentally studied the flow and heat transfer properties of Cu-water nanofluids and reported that if volume fraction of nanoparticles is low, the friction factor of the nanofluids is as same as the base fluid.

In this work, the numerical simulation of flow of nano fluid in channel containing submerged bodies was carried out at different Reynolds number in laminar region. The heat transfer performance of channel in terms of Nusselt number and entropy generation is studied for different concentration of the nanofluids.

2 Computational Geometry

The computational geometry for simulation is shown in Fig. 1.

Five different geometry shapes of submerged tubes were considered: circular, flat, oval, and diamond, and rhombus. The space between two successive tubes was kept constant. The length (L) of the module is taken as 1 cm. The ratio $L/D = 3$, $H/D = 2$, and $l/D = 0.1$ was kept constant for all the tubes. The entry length of channel is considered equal to the length of one module length and exit length of channel is considered equal to three time that of a module length.

3 Governing Equations and Boundary Conditions

The general conservation equations of continuity, momentum, and energy for two dimensional, steady and incompressible fluid flowing in a channels are (Bird et al. 2002)

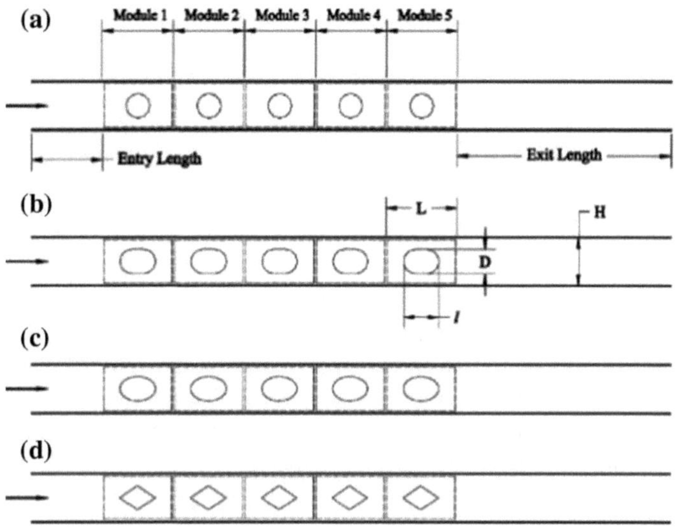

Fig. 1 Computational geometry (Bahaidarah et al. 2006)

$$\nabla \cdot \bar{V} = 0 \tag{1}$$

$$\bar{V} \cdot \nabla \bar{V} = -\frac{1}{\rho}\nabla P + \frac{1}{\rho}\nabla \cdot (\mu \nabla \bar{V}) \tag{2}$$

$$\rho C_P \bar{V} \cdot \nabla \bar{V} = \lambda \Delta^2 T - \tau : \nabla \bar{V} \tag{3}$$

where, V is the velocity vector. The set of equations were simulated at different values of Reynolds number.

The entropy generation in channel is given as (Khapre and Munshi 2015)

$$\dot{S}_{gen} = \dot{S}_{HT} + \dot{S}_{VD} \tag{4}$$

where

$$\dot{S}_{HT} = \frac{\lambda}{\bar{T}^2}\left[\left(\frac{\partial \bar{T}}{\partial x}\right)^2 + \left(\frac{\partial \bar{T}}{\partial y}\right)^2 + \left(\frac{\partial \bar{T}}{\partial z}\right)^2\right] \tag{5}$$

$$\dot{S}_{VD} = \frac{\mu}{\bar{T}}\left[2\left\{\left(\frac{\partial \bar{u}}{\partial x}\right)^2 + \left(\frac{\partial \bar{v}}{\partial y}\right)^2 + \left(\frac{\partial \bar{w}}{\partial z}\right)^2\right\}\right] + \left(\frac{\partial \bar{u}}{\partial y} + \frac{\partial \bar{v}}{\partial x}\right)^2 + \left(\frac{\partial \bar{u}}{\partial z} + \frac{\partial \bar{w}}{\partial x}\right) + \left(\frac{\partial \bar{v}}{\partial z} + \frac{\partial \bar{w}}{\partial y}\right) \tag{6}$$

For simulation Al_2O_3–water nanofluids are used which is the mixture of base fluid, water, and Al_2O_3 nanoparticle with 30 nm diameter. The physical properties

Table 1 Physical properties of Al_2O_3–water nanofluids

Concentration (Φ)	Thermal conductivity (K)	Specific heat (C_p)	Density (ρ)	Viscosity (μ)
0.0	0.6	4182.00	998.2	9.93e-4
1.0	0.62	4053.21	1027.02	1.08e-3
2.0	0.63	3931.5	1055.84	1.19e-3
3.0	0.64	3816.5	1084.65	1.32e-3

of base fluid and a nanoparticle are taken from literature (Chun et al. 2008; Manca et al. 2012), and shown in Table 1.

The velocity inlet boundary condition is applied at the entry of channel and at the outlet pressure boundary condition is applied. The inlet temperature of nanofluid is kept constant at 293 K for all simulation cases. While the temperature of walls of channel and the tubes is kept constant at 350 K.

4 Results and Discussions

The numerical simulations of different concentration of nanoparticles were carried out at laminar region (*Re* range: 10–2500). Figure 2a–d shows the effect of Reynolds Number on Nusselt number for different concentrations of nano fluids and for various geometries of submerged bodies. It can be seen that, as the particle concentration increases, the Nusselt number increases and hence the heat transfer coefficient increase. The increase in heat transport in the presence of nano particles is may be due to the increase in solid-solid interactions. The detail mechanism is explored and given by Keblinski et al. (2002). From Fig. 2, it can be observed that at high *Re* (1000–2500), the Nusselt number for oval and obround tubes are relatively high as compared to circular, diamond and rhombus tubes, respectively. However, at low *Re* (10–1000), the Nusselt number is almost equal for all shape of tubes.

The pressure drop (ΔP) across the channel, at $Re = 2500$, with different shape of tubes at different nanofluid concentration (Φ) is tabulated in Table 2. From Table 2, it is inferred that the rhombus tube has higher pressure drop than the other shape of tubes. The order of pressure drop for all concentrations ($\Phi = 0$–2) is: $\Delta P_{oval} < \Delta P_{obround} < \Delta P_{\mathrm{circle}} < \Delta P_{diamond} < \Delta P_{rhombus}$. From this order, it can be conclude that the pressure drop across the channel is inversely proportional to the Nusselt number that is heat transfer characteristics. Thus the rhombus tube has least heat transfer characteristic than other shapes of tubes and oval, and obround tube has high heat transfer characteristics. The results are in agreement with the result of the effect of *Re* on Nusselt numbers (Fig. 2).

Figure 3 shows the volumetric entropy generation in the channel. The entropy of system increases with the increase in Re and volume concentration of nanoparticles.

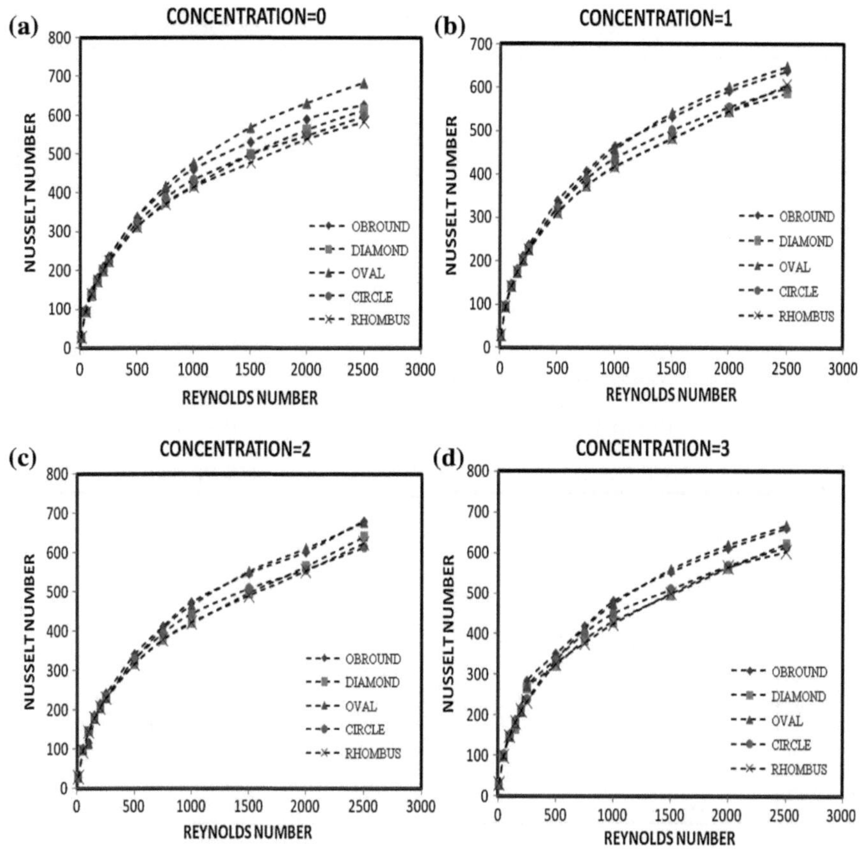

Fig. 2 *Re* versus *Nu*, at different nanofluid concentrations for different shapes of sub-merged tubes

Table 2 Pressure drop across the channel with different shapes of tubes at *Re* = 2500 and different nanofluid concentration

Shape of tubes	$\Phi = 0$	$\Phi = 1$	$\Phi = 2$
Oval	625.4	722	996
Obround	649.3	744	1020.1
Circle	711.1	835.3	1135.1
Diamond	770.4	1066.7	1244.7
Rhombus	952.3	1083.5	1402.9

From Fig. 3, the order of entropy generation for different concentration is: $S_{gen,oval} > S_{gen,obround} > S_{gen,circle} > S_{gen,dimond} > S_{gen,rhombus}$. Thus, the entropy generation in the channel is relatively proportional to Nusselt number that is heat transfer characteristics.

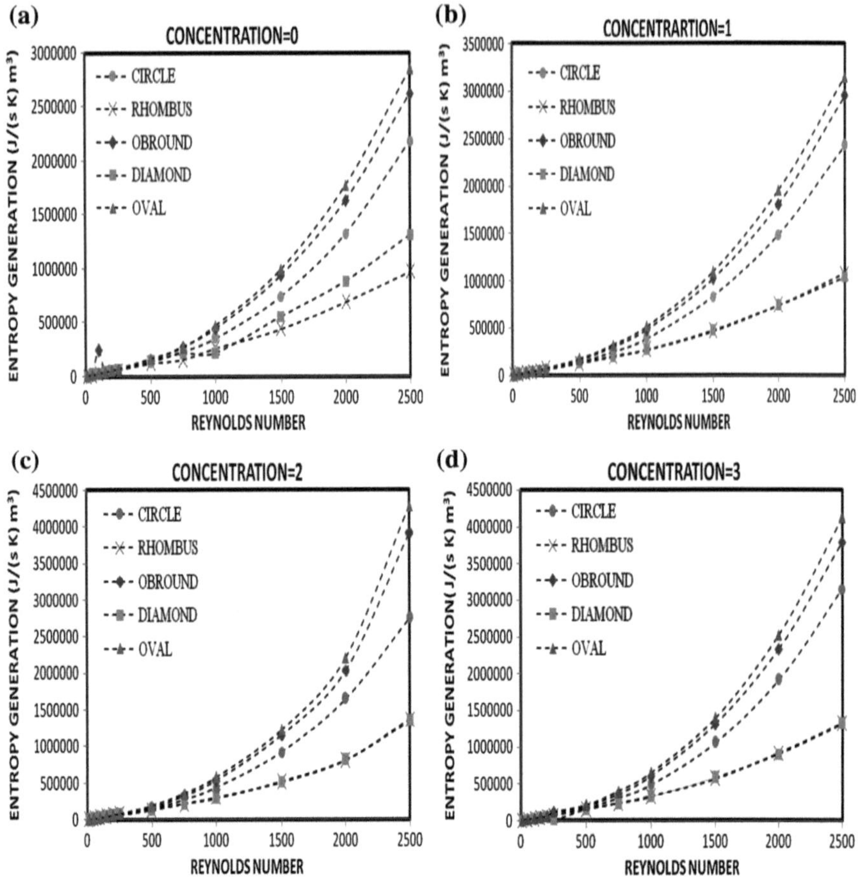

Fig. 3 Entropy generation at different Re and concentrations of nanofluids for different shapes of tube

5 Conclusions

The heat transfer characteristics in terms of Nusselt number and entropy generation is studied numerically with Ansys Fluent for the channel confined with different shapes of tubes. The simulations are carried out for different concentration of nanofluids at different *Re*. The Nusselt number, and entropy generation is predicted and found that they are increased with *Re* and concentration of nanofluids. Out of five different shapes tubes, the oval shape tube has shown the higher heat transfer characteristics that is Nusselt number and entropy generation than other shapes of tube.

References

Bahaidarah, H.M., Anand, N.K., Chen, H.C.: Numerical study of fluid flow and heat transfer over a series of in-line noncircular tubes confined in a parallel-plate channel. Numer. Heat Transf. Part B **50**, 97–119 (2006)

Bird, R.B., Stewart, W.E., Lightfoot, E.N.: Transport Phenomena, 2nd edn. John Willy & Sons Inc, New York (2002)

Choi, S.U.S.: Enhancing thermal conductivity of fluids with nanoparticle. In: Siginer, D.A., Wang, H.P. (eds.) Developments and Applications of Non-Newtonian Flows, ASME MD 231 and FED 66, pp. 99–105 (1995)

Chun, B.H., Kang, H.U., Kim, S.H.: Effect of alumina nanoparticles in the fluid on heat transfer in double-pipe heat exchanger system. Korean J. Chem. Eng. **25**, 966–971 (2008)

Keblinski, P., Philpot, S.R., Choi, S.U.S., Eastman, J.A.: Mechanisms of heat flow in suspensions of nano-sized particles (nanofluids). Int. J. Heat Mass Transf. **45**, 855–863 (2002)

Khapre, A., Munshi, B.: Numerical investigation of hydrodynamic behavior of shear thinning fluids in stirred tank. J. Taiwan Inst. Chem. Eng. **56**, 16–27 (2015)

Li, Q., Xuan, Y.: Experimental investigation on convective heat transfer of nanofluids. J. Eng. Thermophys. **23**, 721–723 (2002)

Manca, O., Mesolella, P., Nardini, S., Ricci, D.: Numerical study of a confined slot impinging jet with nanofluids. Nanoscale Res. Lett. **6**, 1–16 (2011)

Manca, O., Nardini, S., Ricci, D., Tamburrino, S.: Numerical investigation on mixed convection in triangular cross-section ducts with nanofluids. Adv. Mech. Eng. Article ID 139370 (2012)

Masuda, H., Ebata, A., Teramae, K., Hishinuma, N.: Alteration of thermal conductivity and viscosity of liquid by dispersing ultra-fine particles. Netsu Bussei **7**, 227–233 (1993)

Wang, X.Q., Mujumdar, A.S.: Heat transfer characteristics of nanofluids: a review. Int. J. Therm. Sci. **46**, 1–19 (2007)

A Numerical Investigation Based on Heat Transfer and Fluid Flow Characteristics of Air in a Circular Tube Heat Exchanger with Inclined Ribs

Suvanjan Bhattacharyya, Anirban Roy, Himadri Chattopadhyay and Aritra Rakshit

1 Introduction

Saving energy has been the most important engineering concern in the last decade. This can be attributed to the exponential increase in population all over the world and the ever-increasing demand of energy. Innovative design solutions and applications to effectively utilize energy have been used in the industry, with the last few decades having seen the wide use of ribbed wall surfaces. Several industrial processes and applications face thermal energy exchange in fluids with laminar flow. In the case of a laminar flow, there is major thermal resistance in the bulk flow. There also lies a dominant thermal resistance within the thin boundary layer adjacent to the flow. The roughness in the form of continuous or gapped ribs could generate turbulence and disintegrate the viscous sub layer near the heat surface, thereby reducing the thermal resistance. The added advantage of increased heat transfer area is achieved by the number of ribs mounted on the wall (Peng et al. 2011).

In an effort to find viable alternative energy sources people have turned towards the Sun. However a solar air heater's thermal performance is not good because of the poor heat transfer rate observed between the working fluid and absorber plate (Duffie and Beckman 1980). Bhattacharyya et al. (2014) studied low thermo-fluids properties of the fluid in part, and to the viscous sub-layer that appears in the vicinity of the absorber and which is resistant to the heat transfer in the other part. Bhattacharyya et al. (2015) presented the idea that in order to make solar air heaters efficient their thermal performance needs to be improved.

S. Bhattacharyya (✉) · H. Chattopadhyay
Mechanical Engineering Department, Jadavpur University, Kolkata 700032, India
e-mail: suva.raindrops@gmail.com; suvanjanr@gmail.com

A. Roy · A. Rakshit
Mechanical Engineering Department, MCKV Institute of Engineering, Liluah, Howrah 71120, India

I. Regupathi et al. (eds.), *Recent Advances in Chemical Engineering*,
DOI 10.1007/978-981-10-1633-2_2

Many techniques based on both active and passive methods have been proposed to enhance heat transfer in these applications (Boulemtafes 2010). One can find numerous studies on solar air heat enhancement techniques. For example, Varun et al. 2007) reviewed the geometry of roughness used in heat exchangers and reported the optimum geometry of roughness that is adapted in case of solar air collectors. (Hans et al. 2009) studied the basic geometry of roughness element employed by various researchers to enhance the thermal efficiency of solar air heaters. In view of the search for optimal roughness pattern, few decent roughness geometries have been compared on the basis of thermo-hydraulic performance. (Bhushan et al. 2010) presented their attempt to classify and examine the geometry of artificial roughness used in the ducts of solar air collectors. (Chandra et al. 2003) explained the thermal characteristics in a square channel with continuous ribs on one, two, three, and four walls and found that the heat transfer and pressure drop increase with the rise in the number of ribbed walls. Promvonge et al. (2008) represented an approximate idea of the effect of various rib shapes and their thermal behaviours in a tube and showed that the staggered triangular rib performs the best. Currently work on nano-fluids is also becoming popular for enhancing heat transfer. Smithberg and Landis (1964), (Lopina and Bergles 1969), (Manglik and Bergles 1993), (Sarma et al. 2002, 2003), and (Bhattacharyya and Chattopadhyay 2015) have reported heat transfer enhancements with swirl generator and twisted tape inserts of single-phase fluids in a tube.

However, all the studies mentioned above, restricted themselves to just one flow regime. This study has tried to bridge the gap and represent the findings of heat transfer enhancement in all flow regimes—laminar, transition, and turbulent. Furthermore, the arrangement of inclined ribs inside a circular duct could provide better mixing and enhanced heat transfer from the wall that can be easily incorporated within a solar heater. The aim of this work is to put forth the findings of heat transfer and thermal performance of the inclined ribs attachment as shown below, at various attack angles (θ) of 15°, 30°, 45° and 60° studied under a very wide Reynolds number regime (200–20,000).

2 Mathematical Model

In the present investigation, circular tubes with four different attack angles were considered, i.e. 15°, 30°, 45°, and 60°. The geometrical configuration of the circular tube with attack angle discussed in the present work is shown in Fig. 1.

The length of the tube is considered as 20 times the diameter of the tube. It is assumed to be a part of a longer pipe section—one that is typically used in industrial applications. Therefore, a fluid with a fully developed velocity profile is assumed to enter through the inlet at a fixed temperature, and flows through the circular tube fitted with ribs with different attack angles (θ), while a pressure outlet condition was applied at the exit. Air inlet temperature with 300 K was assumed in the flow direction. The wall temperature is 500 K and is constant throughout the experiment. Initially the physical properties of the air at the inlet temperature were

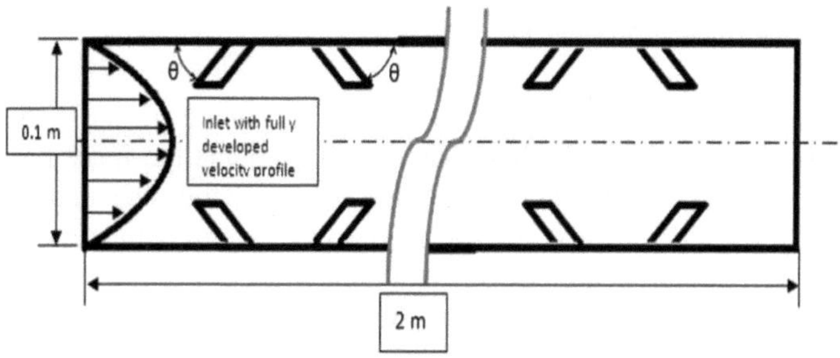

Fig. 1 Circular tube geometry with artificial ribs

used to calculate the mean bulk temperature for each case; after which the correct physical properties at the mean bulk temperature for each individual case was incorporated and all cases were reiterated. This was done until very little variation (less than 1.0 %), i.e. convergence, in the result values was achieved.

The mathematical modelling for this simulation was based on the following assumptions:

- Steady axis-symmetric fluid flow and heat transfer.
- The flow is incompressible.
- Heat transfer by radiation, body forces, and viscous dissipation has been neglected.

The three dimensional governing equations of continuity, momentum, and energy equations are solved using Transient SST model modified and proposed by (Abraham et al. 2009). Ansys Fluent 15 is used to solve the following governing equations.

$$\frac{\partial u_i}{\partial x_i} = 0 \tag{1}$$

$$\rho\left(u_i\frac{\partial u_j}{\partial x_i}\right) = -\frac{\partial p}{\partial x_i} + \frac{\partial}{\partial x_i}\left((\mu+\mu_{turb})\frac{\partial u_i}{\partial x_i}\right), \quad j = 1, 2, 3 \tag{2}$$

$$\left(u_i\frac{\partial \Theta}{\partial x_i}\right) = \frac{\partial}{\partial x_i}\left(\left(\alpha + \frac{\nu_{turb}}{\mathrm{Pr}_{turb}}\right)\frac{\partial \Theta}{\partial x_i}\right) \tag{3}$$

The values of turbulent Prandtl number ($\mathrm{Pr}_{\mathrm{turb}}$) are taken from Abraham et al. (2009).

$$\frac{\partial(\rho u_i \kappa)}{\partial x_i} = \gamma \cdot P_\kappa - \beta_1 \rho \kappa \omega + \frac{\partial}{\partial x_i}\left(\left(\mu + \frac{\mu_{turb}}{\sigma_\kappa}\right)\frac{\partial \kappa}{\partial x_i}\right) \tag{4}$$

$$\frac{\partial(\rho u_i \omega)}{\partial x_i} = A\rho S^2 - \beta_2 \rho \omega^2 + \frac{\partial}{\partial x_i}\left(\left(\mu + \frac{\mu_{turb}}{\sigma_\omega}\right)\frac{\partial \omega}{\partial x_i}\right) + 2(1-F_1)\rho\frac{1}{\sigma_{\omega 2}\omega}\frac{\partial \kappa}{\partial x_i}\frac{\partial \omega}{\partial x_i} \tag{5}$$

3 Meshing Arrangement

Computations were carried out for θ = 15°, 30°, 45°, and 60°. The diameter of the tube was 100 mm. The height of each rib was 30 mm and the space between each rib was 100 mm. Each rib was 5 mm thick. For reducing calculations we have assumed only the top half and run the model in the axis-symmetric setting. The work has been validated with the computational work of (Tanda 2011) on angular ribs. A preliminary mesh independence study is done. Grid containing 125,807 numbers of elements was sufficient to arrive at results of good resolution with only 1 % change in result parameters if the grid as further increased.

4 Results and Discussion

All the results obtained from computation for plain tube, heat transfer and pressure penalty characteristics are verified in terms of Nusselt number and friction factor. The Nusselt number and friction factor obtained from the present numerical computation compared with those from the proposed correlations by (Ozisik 1985) and proposed correlation by (Ozisik 1985) respectively.

The data obtained from the computational analysis for the plain duct are in good agreement with the predicted results from the proposed correlations with data range of +3.02 % to +4.12 % and +0.8 % to +1.6 % for the Nusselt number and friction factor, respectively as shown in Figs. 2 and 3.

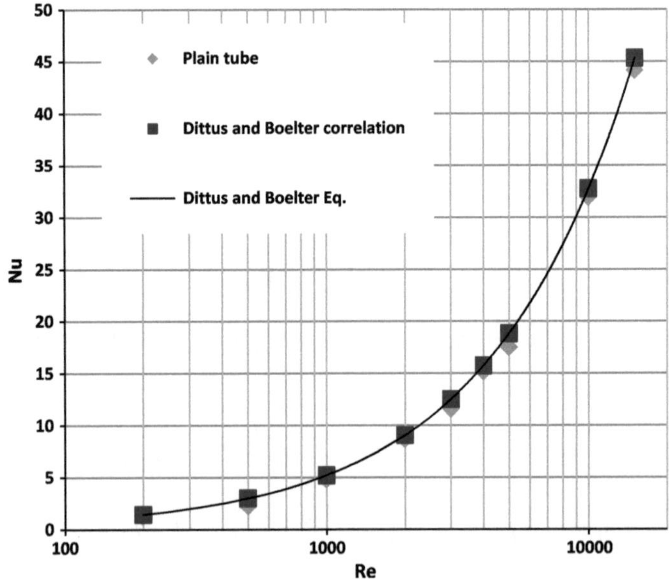

Fig. 2 Correlation of Nu_0 for plain tube with Dittus-Boelter equation

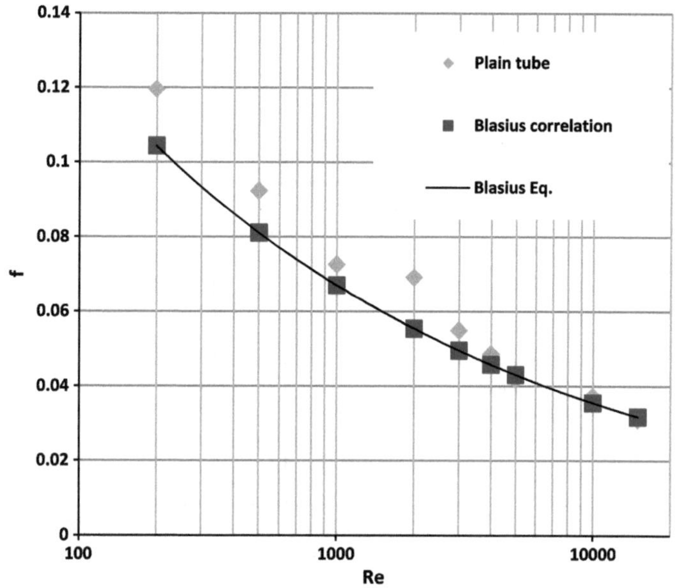

Fig. 3 Correlation of f_0 for plain tube with Blasius equation

It can be observed in Fig. 4 that the Nusselt number increases with the increase of Reynolds number. From Fig. 4, one can notice that the attack angle $\theta = 60°$ yields higher heat transfer than the other attack angles. The rate of heat transfer increases with the increase of attack angles. The better performance of the inserted artificial inclined ribbed tube can be attributed to the swirl flow created by the artificial ribs, leading to more effectiveness in disruption of the boundary layer compared to those caused by the individual device. Friction factor results are presented in Fig. 5. It is seen from Fig. 5 that as the Reynolds number is increased, with the decrease of friction factor. The friction factor values for all the artificial inclined ribs have higher value than the base case. Friction factors are higher at lower Reynolds number and then decreases slowly at higher Reynolds numbers. This is because the attributed to the use of inclined ribs with a higher attack angle which led to a higher viscous loss near the duct wall regions caused by a stronger swirl flow or turbulence flow and long residence time in the tube.

From Fig. 6 it can be observed that Nu/Nu_0 shows a slightly opposite trend for all cases as Nusselt number shows—it shows an increase in the transition region with Reynolds number ranging between 2000 and 4000 and then decreases at the downstream end of the graph. This enhancement is due to a secondary flow induced by the artificial inclined ribs, which promotes radial mixing of bulk flow with

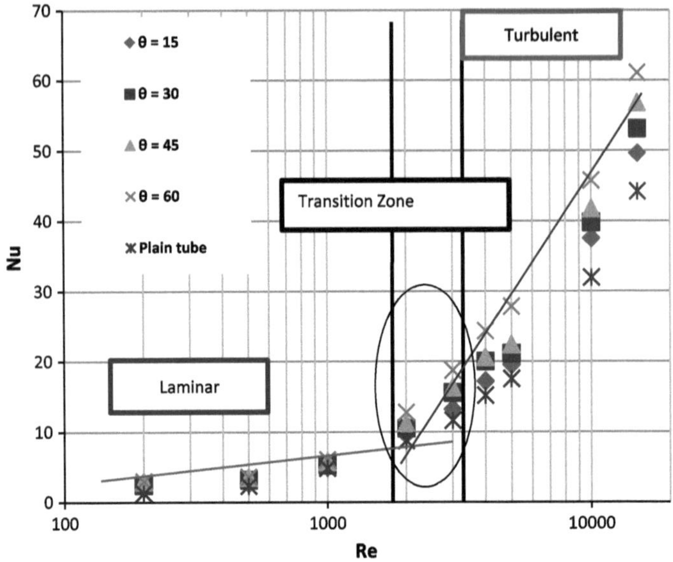

Fig. 4 Variation of Nusselt number with Reynolds number for different angle of attacks

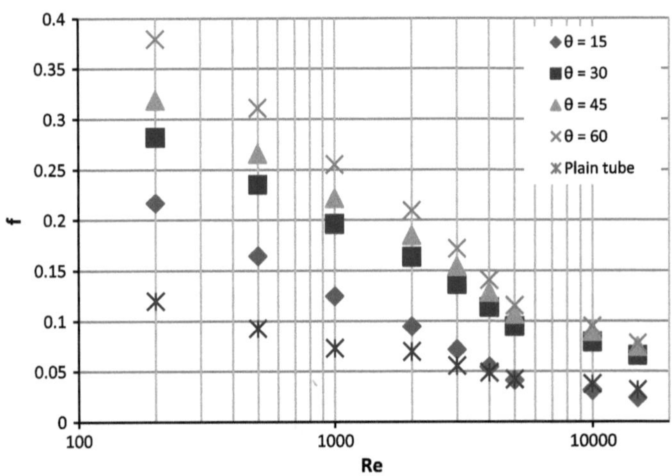

Fig. 5 Variation of friction factor with Reynolds number for different angle of attacks

near-wall flow. At similar operating condition, the circular tube with artificial ribs of different attack angle yields higher heat transfer rate than the plain tube acting alone.

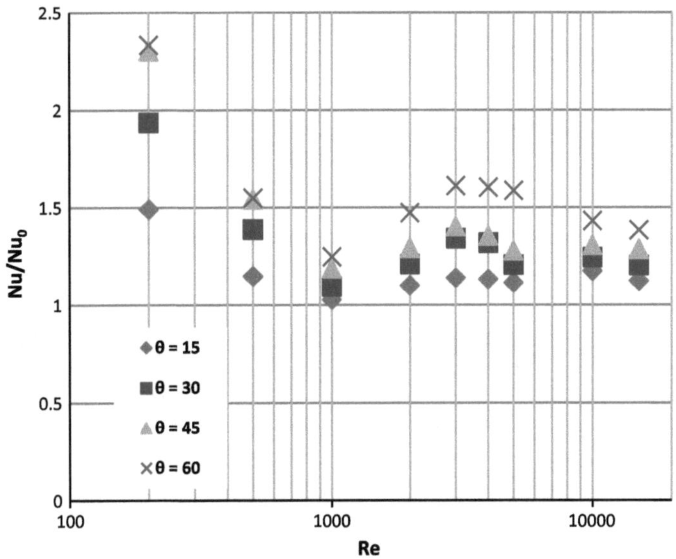

Fig. 6 Variation of Nu/Nu$_0$ with Reynolds number for different angle of attacks

Figure 7 shows the turbulent intensity contour for different attack angles of the ribbed tube for (Re = 3000). The vortices are formed in the section due to the flow separation, recirculation and rib-induced flow. From Fig. 4 one can see that a number of vortices are formed when the attack angle is highest (θ = 60°). The vector distributions around the artificial roughness are for observing more closely the main recirculation zones that appear when the flow is separated. When the fluid turbulent intensity is about 9–10 % more than the normal inlet fluid velocity, the second recirculation zone generates. Near the wall, there is a recirculation zone which is a reattachment area.

Overall performance evaluation is important for comparing the performances of different rib configurations. Figure 8 shows the thermal enhancement factor (PEC) with variations of rib attack angles.

This parameter is called the thermal performance factor, which means that the comparison is made based on constant pumping power. The computational analysis reveals that the PEC decreases at laminar region and increase with an increase in Reynolds number at turbulent region. This shows the significant role of the ribs at different attack angles in increasing the turbulence intensity at higher velocities. Also, the values of PEC vary between 0.8 and 1.75 for all cases. From Fig. 8, it is evident that the attack angle θ = 15° and θ = 60° shows the most-promising results in the transition zone and turbulent region, while θ = 30° has the lowest performance.

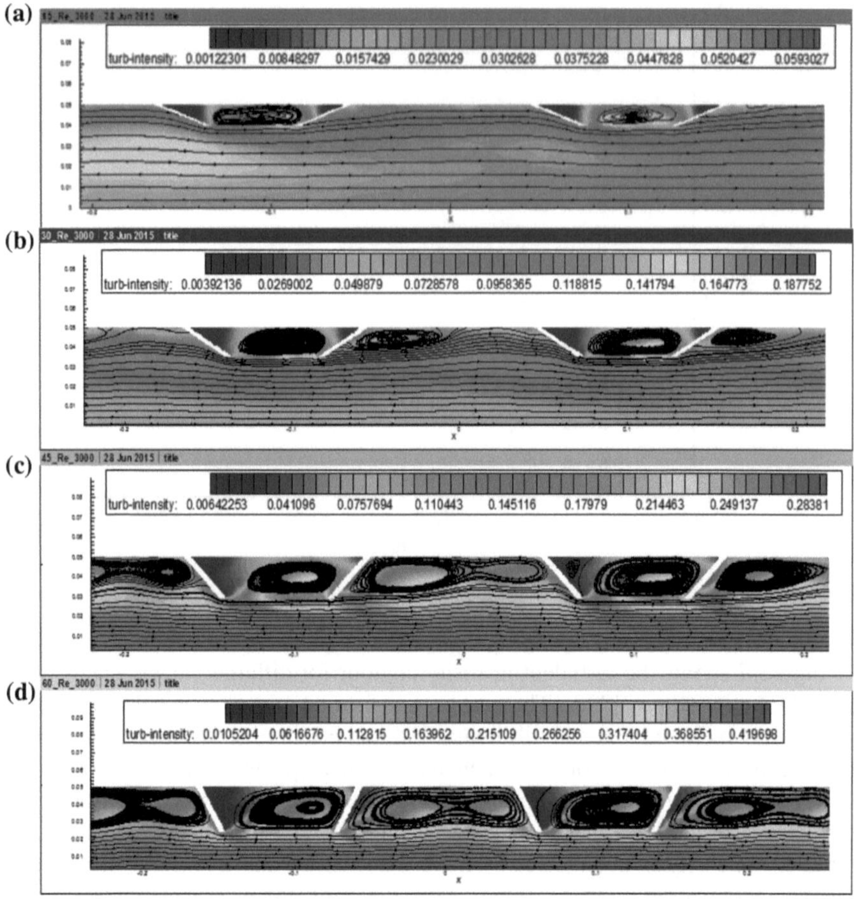

Fig. 7 Turbulent Intensity contours of u-components for different attack angles

5 Conclusion

In the present study, the thermal performance factor of different artificial attack angles has been compared with plain smooth tube. The studies of artificial inclined angles of 15°, 30°, 45°, and 60°, were placed in a circular tube of the solar air heater/heat exchanger. Effects of the artificial attack angles (θ) on local convective heat transfer, friction factor, and thermal performance characteristics are also reported. Smooth, plain tube consistently provides poorer heat transfer than tube with ribs inserts over the range studied. The attack angle of θ = 15° and θ = 60° yields better thermal performance than the other two attack angles (θ = 30° and 45°)

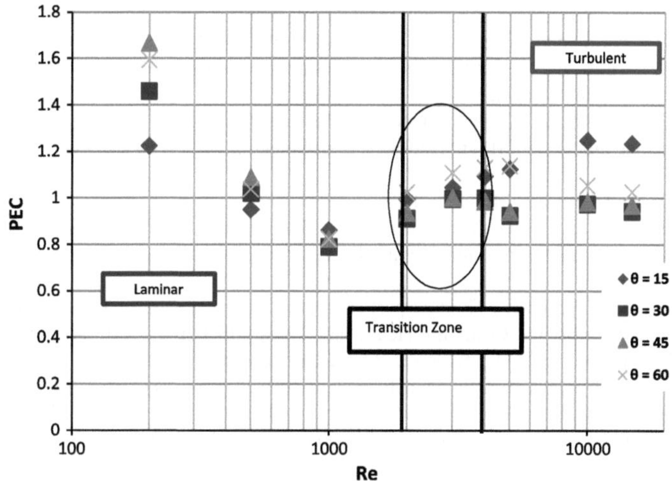

Fig. 8 Comparison of thermal enhancement factor for different rib attack angles

in varied Reynolds number range. In addition to those good performances, the artificial inclined ribs are also easy to fabricate. Thus, it is a promising work which can be widely used in heat transfer enhancement of turbulent flow.

References

Abraham, J.P., Sparrow, E.M., Tong, J.C.K.: Heat transfer in all pipe flow regimes: laminar, transitional/intermittent, and turbulent. Int. J. Heat Mass Transf. **52**, 557–563 (2009)

Bhattacharyya, S., Chattopadhyay, H., Saha, S.K.: Numerical Study on heat transfer enhancement of laminar flow through a circular tube with artificial rib roughness. J. Refrig, Air Conditioning, Heat. Ventilation **1**(3), 14–19 (2014)

Bhattacharyya, S., Chattopadhyay, H.: Computational of studies of heat transfer enhancement in turbulent channel flow with twisted strip inserts. In: Proceedings of CHT-15, ICHMT International Symposium on Advances in Computational Heat Transfer, Rutgers University, Piscataway, USA (2015)

Bhattacharyya, S., Roy, A., Bhattacharyya, A.: Computational heat transfer analysis of a counter-flow heat exchanger with fins. In: Proceedings of Recent Developments in Mechanical Engineering, Pune (2015)

Bhushan., B., Singh, R.: A review on methodology of artificial roughness used in duct of solar air heaters. Energy **35**, 202–212 (2010)

Boulemtafes, A.: Numerical simulation of heat transfer enhancement in solar air heaters. Magister Thesis, USTHB (2010)

Chandra, P.R., Alexander, V.R., Han, J.C.: Heat transfer and friction behavior in rectangular channels with varying number of ribbed walls. Int. J. Heat and Mass Transf. **46**, 481–495 (2003)

Duffie, J.A., Beckman, W.A.: Solar engineering of thermal processes. Wiley Inter-science publications, NY, New York (1980)

Hans, V.S., Saini, R.P., Saini, J.S.: Performance of artificially roughened solar air heaters—a review. Renew. Sustain. Energy Rev. **13**, 1854–1869 (2009)

Lopina, R.F., Bergles, A.E.: Heat transfer and pressure drop in tape-generated swirl flow of single phase water. J. Heat Transf. **91**, 434–442 (1969)

Manglik, R.M., Bergles, A.E.: Heat transfer and pressure drop correlations for twisted-tape inserts in isothermal tubes: part II—transition and turbulent flows. J. Heat Transf. **115**, 890–896 (1993)

Ozisik, M.N: Heat Transfer. Mc-Graw-Hill (1985)

Peng, W., Jiang, P.X., Wang, Y.P., Wei, B.Y.: Experimental and numerical investigation of convection heat transfer in channels with different types of ribs. Appl. Therm. Eng. **31**, 2702–2708 (2011)

Promvonge, P., Thianpong, C.: Thermal performance assessment of turbulent channel flows over different shaped ribs. Int. Commun. Heat Mass Transf. **35**(10), 1327–1334 (2008)

Sarma, P.K., Subramanyam, T., Kishore, P.S., Dharma Rao, V., Kakac, S.: A new method to predict convective heat transfer in a tube with twisted tape inserts for turbulent flow. Int. J. Therm. Sci. **41**, 955–960 (2002)

Sarma, P.K., Subramanyam, T., Kishore, P.S., Dharma Rao, V., Kakac, S.: Laminar convective heat transfer with twisted tape inserts in a tube. Int. J. Therm. Sci. **42**, 821–828 (2003)

Smithberg, E., Landis, F.: Friction and forced convective heat transfer characteristics in tube with twisted-tape swirl generators. J. Heat Transf. **86**, 39–49 (1964)

Tanda, G.: Effect of rib spacing on heat transfer and friction in a rectangular channel with 45° angled rib turbulator on one/two walls. Int. J. Heat Mass Transf. **54**, 1081–1090 (2011)

Varun, S.R.P., Singal, S.K.: A review on roughness geometry used in solar air heaters. Solar Energy **81**, 1340–50 (2007)

Experimental Studies Involving Flow Visualization Over Non-circular Geometries

Saisarath Kruthiventi, Subbarao Rayapati, Sai Nikhil and Manideep

1 Introduction

Reduction of drag force and understanding the physics behind any wake zone plays vital role in both low and high speed flows. Flow visualization technique is used to understand various flow parameters behind the wake zone. In the present work, same technique is used to study flow over non-circular geometries. Maintaining the recirculation zone size within a desired limit is an important task in the mixing process. Earlier, Nezu and Rodi (1986) made flow measurements in an open-channel using a laser Doppler anemometer and provided the data of turbulence intensities that would offer valuable information for the further understanding of open-channel flows and the development of calculation methods. Prasad and Williamson (1997) studied the shear layer instability in case of a bluff body. Investigation of vortex formation in case of similar bodies was done by Rathakrishnan (2012) and various clarifications for vortex length were discussed. It was also observed that management of size of vortices would be essential for mixing processes and pressure hill was the main cause for bluff body drag. Konstantinidis and Balabani (2007) reported that formation of symmetric vortex shedding near wake zone is due to stream wise separation. Takama et al. (2008) concluded from his set of experiments that modification of the geometry leads to change in the size of the vortices. With this aim, present work investigates the flow visualization over forward facing, backward facing and combined steps with different inclinations in addition to the square blocks. Comparison has been made for all the objects with respect to variation in geometrical dimensions, angle of attack,

S. Kruthiventi (✉) · S. Nikhil · Manideep
K.L. University, Guntur, Andhra Pradesh, India
e-mail: satyasai222@gmail.com

S. Rayapati
NITTTR Kolkata, Kolkata, West Bengal, India

I. Regupathi et al. (eds.), *Recent Advances in Chemical Engineering*,
DOI 10.1007/978-981-10-1633-2_3

and Reynolds numbers. Nezu and Sanjou (2011) successfully applied PIV/PTV to open-channel flows with sediments and shown the fluid and particle interaction along with the association of coherent structures and sediment concentration. Synchronized velocity measurements and dye concentration with the combination of PIV and LIF in a stagnated open-channel flow enables us to study turbulent scalar flux in an inactive contaminant. Recently, Chen et al. (2014) found significant contribution of the induced motion of the span wise vortex filaments. In such a scenario, present work finds its significance, which is involved in studying about the formation of eddies, location of flow separation, origination of reverse flow behind the objects. Different sets of Reynolds numbers ranging from 1000 to 7000 are considered.

2 Experimental Facility

Flow analysis over various geometries is visualized using a rectangular water flow channel. Schematic view and fabricated experimentation unit are as shown in Fig. 1. Visualization studies are carried out in the test section of 300 mm width and 500 mm length with a uniform water stream of 5 mm deep. Initially, water is allowed to the stagnation chamber and passes over a wedge that is placed after the test section. It is ensured that water attains uniform flow. Three wire screens of different hole diameters are placed between the test section and the wedge leading to uniform velocity. Square blocks of thickness 4, 6 and 12 mm are placed perpendicular to the direction of the flow. Each block is visualized individually and it is ensured that there is no flow taking place at the bottom face of the model. Floating particle method is used to measure the flow velocity. For clear visibility, a die is injected along with water just before the last screen such that it is easy to visualize the flow pattern over various geometries. Entire flow structure over various geometries is video recorded with the view from the measuring scale, drawn within the test section, ensuring no disturbance to the flow. Also, it would be easy to measure the distance of reverse flow and the size of wake zone. Parameters such as drag force, coefficient of drag, flow separation point, flow attachment point, recirculation zone size and formation of eddies are measured and calculated

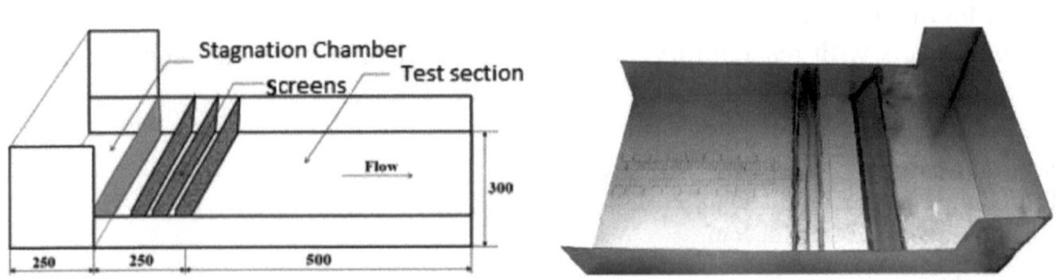

Fig. 1 Schematic view and fabrication of the experimentation unit

experiments for different Reynolds numbers in the range between 1000 and 7000. After that, the work is extended for flow visualization over steps of forward facing, backward facing and combined steps of various inclinations of 90° and 30°.

3 Results and Discussion

Analysis of flow over different non-circular geometries plays an important role the reduction of drag. In the present analysis, different square blocks with thickness of 4, 6 and 12 mm with forward, backward and combined facing steps five Reynolds numbers and inclinations of 90° and 30° as shown in Fig. 2. The uncertainty in the measurements of separation zone, reverse flow location, pressure uphill, and coefficient of drag is about ±5 % based on the values obtained from the images. Figure 3 shows the variation of flow separation zone size and width of pressure uphill with respect to Reynolds number. For Re = 1000, it is observed that there is a separation zone of size 3.5 cm, but it gradually increases as Re increases. This is due to the increment of suction force and formation of vortices behind the plate. These vortices create the low pressure zone as it becomes strengthened up to 4000

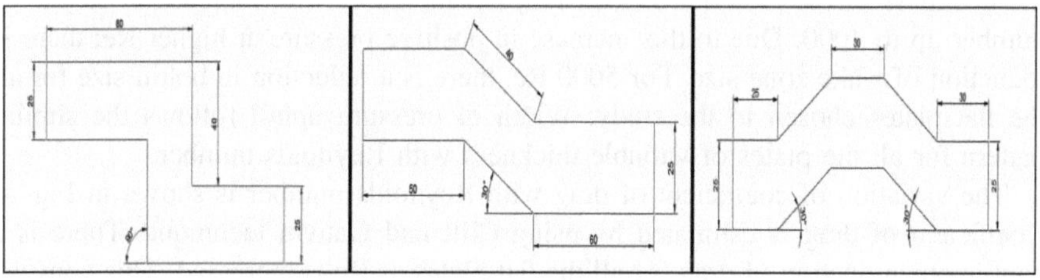

Fig. 2 Backward and combined facing steps with 90° and 30° with all dimensions in mm

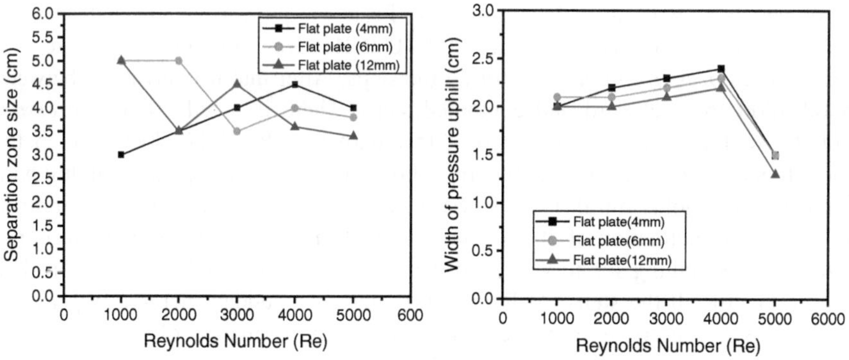

Fig. 3 Variation flow separation zone size and width of pressure uphill with Re

Fig. 4 Variation of coefficient of drag with Re number

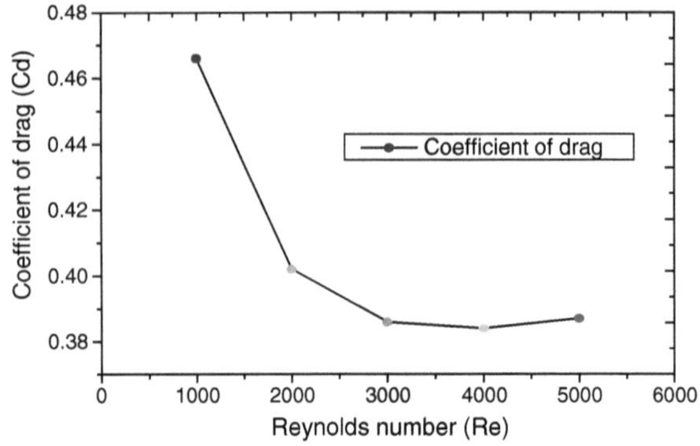

Re. Further it becomes invariant and reduces in size from Re of 5000 due to the effect of positive pressure known as *pressure uphill*. But, for flat plate of thickness 6 mm, it is observed that at 1000 Re, size of separation zone is 5 cm and it is not varying much up to 2000 Re. Afterwards, there is a continuous reduction in size, which is due to reduction in the formation of vortices. For 12 mm thickness, there is a significant change in separation zone size for each and every Reynolds number. This means that there is drastic variation in size of recirculation zone with respect to the thickness of the material. *Pressure uphill* increases with increase in Reynolds number up to 4000. Due to this increase in positive pressure at higher Re, there is reduction of wake zone size. For 5000 Re, there is a reduction in uphill size for all the flat plates chosen in the study. Width of pressure uphill follows the similar pattern for all the plates of variable thickness with Reynolds number.

The variation of coefficient of drag with Reynolds number is shown in Fig. 4. Coefficient of drag is estimated by using Clift and Gauvin technique. There is a continuous reduction of drag for all the flat plates as Re is increased. This conveys that one should prefer higher Reynolds numbers, irrespective of the geometry of the material. Figure 5 describes the flow visualization over flat plates. The formation of vortices and size of recirculation zone are identified and analyzed. Figure 6 shows the variation of pressure uphill with Re for forward facing, backward facing and combined steps with an inclination of 90º. From the plot, it is clear that *pressure uphill* is almost uniform between 3000–4000 Re. But, for 5000 Re, there is a reduction in *pressure uphill* size for all the steps. After that it is invariant between 5000–7000 for backward and combined steps. There is a reduction in *pressure uphill* size for forward facing step at that particular Re 5000–7000 due to the formation of recirculation zone. Figure 7 shows the variation of pressure uphill with respect to Reynolds number for forward facing, backward facing and combined steps with an inclination of 30°. Is observed that there is a raise of pressure uphill size at the beginning for forward and combined steps, but at Re 5000, there is a drastic fall for all the steps. At Re = 6000, there is a sudden increase in *pressure uphill*, due to the pressure recovery. This causes the formation of second recirculation zone and gradual pressure drop is observed at 7000 Re. The variation of

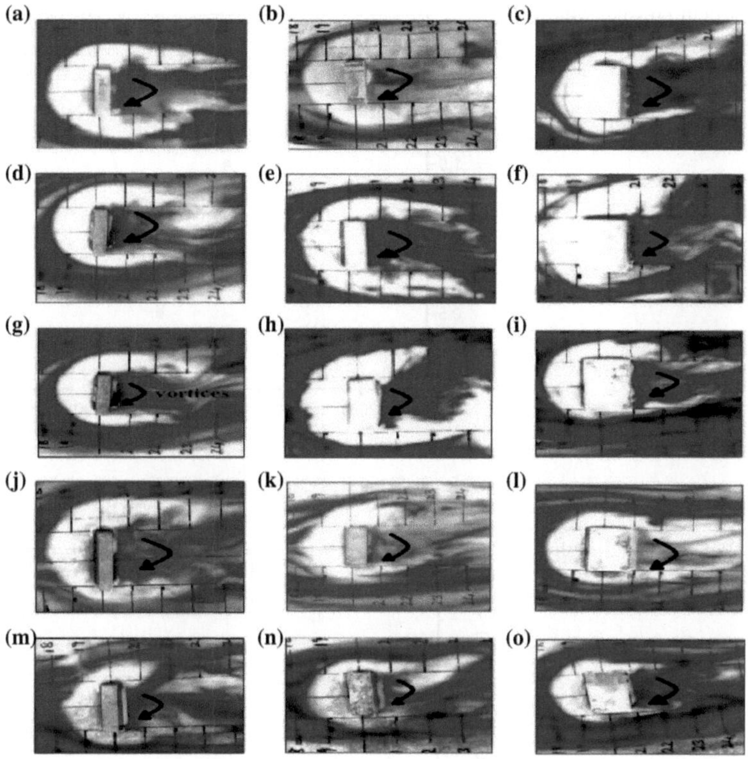

Fig. 5 Flow visualisation over flat plates

Fig. 6 Uphill pressure with Re for 90°

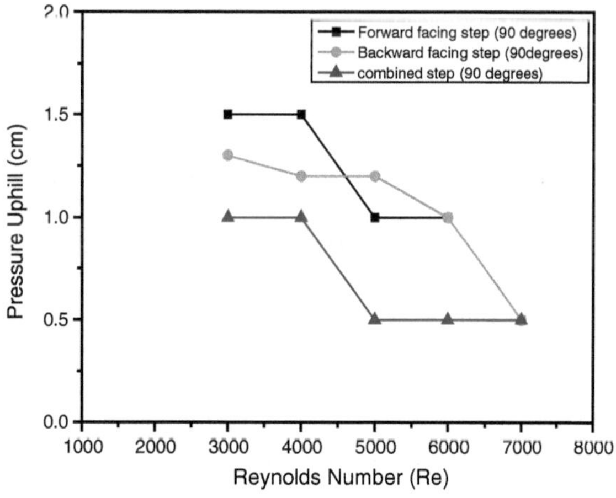

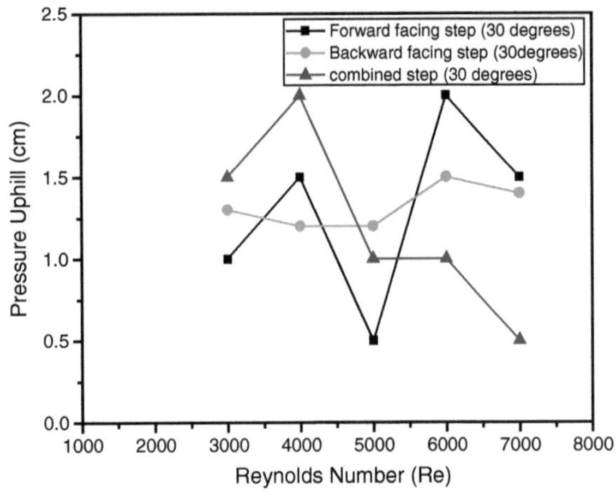

Fig. 7 Uphill pressure with Re for 30°

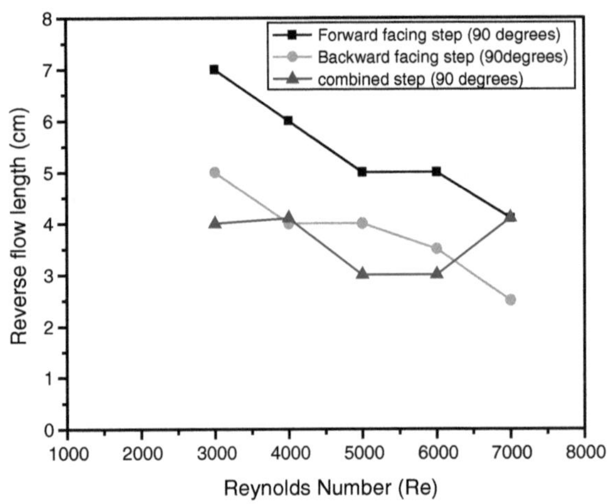

Fig. 8 Reverse flow length with Re for 90°

reverse flow length with Re for forward facing, backward facing and combined steps with an inclination of 90° is shown in Fig. 8. There is decrement in reverse flow length between 3000–4000 Re for all the steps and it is invariant between 4000–6000. Afterwards, there is a sudden fall in the reverse flow length. Figure 9 shows the change of reverse flow length with respect to Re with an inclination of 30º. There is a continuous decrement in reverse flow length for forward and backward steps. But for the combined step, there is uniformity between 3000–4000 Re and 5000–6000 Re due to the formation of twin vortices. Figure 10 shows the actual images obtained for all the configurations.

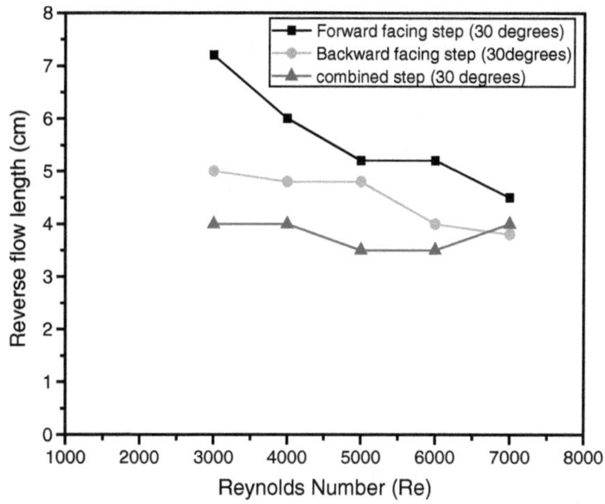

Fig. 9 Reverse flow length with Re for 30°

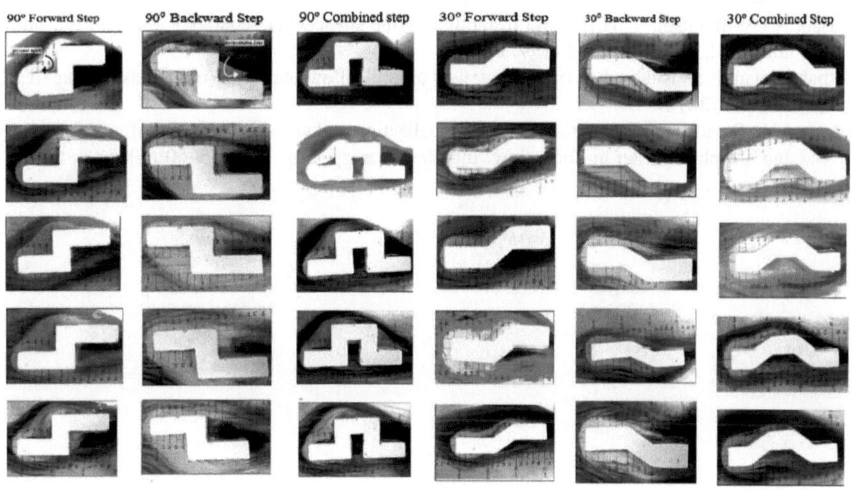

Fig. 10 Flow visualization over steps

4 Conclusions

Flow visualization over flat plate of different thickness is done for various Reynolds numbers ranging from 1000 to 7000. *Pressure uphill* and separation zone size increase up to 4000 Re and at 5000 Re, there is a reduction in the width of the same. For the flat plate of 6 mm, there is reduction in the separation zone size at 3000 Re. For the 12 mm flat plate, there is increase in the width of the separation zone. Coefficient of drag also reduces with increase in Re. This gives encouraging information regarding the increase in the thickness of the plate. Flow visualization over steps with different inclinations is made and parameters such as *pressure uphill* size, reverse flow length variation with respect to Re are compared. Reduction in recirculation zone size leads to formation of less number of wakes behind the object which decreases the drag force and increases the efficiency. Present study will be helpful for better flow visualization over similar objects for different applications.

References

Chen, Q., Ronald, J.A., Zhong, Q, Li, D., Wang, Xi.: Experimental study on the role of spanwise vorticity and vortex filaments in the outer region of open-channel flow. J. Hydraul. Res. **52**(4), 476–489 (2014)

Konstantinidis, E., Balabani, S.: Symmetric vortex Shedding in the near wake of a circular cylinder due to stream wise perturbations. Journal of Fluid Structure **23**, 1047–1063 (2007)

Nezu, I., Rodi, W.: Open-channel flow measurements with a laser Doppler anemometer. J. Hydraul. Eng. **112**(5), 335–355 (1986)

Nezu, I., Sanjou, M.: PIV and PTV measurements in hydro-sciences with focus on turbulent open-channel flows. J. Hydro-environ. Res. **5**(4), 215–230 (2011)

Prasad, A., Williamson, C.H.K.: The instability of the shear layer separating form a bluff body. J. Fluid Mech. **333**, 375–402 (1997)

Rathakrishnan, E.: Visualization of the flow field around a flow plate. Instrum. Measur. Mag. IEEE **15**(6), 8–12 (2012)

Takama, Y., Suzuki, K., Rathakrishnan, E.: Visualization and size measurement of vortex shed by flat and arc plates in an uniform flow. Int. Rev. Aerosp. Eng. **1**(1), 55–60 (2008)

Thermohydraulic Transport Characteristics of Micro Mixer in Micro Channel

Suvanjan Bhattacharyya, Himadri Chattopadhyay, Anirban Roy, Aritra Rakshit and Indraneel Roy Chowdhury

1 Introduction

The discipline of Micro fluidics has been developed as a major area of research due to its various applications in various forms. The recent technologies developed in the field of micro-fluidics have application ranging from pharmaceuticals and biomedical industries (such as drug design, delivery and detection, diagnostic devices etc.) to electronics industry (as heat exchangers for integrated cooling of electronic circuits). Also, in daily lives one can see extensive application like print head of an inkjet printer or nib of a fountain pen etc. Other areas of applications are in fuel cells, pumps and in gas turbines. The continuous enhancement in the functionality and reduction in size of micro-electronics has opened the requirement for new ways for effective removal or dissipation of high heat fluxes in order to enhance its performance. The minute dimensions of the heat removal devices and very strict operational temperature conditions for optimum performance of electronic devices, make heat management of micro-electronic devices a challenging problem. A number of different heat removal techniques such as impinging jets and heat pipes have been applied to achieve optimum cooling. Use of micro-mixers on the channel wall can help to achieve effective removal of heat fluxes. The main objectives of this study are minimization of maximum device temperature and minimization of temperature gradients, which are possible by configuring the wall of circular shape micro-mixers.

S. Bhattacharyya (✉) · H. Chattopadhyay
Mechanical Engineering Department, Jadavpur University, Kolkata 700032, India
e mail: suvanjanr@gmail.com; suva.raindrops@gmail.com

A. Roy · A. Rakshit · I.R. Chowdhury
Mechanical Engineering Department, MCKV Institute of Engineering,
Liluah, Howrah 71120, India

I. Regupathi et al. (eds.), *Recent Advances in Chemical Engineering*,
DOI 10.1007/978-981-10-1633-2_4

The initiation of study of micro channels happened in 1981 when (Tuckerman and Pease 1981) published a work that dealt with the advantageous effects of using channels which are small in diameter for heat dissipation purposes of very large scale integrated circuits. It was shown that the hydraulic diameter of the channel varies inversely to the heat transfer coefficient. Some of the results obtained by means of experiment for fluid flow in micro channel using gas traces its source to (Wu and Little 1983). In this experiment the friction factors in the laminar region were more than expected, and they found that 350–900 was the transition regime. (Peiyi and Little 1983) performed an investigation on the measurement of the friction factors for gas flow in channels of fine dimensions, of, 130–200 μm width. The results obtained by them exhibited dissimilar characteristics from that indicated by conventional models of fluid flow. The reason for this cause was attributed widely to the surface roughness of the micro channels. An experimentation was conducted by (Pfahler et al. 1990) on flow of fluid in micro channels. It was found that for large flow channels the observations obtained by experimentation were in rough agreement with the conventional model whereas for small channels the magnitude of difference augmented. In addition to the previous work, (Pfahler et al. 1991) extracted measurements, through experimentation of friction factor or apparent viscosity of silicon oil and isopropyl alcohol by virtue of its flow in micro channels. It was observed by them that for channels of greater dimensions the results of experimentation showed a negligible disagreement with the estimated values of conventional theory. However, the apparent viscosity began to decline from the theoretical value for a constant pressure drop, as the channel dimensions decreased, though noticeably dissimilar behaviours were detected between the non-polar silicon oil and the polar isopropyl alcohol. Investigation by (Choi et al. 1991) on the friction factor, coefficient of convective heat transfer and the consequential effects of inner wall surface roughness for flow in laminar and turbulent regime in micro channels. The results of experimentation obtained by them were significantly dissimilar from the correlations in the classical theories. There was a certain disparity found by (Peng et al. 1994a, b, 1995; Peng and Peterson 1995, 1996) between conventional flow model and results of experimental investigation for micro tubes. Testing of a range of micro channels of varied hydraulic diameters ranging from 133 to 367 μm were undertaken, and they revealed a friction factor dependence on hydraulic diameter and aspect ratio of the channel in which fluid flow takes place. Experimental investigation were undertaken by (Kazmierezak et al. 1999) for single-phase forced convection in micro channels of rectangular cross-section and considerable depth. Testing for two configurations were successfully performed, a multiple channel system and a single channel system. In this particular case of the multiple channel system the channels were 251 μm in width and the channel walls were 119 μm in thickness. In those systems the channels were approximately 1000 μm in depth and a projected area of 2.5 cm × 2.5 cm is defined. The fluid flow was in the form of de-ionized water, of Reynolds number range varying from 173 to 12,900. The local Nusselt number, which was obtained through experimental means matched reasonably well with conventional developing channel flow theory. A theoretical as well as experimental work by (Hetsroni et al. 2005), on single-phase transfer of heat in micro channels, shed light on several facets of flow in micro channels as pressure drop, transition of fluid flow from laminar to turbulent regime etc. Transfer of heat pertaining to small Knudsen number was considered as the problem. Analysis of data of heat transfer in micro-channels with hydraulic diameter range between 60 and 2000 μm was performed. The consequential effects from different dimensional variations, axial heat flux due to thermal conduction through the fluid used in the experimental setup and channel walls, as well as the dissipation of energy were discussed. It was found by them that the effect of dissipation of energy on transfer of heat in micro-channels is insignificant under typical condition of flow.

It has been observed from the literature review that heat transfer by virtue of fluid flow in circular micro mixer has not been reported in the past. In this paper, therefore, the laminar flow numerical heat transfer and pressure drop results of circular micro mixer of different diameter in circular micro-channel are presented.

2 Boundary Conditions

The full length micro-channel as shown in Fig. 1, of inlet diameter (D) with circular roughness of diameter, (d) and pitch, (p) is employed. Air velocity was introduced at inlet of the channel and a pressure outlet condition was applied at exit. Air inlet temperature of 300 K was set in the direction of fluid flow. The temperature of wall was kept constant of 500 K throughout the experiment. At mean bulk temperature, the thermal and the physical properties of air were taken to be invariant. The channel walls were assumed to be impervious and no-slip condition was implemented.

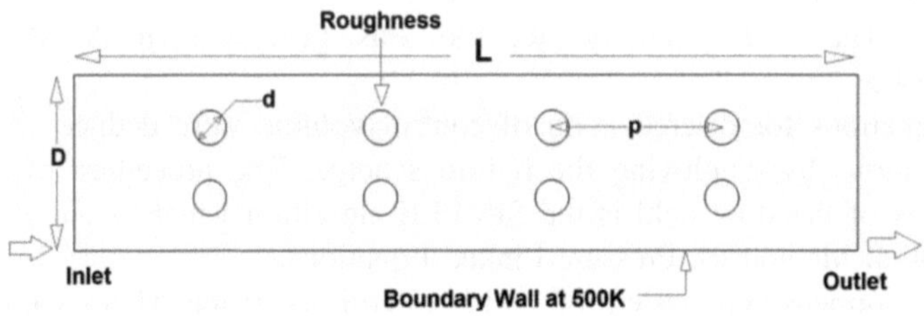

Fig. 1 Computational domain

3 Mathematical Model

The basic form of continuity, energy and momentum equations for a two dimensional, incompressible, steady state flow and laminar forced convection of transfer of heat minus viscous dissipation are as follows:

Continuity Equation

$$\frac{\partial u}{\partial x} + \frac{\partial v}{\partial y} = 0 \tag{1}$$

Momentum Equation

$$\rho\left(u\frac{\partial u}{\partial x} + v\frac{\partial u}{\partial x}\right) = -\frac{\partial p}{\partial x} + \mu\left(\frac{\partial^2 u}{\partial x^2} + \frac{\partial^2 u}{\partial y^2}\right) \tag{2}$$

$$\rho\left(u\frac{\partial v}{\partial y} + v\frac{\partial v}{\partial y}\right) = -\frac{\partial p}{\partial y} + \mu\left(\frac{\partial^2 v}{\partial x^2} + \frac{\partial^2 v}{\partial y^2}\right) \tag{3}$$

Energy Equation

$$u\frac{\partial T}{\partial y} + v\frac{\partial T}{\partial y} = \alpha\left(\frac{\partial^2 T}{\partial x^2} + \frac{\partial^2 T}{\partial y^2}\right) \tag{4}$$

4 Numerical Solution

Grid sizes of various dimensions were tested as a part of study for grid independence. After a rigorous grid independence check, a mesh consisting of 154,656 elements and 78,665 nodes were used for the present computation. Theoretical equations for laminar model in a 2-D geometry of diameter, 0.5 mm and length, 10 mm, were solved numerically for heating of the working fluid which was air (Pr = 0.7). The 2-D geometry-data file was generated in ANSYS Design Modeler 14.5.

The equations for discretization of control volume were deduced from these basic equations by employing the hybrid scheme. The procedure of numerical calculations of the flow field is the SIMPLE algorithm which is an acronym for Semi-Implicit Method for Pressure-Linked Equations.

The two-dimensional system of grid is established using ANSYS Fluent 14.5. Considering air flow in the channel with heat transfer, the mathematical model applied is composed of the conservation equations of mass, momentum and energy for incompressible flow in two dimensions with the following assumptions:

- The flow is two-dimensional, laminar and stationary.
- The properties of the air, which are thermo-physical, are assumed to be invariant.
- The thermal conductivity of the walls is supposed to be constant.

The Reynolds number of flow of air in the duct is found out from the following expression:

$$Re = \frac{\rho v D}{\mu}; \tag{5}$$

The convective heat transfer coefficient is then used to obtain Nusselt number, Nu, from the following expression:

$$Nu = \frac{hD}{k}; \tag{6}$$

The friction factor is found out from the measured values of drop in pressure (ΔP), across the length of test section.

$$f = \frac{2\Delta PD}{\rho LV^2}; \tag{7}$$

To assess the increment of heat transfer under a pumping power of given magnitude, the performance evaluation criteria is expressed below as:

$$\eta = \left(\frac{Nu}{Nu_0}\right) / \left(\frac{f}{f_0}\right)^{0.33}; \tag{8}$$

Where, Nu_0 and Nu are Nusselt numbers for the smooth channel and the augmented channel correspondingly, f and f_0 are friction coefficients for augmented channel and smooth channel correspondingly.

5 Results and Discussion

Correlations provided by (Shah and London 1978) were used to compare and validate the results of the simulation. The values of the data found from CFD simulation of the plain micro-channel are quite close with the predicted results from the proposed correlations with little margin of error (Shah and London 1978) with data range of +6.0 to +8.5% and +2.2 to +2.7% for the Nusselt number, Nu, and friction factor, f, respectively as revealed in Fig. 2a, b.

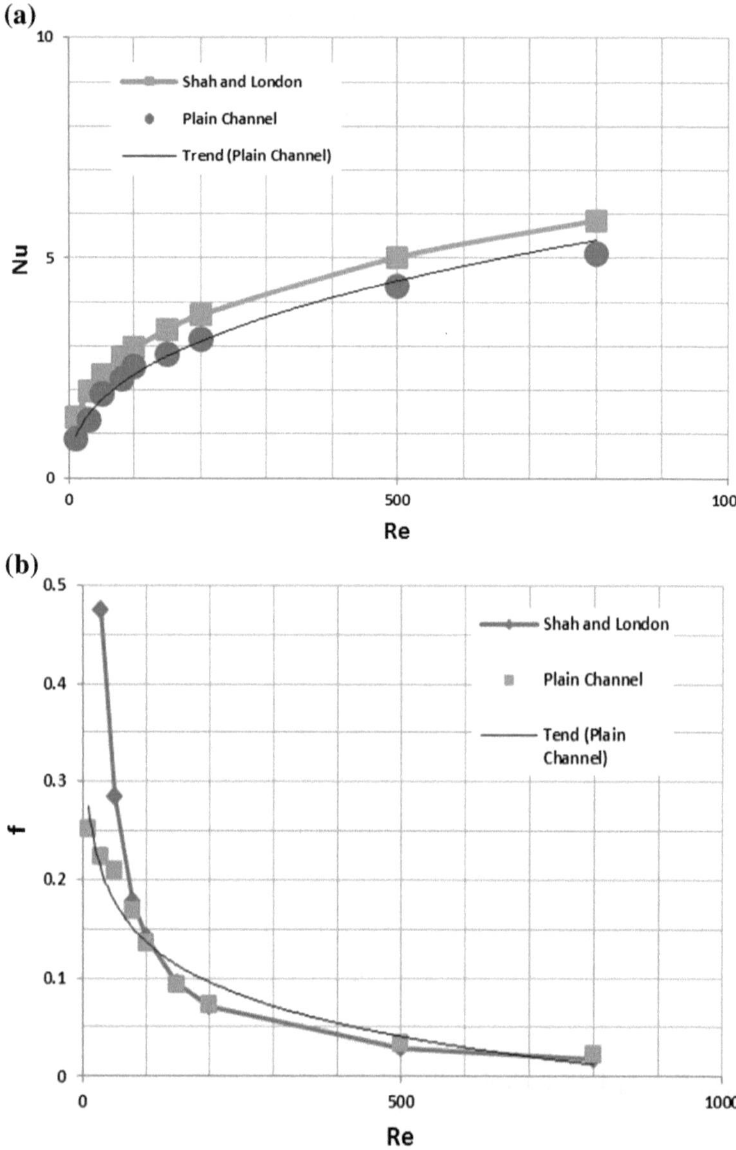

Fig. 2 **a** Verification of the plain micro-channel: Nusselt number. **b** Verification of the plain micro-channel: Friction factor

As predicted by (Shah and London 1978) and also visible from Fig. 3a the Nusselt number does show a definite trend—increasing with rise in Reynolds number. The addition of micro mixer does aid to effectively increase the rate of heat transfer by disrupting the boundary layer development and also by creating local

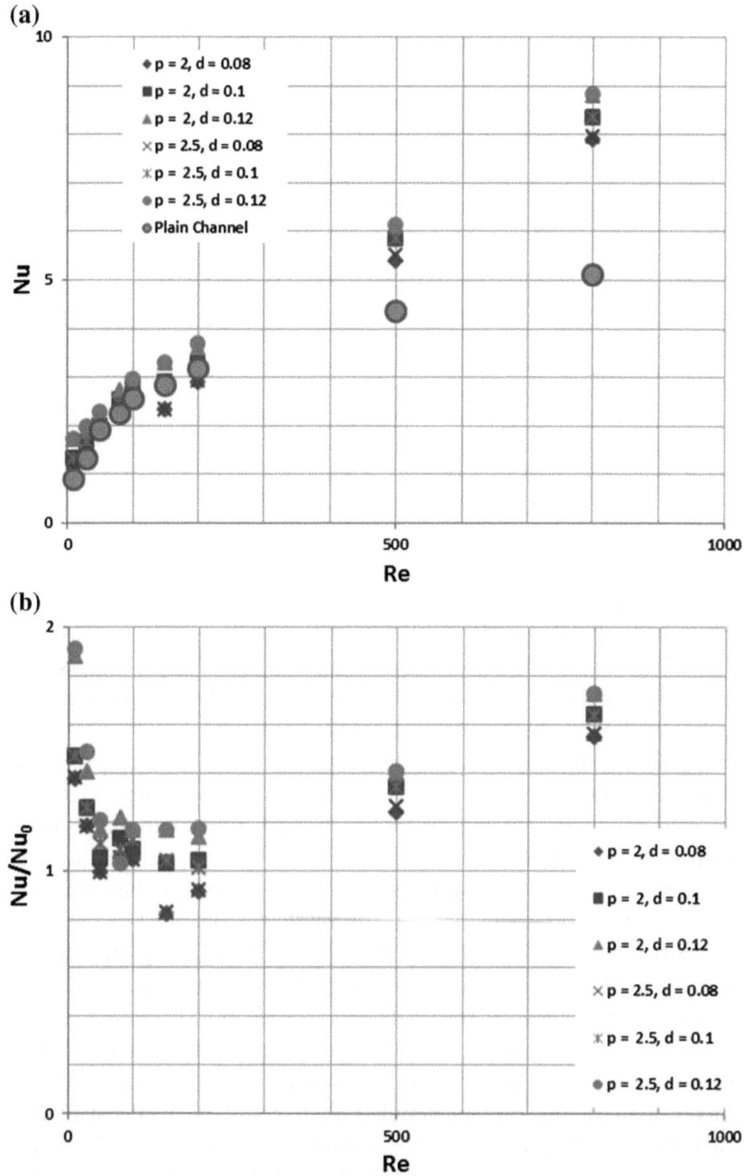

Fig. 3 **a** Variation of Nusselt number with Reynolds number with different (*p*) and (*d*). **b** Variation of Nu/Nu_0 with Reynolds number with different (*p*) and (*d*)

turbulence and swirl flow. This phenomenon accounts for the increase in Nusselt number that is clearly seen in Fig. 3a. It is also evident from Fig. 3b that there is a significant increase in enhancement which is noticed for Reynolds number 500 and above. The dimensionless quantity of Nu/Nu_0 is plot with against Reynolds number

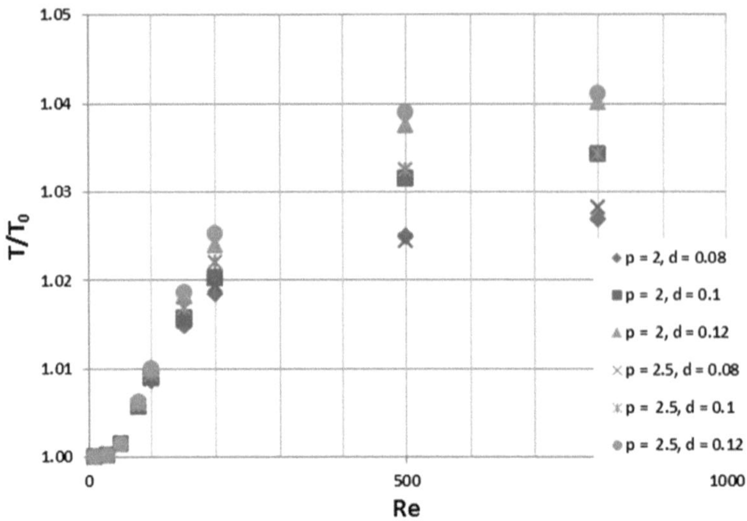

Fig. 4 Relation between the T/T_0 and Reynolds number

and it shows how effective each of the different geometries is in increasing the rate heat transfer. The results show that with increase in size of micro mixers there in an enhancement in transfer of heat. According to the figure, Fig. 3b, the geometry with p = 2.5, d = 0.12 gives the best enhancement.

Further evidence that supports this theory of heat transfer enhancement within a micro-channel by creating obstructions/indentations is presented in Fig. 4. This shows the ratio of outlet temperatures for the channels fitted with micro-mixer to the outlet temperatures of a plain micro-channel. From Fig. 4 the same inference can be made—increase in size of micro-mixers there in an increase in heat transfer which makes it easier to carry more hot fluid through the outlet. The geometry with p = 2.5, d = 0.12 is best in achieving this effect. Fluid having Reynolds number more or equal to 500, a significant increase in the outlet temperature is perceived.

The friction factor characteristics obtained as a result of the effects exhibited by the micro-mixer are shown in Fig. 5a. The figure represents the association between the Reynolds number and the friction factor at various pitch and diameter of the micro mixer used in the present simulation. It could be well established from Fig. 5a that the friction factor was in the analogous trend, both for the channel with micro mixers and also with the plain micro channel. The friction factor of the micro mixer in the micro channel is inversely proportional to Reynolds number. At a certain Reynolds number, the micro channel with micro mixer resulted in greater friction factors over those of the plain micro channel. This was because of the blockage of flow, greater contact area of the surface, the phenomenon caused by the swirl flow as well as the dynamic pressure dissipation of the fluid due to the viscosity loss nearby the tube wall. Moreover, the pressure drop had a greater possibility to occur by the interaction of the inertia forces with pressure forces in the

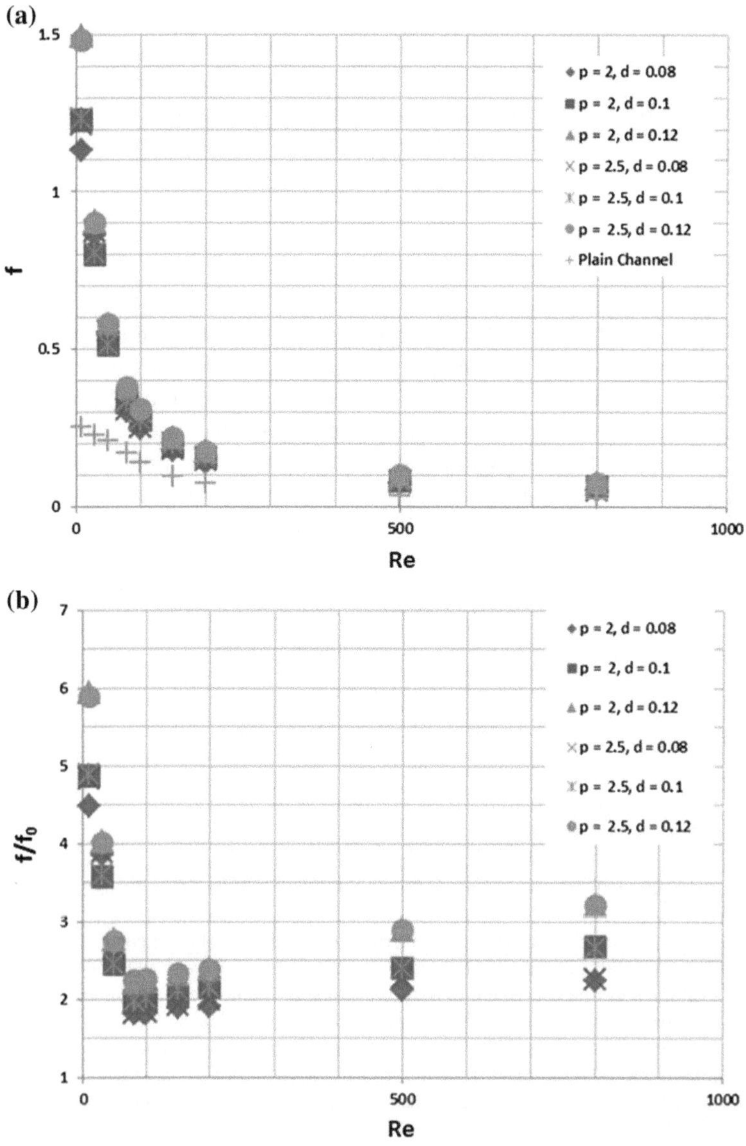

Fig. 5 **a** Variation between the friction factor and Reynolds number for (*p*) and (*d*). **b** Variation between the f/f_0 and Reynolds number for (*p*) and (*d*)

boundary layer. Figure 5a shows that at lower Reynolds number, surge in the friction factor was relatively high and vice versa. This particular occurrence might be elucidated by the fact that at lower values of Reynolds number, corresponding to lower flow rates, air could pass all over the micro mixer and created high frictional forces due to the presence of small vortices behind the micro mixer. Figure 5b depicts the variation of friction factor ratio (f/f_0) with Reynolds number for different pitch and diameter. It was shown from Fig. 5b that the friction factor ratio tended to decline with raising the Reynolds number for all the examined cases. In higher Reynolds number region, the friction factor values for the micro channel with micro mixer were 2.56–2.88 times higher than those of the plain micro channel.

6 Conclusion

Thermo hydraulic transport characteristics of micro mixer in micro channel were investigated numerically. The simulation was conducted for the micro channel with micro mixer of different pitch and diameter. The use of micro mixer produced substantial increase of heat transfer with a considerable increase in friction factor. In general observations, it was established that the heat transfer, outlet temperature and friction factor increased at $p = 2.5$, $d = 0.12$. It is also evident that, the Nusselt number is directly varies with the Reynolds number while the opposite was found for the instance of friction factor. The Nusselt number for the micro channel with micro mixes has a reasonable enhancement than those of the plain micro channel values.

References

Choi, S.B., Baron, R.R., Warrington, R.O.: Fluid flow and heat transfer in micro tubes. ASME DSC **40**, 89–93 (1991)

Hetsroni, G., Mosyak, A., Pogrebnyak, E., Yarin, L.P.: Heat transfer in micro-channel: comparison of experiments with theory and numerical results. Int. J. Heat Mass Transf. **48**, 5580–5601 (2005)

Harms Todd, M., Kazmierczak Michael, J., Gerner Frank, M.: Developing convective heat transfer in deep rectangular micro channels. Int. J. Heat Fluid Flow, 149–157 (1999)

Peiyi, W., Little, W.A.: Measurement of friction factors for the flow of gases in very fine channels used for micro miniature Joule–Thomson refrigerators. Cryogenics **23**, 273–277 (1983)

Pfahler, J., Harley, J., Bau, H.H., Zemel, J.: Liquid and gas transport in small channels. AMSE DSC **19**, 149–157 (1990)

Pfahler, J., Harley, J., Bau, H., Zemel, J.: Gas and liquid flow in small channels. ASME DSC. **32**, 49–60 (1991)

Peng, X.F., Peterson, G.P., Wang, B.X.: Frictional flow characteristics of water flowing through micro channels. Experimental Heat Transfer **7**, 249–264 (1994a)

Peng, X.F., Peterson, G.P., Wang, B.X.: Heat transfer characteristics of water flowing through micro channels. Experimental Heat Transfer. **7**, 265–283 (1994b)

Peng, X.F., Wang, B.X., Peterson, G.P., Ma, H.B.: Experimental investigation of heat transfer in flat plates with rectangular micro channels. Int. J. Heat Mass Transf. **38**(1), 127–137 (1995)

Peng, X.F., Peterson, G.P.: Effect of thermo fluid and geometrical parameters on convection of liquids through rectangular micro channels. Int. J. Heat Mass Transf. **38**(4), 755–758 (1995)

Peng, X.F., Peterson, G.P.: Convective heat transfer and flow friction for water flow in micro channel structures. Int. J. Heat Mass Transf. **39**(12), 2599–2608 (1996)

Shah, R.K., London, A.L.: Laminar Flow Forced Convection in Ducts. Academic Press, New York, Advanced Heat transfer (1978)
Tuckerman, D.B., Pease, R.F.W.: High-performance heat sinking for VLSI. IEEE Electron Device Lett. **EDL**(5), 126–129 (1981)
Wu, P., Little, W.A.: Measurement of friction factors for the flow of gases in very fine channels used for microminiature Joule-Thomson refrigerators. Cryogenics **23**(5), 273–277 (1983)

Esterification of Ethanol and Maleic Acid in Packed Bed Reactor Catalyzed by Sulfonic Acid Functionalized Silica (SAFS)

Rajkumar Sirsam and Ghayas Usmani

1 Introduction

Organic esters were widely used in the synthesis of various chemicals (Ogliaruso and Wolfe 1991). These are commonly used as flavouring and fragrancing agents, intermediate in drugs, plasticizers in polymer synthesis, solvents, perfumery and in other various fine chemicals. These are also used in verity of applications in pharmaceutical and agrochemical industries (Zaidi et al. 1995).

Conventionally organic esters were extensively produced in a batch reactor catalyzed by either homogeneous acid catalysts or with heterogeneous acid catalysts. Solid supported acid catalyst were used in fixed bed reactor to achieve the maximum reaction rates. Homogeneous catalysts shows better activity but requires more effective separation and purification downstream processes. Therefore eco-friendly solid catalyst are extensive used in industry, as these reduces processing steps and offers more advantages over homogeneous catalyst (Kirumakki et al. 2006).

Difficulties encountered while using homogeneous catalysts are overcome by using heterogeneous catalyst like zeolites, metal oxides, acid treated clays, ion exchange resin (Chakrabati and Sharma 1993; Yadav and Thathagar 2002). These solid catalysts become attractive alternative because of their eco-friendly nature, easily recovered and reused (Hass et al. 2003; Zhang et al. 2003; Kulkarni and Dalai 2006; Phan and Phan 2008). Considering the economical aspect, hybrid acid solid catalyst can be synthesized by anchoring reactive centres of traditional acid catalyst onto the inorganic solid material such as silica gel (Shylesh et al. 2004;

R. Sirsam (✉) · G. Usmani
University Institute of Chemical Technology, North Maharashtra University, Jalgaon, India
e-mail: rajkumarsirsam@gmail.com

I. Regupathi et al. (eds.), *Recent Advances in Chemical Engineering*,
DOI 10.1007/978-981-10-1633-2_5

Karimi and Khalkhali 2005; Adam et al. 2012; Shah et al. 2013; Ziarani et al. 2013), clays (Vijaykumar et al. 2012; Baura et al. 2014), titanium dioxide (Atghia and Beigbaghlou 2013), carbon (Chang et al. 2013).

In the current work, SAFS has been synthesized and characterized previously in our laboratory (Sirsam and Usmani 2015) was used as catalyst for the esterification of ethanol with maleic acid in packed bed reactor (PBR). An objective of this work was to optimize the process parameters in PBR by using Response surface methodology (RSM) (Aziz et al. 2008).

2 Material and Method

2.1 Chemicals

All the chemicals required for the experimentation were of analytical grade and purchased from reputed companies (s. d. fine Chemicals ltd., Sisco Chem, Sigma Aldrich, Merck, Fisher Scientific). Silica gel, hydrochloric acid, methanol, toluene, 3-mercaptopropyltrimethoxy-silane (3MPTMS), hydrogen peroxide, sulfuric acid were used for the synthesis of sulfonic acid functionalized silica (SAFS). Maleic acid, ethanol and potassium hydroxide were used for esterification and analysis purpose.

2.2 Analysis

Chemical analysis was carried out for the samples using alcoholic KOH (Yadav and Thathagar 2002). Randomly samples were analyzed by GC (Agilent 7890B) for confirmation as discussed by Sirsam and Usmani (2015).

2.3 Catalyst Preapration and Characterization

The functionalization of commercially available silica gel was carried out as discussed in literature (Karimi and Khalkhali 2005; Adam et al. 2012; Shah et al. 2013; Ziarani et al. 2013), Detail procedure for the synthesis and characterization by FTIR and TGA of the catalyst (SAFS) used in this study was discussed by Sirsam and Usmani (2015). Figure 1 represents the process of SAFS synthesis.

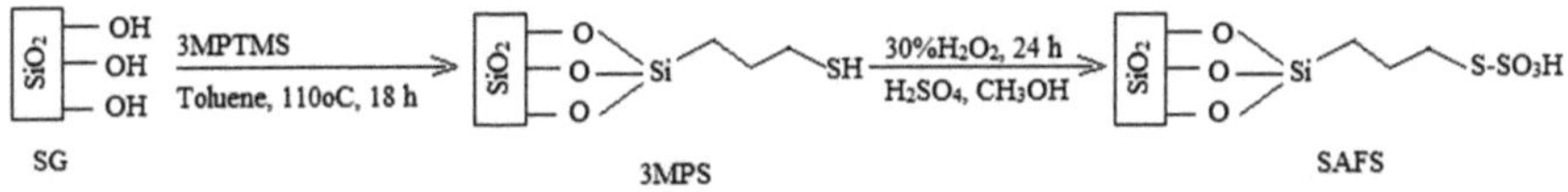

Fig. 1 Process of SAFS synthesis

2.4 *Experimental Setup and Procedure*

Packed bed reactor was comprised of a glass column of ID 3 cm and height 46 cm. Water jacket was provided for hot water circulation to maintain reaction temperature. The desired amount of catalyst confined in stainless steel wire mesh bags (2 × 1 cm) was randomly placed in reactor. As the catalyst is of 60–200 mesh size may develop pressure drop across the column and can be eliminated by the use of wire mesh bags (Chopade and Sharma 1997). Peristaltic pump was used to pump the reaction mixture of ethanol and maleic acid through the column. Raschig ring are placed at the bottom of column to offer the space to reaction mixture to attend required temperature. Condenser was provided at the top to recover the any loss of volatile components. Schematic representation of experimental setup is shown in Fig. 2.

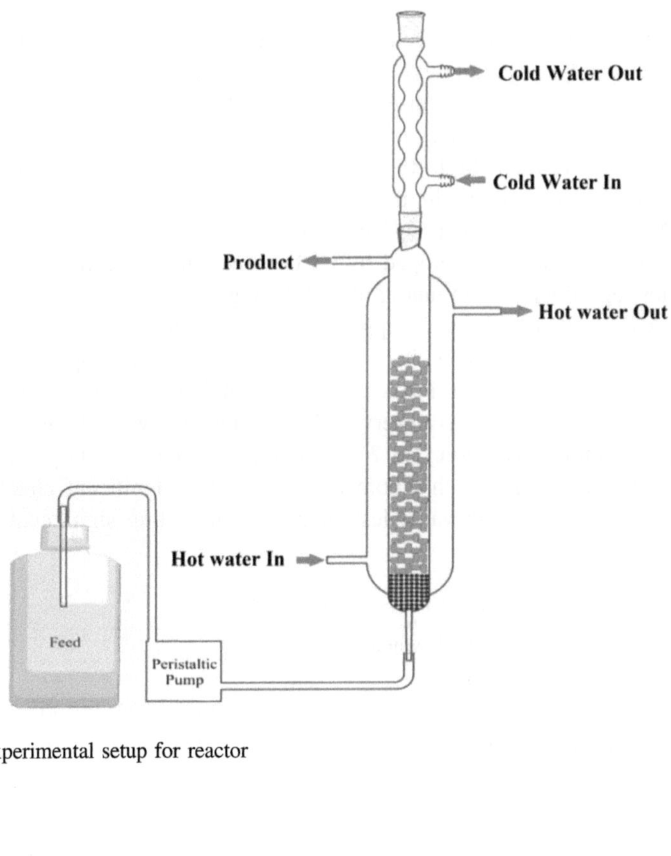

Fig. 2 Experimental setup for reactor

3 Design of Experiments

A five level two factorial, CCD was used to optimize the esterification study. The factors to be optimized in the esterification study are Catalyst amount and reactant flow rate. The percent conversion of maleic acid was considered as the response of the design experiments. Table 1 shows the coded and actual values of the independent variables used in the study. Thirteen experiments were carried out with five replications to evaluate the pure error at the centre point.

Experiments were conducted in PBR as shown in Fig. 2. Operating condition for packed bed reactor i.e. mole ratio of maleic acid to ethanol and temperature was maintained as previously discussed by Sirsam and Usmani (2015).

3.1 *Statistical Analysis*

A quadratic model suggested for the esterification of maleic acid, in terms of coded values, by RSM-CCD is as follows:

$$\%Conv. = 43.64 + 9.83\,A - 7.3\,B - 2.95 - 10.05\,A^2 - 10.41\,B^2 \quad (1)$$

The relationship between percent conversion of maleic acid and independent variables i.e. catalyst weight and feed flow rate were examined. The plot of experimental and predicted response shows the inclination of actual and predicted values of conversion of maleic acid with $R^2 = 0.9964$. The value of adjusted $R^2(0.9939)$ is sufficiently high to support for the significance of the model. However, correlation between the independent variables validated by predicted R^2 (0.9806). Model to be reproducible Coefficient of variation (CV) value must be less than 10 % and the model developed by CCD shows the value (CV = 3.99) indicating better reproducibility (Yuan et al. 2008; Beg et al. 2003).

ANOVA illustrated in Table 2 supports to evaluate the reliability of the model. Model *F*-value (390.47) implies that the regression model developed is reliable to predict the percent conversion. From the *F*-value of Catalyst weight (597.83) and Feed flow rate (330.18), it was observed that the catalyst weight show more significant effect on the conversion. The *F*-value of quadratic term catalyst weight (A^2) feed flow rate (B^2) indicates that both the terms have significant effect. All the model terms show Prob > *F* value less than 0.05 indicating significant effect on

Table 1 Coded symbol and actual values of factor for PBR

Factor	Coded symbol	Level				
		$-\alpha$	-1	0	+1	$+\alpha$
Catalyst weight (gm)	A	5.86	10	20	30	34.14
Feed flow rate (ml/min)	B	1.17	2	4	6	6.83

Table 2 Coded symbol and actual values of factor for PBR

Source	Sum of squares	DF	Mean square	*F* value	Prob > *F*
Model	2522.16	5	504.43	390.47	<0.0001
A-Catalyst weight	772.32	1	772.32	597.83	<0.0001
B-feed flow rate	426.55	1	426.55	330.18	<0.0001
A^2	702.52	1	702.52	543.81	<0.0001
B^2	753.76	1	753.76	583.47	<0.0001
AB	34.93	1	34.93	27.04	0.0013
Residual	9.04	7	1.29		
Lack of fit	6.32	3	2.11	3.10	0.1518
Pure error	2.72	4	0.68		
Cor. total	2531.20	12			
Std. dev.	1.14		R-squared		0.9964
Mean	31.05		Adj R-squared		0.9939
C.V.	3.99		Pred R-squared		0.9806
PRESS	49.2		Adeq precision		44.849

response. The interaction between catalyst weight and feed flow rate also shows insignificant effect on the conversion of maleic acid. The "Lack of Fit *F*-value" of 3.1 indicated that is not significant and non-significant lack-of-fit implies that the model is valid for present study.

Figure 3 shows 2D contour and 3D surface plot for the combined effect of catalyst weight and feed flow rate on the response. It was observed that, percent conversion increases sharply with increasing catalyst weight than feed flow rate. An optimum point was achieved around catalyst weight 25 gm and feed flow rate 3 ml/min. Further increase in feed flow rate does not show significant effect whereas at higher value of flow rate and catalyst weight, conversion is reduced.

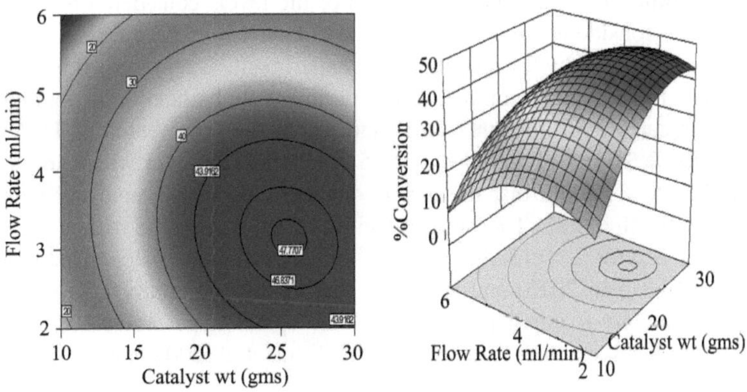

Fig. 3 Response surface plot showing the effect of flow rate and catalyst weight and their mutual interaction on the conversion in PBR catalyzed by SAFS

Fig. 4 Response surface plot for optimized conditions for PBR using SAFS

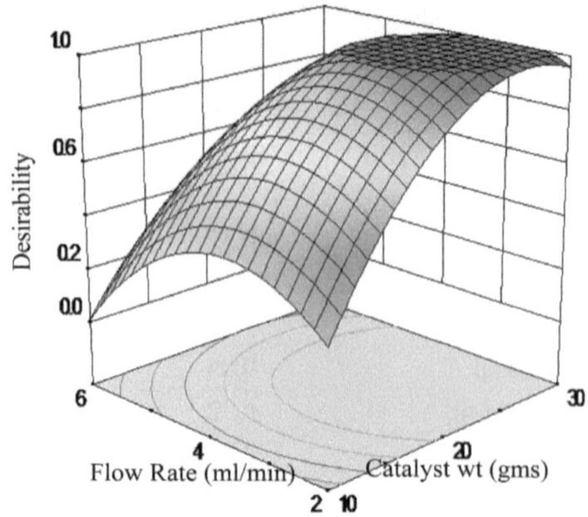

Table 3 Optimum condition suggested by DOE for PBR using SAFS

Run	Factors		Response		Std. devi. σ	Std. error α	95 % CL
	Cat. wt (gm)	Feed flow rate (ml/min)	Observed	Predicted			
1	26.42	3.11	46.67	47.83	0.347	0.115	0.22
2	26.42	3.11	46.89	47.83			
3	26.42	3.11	47.35	47.83			

$\sigma = \sqrt{(\sum x^2)/(n-1)}$, $\alpha = (\sqrt{(\sum x^2)/(n-1)})/n$, where x represents deviation from mean and n represents the sample size

3.2 *Optimization Analysis*

Numerial optimization aspect was considered to optimize the independent factor for PBR. Fourteen set of experiments were recommended to get maximum possible response within the lower and upper limits of the levels considered for study and response was considered at maximum value obtained in experiments.

Predicted response by the recommended experiments is represented in Fig. 4. To validate the model, the recommended experiment with maximum possible conversion was carried out for three times and results are represented in Table 3. An error of ±0.22 % was observed under 95 % confidence level which implies that one can be 95 % certain that the mean conversion of maleic acid fall within the range of 46.97 % ±0.22. The ratio of weight of catalyst to the maleic acid in the packed bed reactor is 0.13, however in the conventional batch reactor is 0.38 (Sirsam and Usmani 2015) and it implies that in PBR at optimized condition the amount of catalyst required per gm of maleic acid fed per cycle is less than in conventional batch reactor (Çiftçi et al. 2009; Tepe et al. 2008; Chang et al. 2007).

4 Conclusion

This article emphasizes on the statistical modeling and optimization of process parameters by Design-Expert software (Version-8.0.7.1) for PBR. CCD methodology was applied successfully to optimize variables for packed bed column. Statistical model recommended by the Design-Expert for the system is suitable to predict the response for PBR. Accuracy of the models was further supported by ANOVA.

Acknowledgments Authors are thankful to TEQIP-II, funded by Ministry of HRD, Govt. of INDIA, New Delhi for the financial support.

References

Adam, F., Batagarawa, M.S., Hello, K.M., Al-Juaid, S.S.: One-step synthesis of solid sulfonic acid catalyst and its application in the acetalization of glycerol: crystal structure of cis-5-hydroxy-2-phenyl-1-3-dioxane trimer. Chem. Pap. 66–1048 (2012)

Atghia, S.V., Beigbaghlou, S.S.: Nanocrystalline titania-based sulfonic acid (TiO_2-Pr-SO_3H) as a new, highly efficient, and recyclable solid acid catalyst for preparation of quinoxaline derivatives. J. Nanostruct. Chem. **3**(38), 1 (2013)

Aziz, H.A., Kamaruddin, A.H., Bakar, M.Z.A.: Process optimization studies on solvent extraction with naphthalene-2-boronic acid ion-pairing with trioctylmethylammonium chloride in sugar purification using design of experiments. Separ. Purif. Technol. **60**(2), 190 (2008)

Baura, S., Dutta, N., Karmarkar, S., Chattopadhyay, P., Aidew, L., Buragohain, A.K., Karak, N.: Biocompatible high performance hyperbranched epoxy/clay nanocomposite as an implantable material. Biomed. Mater. **9**, 1 (2014)

Beg, Q., Sahai, V., Gupta, R.: Statistical media optimization and alkaline protease production from Bacillus mojavensis in bioreactor. Proc Biochem. **39**, 203 (2003)

Chakrabati, A., Sharma, M.M.: Cationic ion exchange resins as catalyst. React. Polym. **20**, 1 (1993)

Chang, B., Fu, J., Tian, Y., Dong, X.: Soft-template synthesis of sulfonated mesoporous carbon with high catalytic activity for biodiesel production. RSC Adv. **3**, 187 (2013)

Chang, S.-W., Shaw, F.-F., Yang, C.-K., Shieh, C.-J.: Optimal continuous biosynthesis of hexyl laurate by packed bed bioreactor. Proc Biochem. **24**, 1362–1366 (2007)

Chopade, S.P., Sharma, M.M.: Reaction of ethanol and formaldehyde: use of versatile cation-exchange resin as catalyst in batch reactor and reactive distillation columns. React. Funct. Polym. **32**, 53 (1997)

Çiftçi, O.N., FadIloglu, S., Gögüs, F.: Conversion of olive pomace oil to cocoa butter-like fat in a packed-bed enzyme reactor. Bioresour. Technol. **100**(1), 324 (2009)

Hass, M.J., Michalski, P., Runyon, S., Nunez, A., Scott, K.M.: Production of FAME from acid oil, a by-product of vegetable oil refining. J. Am. Oil Chem. Soc. **80**(1), 97 (2003)

Karimi, B., Khalkhali, M.: Solid silica-based sulfonic acid as an efficient and recoverable interphase catalyst for selective tetrahydropyranylation of alcohols and phenols. J. Molec. Cata A. Chem. **232**, 113 (2005)

Kirumakki, S.R., Nagaraju, N., Komandur, V.R.: Esterification of alcohols with acetic acid over zeolites Hβ, HY and HZSM5. Appl. Catal. A **299**, 185 (2006)

Kulkarni, M.G., Dalai, A.K.: Waste cooking oil an economical source for biodiesel: a review. Ind. Eng. Chem. Res. **45**, 2901 (2006)

Ogliaruso, M.A., Wolfe, J.F.: Synthesis of Carboxylic Acids, Esters and Their Derivatives. Wiley, New York (1991)

Phan, A.N., Phan, T.M.: Biodiesel production from waste cooking oils. Fuel **87**, 3490 (2008)

Shah, K.A., Parikh, J.K., Maheria, K.C.: Use of sulfonic acid-functionalized silica as catalyst for esterification of free fatty acids (FFA) in acid oil for biodiesel production: an optimization study. Res. Chem. Intermed. (2013). doi:10.1007/s11164-013-1253-6

Shylesh, S., Sharma, S., Mirajkar, S.P., Singh, A.P.: Silica functionalised sulphonic acid groups: synthesis, characterization and catalytic activity in acetalization and acetylation reactions. J. Molec. Cat A. Chem. **212**, 219 (2004)

Sirsam, R.S., Usmani, G.A.: Preparation and Characterization of Sulfonic acid Functionalized Silica and its application for the Esterification of Ethanol and Maleic acid. J. Inst. Eng. India Ser. E. (2015). doi:10.1007/s40034-015-0068-y

Tepe, O., Dursun, A.Y.: Combined effects of external mass transfer and biodegradation rates on removal of phenol by immobilized Ralstonia eutropha in a packed bed reactor. J. Hazard. Mater. **151**, 9 (2008)

Vijaykumar, B., Mahadevaiah, N., Nagendrappa, G., Jai Prakash, B.S.: Esterification of stearic acid with p-cresol over modified Indian bentonite clay catalyst. J. Porous Mater. **19**, 201 (2012)

Yadav, G.D., Thathagar, M.B.: Esterification of maleic acid with ethanol over cation-exchange resin catalyst. React. Funct. Polym. **52**, 99 (2002)

Yuan, X., Liu, J., Zeng, G., Shi, J., Tong, J., Huang, G.: Optimization of conversion of waste rapeseed oil with high FFA to biodiesel using response surface methodology. Renew. Energy. **33**, 1678 (2008)

Zaidi, A., Gainer, J.L., Carta, G.: Fatty acid esterification using nylon-immobilized lipase. Biotechnol. Bioeng. **48**(6), 601 (1995)

Zhang, Y., Dube, M.A., McLean, D.D., Kates, M.: Biodiesel production from waste cooking oil: 2 Economic assessment and sensitivity analysis. Bioresour. Technol. **90**(3), 229 (2003)

Ziarani, G.M., Badiei, A., Dashtianeh, Z., Gholamzadeh, P., Mohtasham, N.H.: Application of SiO_2-Pr-SO_3H as an efficient catalyst in the Ritter reaction. Res. Chem. Intermed. **39**, 3157 (2013)

Part II
Separation Processes

Batch Drying of Wheat in a Multiple Porous Draft Tube Spouted Bed

S. Rajashekhara and D.V.R. Murthy

1 Introduction

Mathur and Gishler (1955) developed the spouted bed technique for drying wheat. A spouted bed is known to provide excellent fluid-solid contact while keeping the bed in a continuously agitated state. There exists a maximum spoutable bed depth (MSBD) in the operation of a conventional spouted bed (CSB) having single spout; also when a fairly deep bed is to be operated gas is to be supplied at a higher discharge pressure. These two issues limit the use of CSB for large scale operations. In addition, when large deep beds are to be spouted the contact times between phases can be higher, which may be undesirable in an operation like solids drying.

A multiple spouted bed (MSB) can be considered for large scale operations by overcoming some of the limitations of CSB. The reported investigations related to multiple spouted beds (MSBs) are those of Peterson (1966), Foong et al. (1975), Huang and Chayang (1993), Murthy and Singh (1994), Hu et al. (2008), Chen et al. (2013, 2014). However MSBs also suffer from the drawbacks of MSBD (Murthy 1991) and limited operational regime to maintain stable spouting (Murthy and Singh 1996) at high fluid superficial velocities. We can overcome these drawbacks by providing an axially positioned draft tubes (DTs) in a MSB. DTs can act as vertical baffles thereby avoiding the hydrodynamic instability of the bed at higher gas velocities and also help in better control of gas residence times as well as solids cycle times. Some of the reported works related to solids drying in DT CSBs are

Note: D.V.R. Murthy and S. Rajashekhara have applied for an Indian Patent having the title "MULTIPLE POROUS DRAFT TUBES SPOUTED BED CONTACTOR". The application Number is 2940/CHE/2015 and dated June 11, 2015.

S. Rajashekhara (✉) · D.V.R. Murthy
Department of Chemical Engineering, National Institute of Technology Karnataka, Surathkal 575025, India
e-mail: rajmst@gmail.com

I. Regupathi et al. (eds.), *Recent Advances in Chemical Engineering*,
DOI 10.1007/978-981-10-1633-2_6

those of Ando et al. (2002), Olazar et al. (2011), Berghel and Renstrom (2014). Ishikura et al. (2003) compared the characteristics of CSBs with and without draft tubes. However in an operation like solids drying presence of a DT prevents lateral transport of gas from spout to annulus region thereby limiting the drying to occur essentially within the DT. So, it is desirable to use a porous draft tube to carry out an operation like solids drying. Considering the points discussed above a MSB having three porous draft tubes has been constructed and used in this study. Rajashekhara and Murthy (2015) reported earlier the results obtained for drying the ragi grains in a multiple draft tube spouted bed.

2 Experimental

A rectangular column (225 mm × 75 mm) having three square cells serve as a drier. Each cell is provided with a draft tube as well as an inverted frustum (apex angle = 60°) as base. Fluid inlets having sizes 08, 15 and 21 mm diameter, located at the base of each frustum, are used in the study. Transparent acrylic sheets are used as broader sides of the column for visual observations. The details of various draft tubes used are given in Table 1. It is to be noted here that the DT holes are provided in a rectangular configuration such that the number of holes for a given DT length are same in all the tubes used. A nylon mesh has been tightly wrapped over 4 mm holes on DT to prevent clogging of the holes by solids; also, the mesh prevents movement of solids between spout and annulus. Mesh is not used for 2 mm holes.

Compressed air passing through a constant pressure regulator, main rotameter, electric heater and spout line rotameters enters each spout cell. Air supplied to the main rotameter is at 28 °C and 101.32 kPa(g). Valves are provided appropriately to adjust flow rate to each cell as well as total flow rate. Humidity and temperature sensors are located for measuring air conditions at required locations. A U—tube manometer is used for measuring bed pressure drop. A schematic line diagram of the experimental rig is shown in Fig. 1. Wheat grains (major—6.32 mm/minor—3.26 mm, bulk density—785.8 kg/m^3) are used as solid material for drying studies.

Drying experiments are conducted using grains having initial moisture contents of 20 and 25 % on dry basis; air temperatures of 40, 50 and 60 °C and air flow rate of 36 m^3/h are used. The bed masses 1, 2, 3 and 4 kg used are for wet solids. However the static bed depths reported are for dry solids at ambient conditions. The flow rates, bed pressure drops and MSBD are measured at ambient conditions. Solids of required mass and water of appropriate quantity are thoroughly mixed to obtain wet solids batch needed for a given experimental run. Air flow is adjusted to the required value and it is heated to required temperature and supplied to empty dryer. Wet

Table 1 Details of draft tubes used

NPS	ID (mm)	Pitch (mm)	Length (mm)	No. of holes
¾″	21	15 × 14	380	114
1″	26	15 × 17	380	114

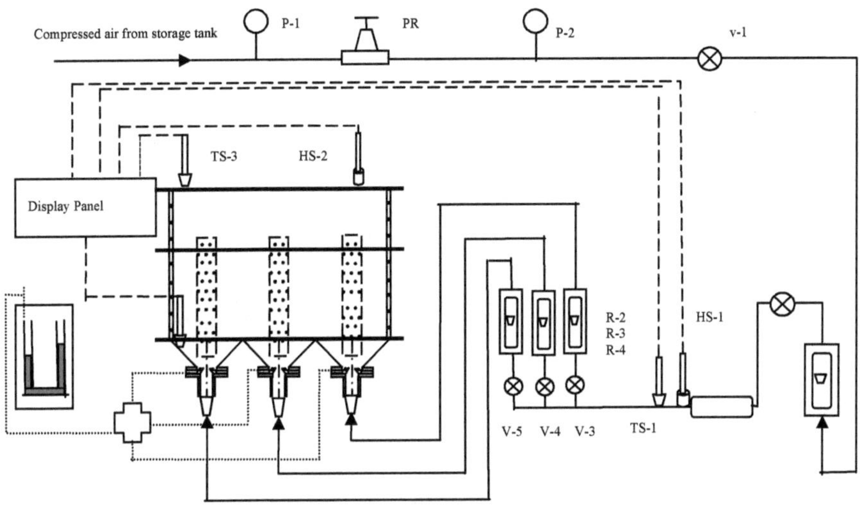

P-1,P-2 =Pressure gauge; PR=Pressure Regulator; V-1, V-2, V-3, V-4, V5, = Globe Valves; R1, R2, R3, R4, = Rotameteres; H = Heater; HS-1,HS-2 = Humidity Sensors; TS-1, TS-2, TS-3 =Temperature Sensors

Fig. 1 Schematic line diagram of experimental rig

solids batch is loaded into the column after it attains steady state. Solid samples are taken from dryer at regular time intervals and kept in desiccators and their masses are noted. These samples are kept in oven maintained at 100 °C for sufficiently long times so that they become bone dry. Using the mass differences the moisture content of each sample at a given time is found. Moisture content versus time plots are prepared for all runs; from these plots batch drying times needed to achieve a final moisture content of 10 % from given initial moisture content are obtained.

3 Results and Discussions

3.1 Influence of Operating Variables on Drying Rate

The batch drying time is influenced by the operating variables like inlet air temperature (T_a), air flow rate (v_0), initial moisture content of solids (X_0), and bed mass (w_b), as well as the design variables like fluid inlet diameter, draft tube diameter, draft tube hole diameter and hole pitch.

The drying of wheat grains has been found to be occurring under falling-rate period for all conditions used in this study. Typical rate curves are shown in Fig. 2. When the solids are amorphous and fibrous in nature the moisture is held as an integral part of the solid structure; the moisture movement through solid structure can be very slow by diffusion. The movement will be much slower when the particles are big. So these solids have not exhibited any constant rate period.

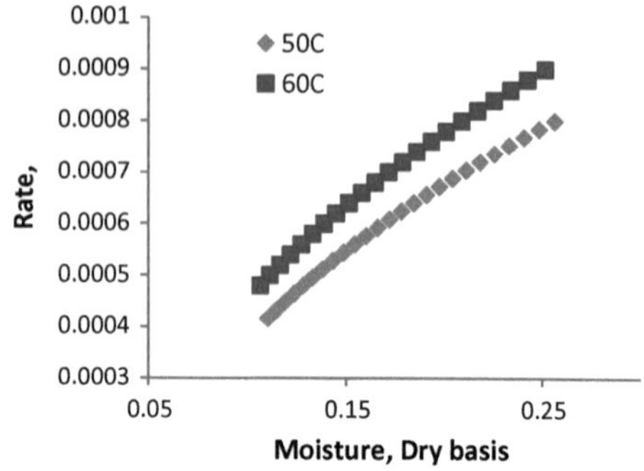

Fig. 2 Effect of inlet air temperature on rate of drying (w_0 = 4 kg, v_0 = 36 m^3/h, X_0 = 0.25 g moisture/g dry solids 26 mm nonporous draft tube)

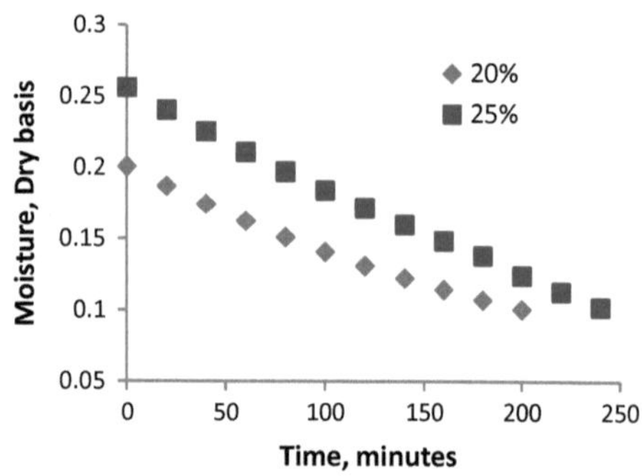

Fig. 3 Effect of initial moisture content on moisture removal (w_0 = 4 kg, v_0 = 36 m^3/h, T_a = 50 °C, 26 mm non-porous draft tube)

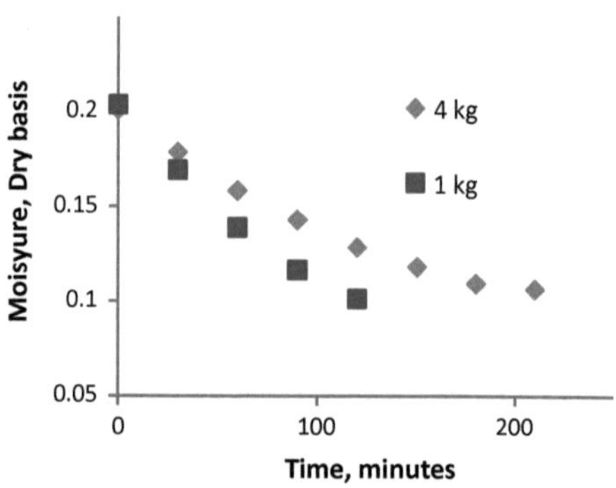

Fig. 4 Effect of bed mass on moisture removal (v_0 = 36 m^3/h, T_a = 60 °C, X_0 = 0.2, 21 mm porous draft tube with 2 mm hole)

The effect of T_a, X_0, and w_b on drying are shown in Figs. 3 and 4. The results indicate that as T_a increases the drying rate (R_d) increased; an increased T_a leads to increased driving forces for both heat and mass transfer leading to higher R_d. The total amount of moisture to be removed will be higher for solids with higher X_0 and hence an increase in X_0 leads to longer batch times to reach given final moisture content of solids. At a given v_0, the solids circulation rate will be lower when w_b is increased; this in turn leads to lower transfer rates of heat and mass, which results in lower moisture removal rate for larger bed mass.

From these results we can say that a higher T_a, lower X_0 and lower w_b would lead to faster drying rates and hence lower batch drying times. However, it should be noted here that higher T_a will increase the operating costs, lower w_b leads to more number of batches to be handled for a given total quantity of solids to be dried. The initial moisture content of fresh agricultural grains is fairly fixed.

3.2 Influence of Design Parameters on Batch Drying Time

In order to understand the influence of DT inside diameter (id) and hole size on batch drying time experiments are conducted using $w_0 = 3$ kg, $v_0 = 36$ m^3/h, $T_0 = 50$ °C, $X_0 = 0.2$ g moisture/g dry solids. Both non-porous and porous DTs are used and the details of DTs are as given in Table 1. The results given in Table 2 indicate that batch drying time for a porous DT is less when compared to a non porous DT; for a given porous DT, the batch drying time decreases with increase in hole size. In the case of non-porous DTs there will not be any lateral movement of hot air into annulus region and as a result drying occurs within the DTs only. However, porous DTs facilitate hot air supply into annular region, the supply of hot air increasing with increasing hole size, leading to better transfer rates and lower drying times. Also, it has been observed that an increased DT id helped to decrease the batch drying times. This could be due to increased available volume of drying zone within the DTs.

The influence of fluid inlet size on batch drying time can be understood from the results presented in Table 3, which indicate that the batch drying time decreases as the fluid inlet size decreases for a given bed mass. As the fluid inlet size decreases the jet velocity would increase leading to an increase in bed recirculation rate and hence reduction in batch drying time.

Table 2 Effect of draft tube parameters on batch drying time ($w_0 = 3$ kg, $v_0 = 36$ m^3/h, $T_a = 50$ °C, $X_0 = 0.2$ g moisture/g dry solids, 21 mm DT, 8 mm fluid inlet)

Hole diameter (mm)	DT inner diameter (mm)	
	21	26
	Batch drying time (min)	
0	250	230
2	220	200
4	200	190

Table 3 Influence of fluid inlet diameter on batch drying time ($v_0 = 36$ m^3/h, $T_a = 50$ °C, $X_0 = 0.2$ g moisture/g dry solids, 26 mm DT with 2 mm holes)

Bed mass (kg)	Fluid inlet diameter (mm)		
	21	15	08
	Batch drying time (min)		
1	130	120	115
4	250	240	210

Table 4 Comparison of performance between MSB without and with draft tubes (v_0 = 36 m^3/h, T_a = 50 °C, X_0 = 0.2 g moisture/g dry solids, DT id = 21 mm, hole diameter = 2 mm, hole pitch = 15 × 14 mm)

Parameters	Case 1	Case 2	Case 3	Case 4	Case 5
Static bed depth (cm)	6	6.5	15	23	31
Bed mass (kg)	1	1	2	3	4
Batch drying time (min)	105	115	170	190	210
Batch drying time per kg (min)	105	115	85	63.33	52.5
Air requirement per kg (m^3)	63	69	51	37.98	31.5

3.3 Comparison of Performance Between MSB Without and with DTs

Batch drying experiments are conducted in a MSB without DT. The static bed depth and the corresponding bed mass chosen are just below MSBD, whereas the bed depths used in MSB with DTs are much higher since no MSBD limitation exists here. The results given in Table 4 indicate much better performance of MSB with porous DTs.

Based on 1 kg solids to be dried, the batch drying time and hot air requirement for MSB with porous DTs is found to be less and both decreasing with increase in batch size. When Case 1 (without DT) and Case 5 (with DT) are compared, it has been found that 50 % reduction in both batch drying time, and hot air requirement is possible in MSB with porous draft tubes.

4 Conclusions

The following conclusions are drawn based on the results obtained in this study.

1. Provision of draft tubes in MSB help in overcoming the limitation of MSBD so that more amount of solids can be dried in the column.
2. An increase in air inlet temperature and air flow rate, decrease in initial moisture content of solids and bed mass will lead to reduction in batch drying time.
3. The batch drying time is found to be more for MSB with non-porous draft tubes.
4. The batch drying time decreases with increase in draft tube diameter and hole diameter.
5. Batch drying time decreases as the fluid inlet size decreases for a given bed mass.
6. The batch drying time and air requirement per kg of solids dried decrease with increase in batch size when porous draft tubes are used.

References

Ando, S., Maki, T., Nakagawa, Y., Namiki, N., Emi, H., Otani, Y.: Analysis of the drying process of seed particles in a spouted bed with a draft tube. Adv. Powder Technol. **13**(1), 73–91 (2002)

Berghel, J., Renström, R.: An experimental study on the influence of using a draft tube in a continuous spouted bed dryer. Dry. Technol. **32**(5), 519–527 (2014)

Chen, C., Lim, J., Grace, J.R.: Stability of slot rectangular spouted beds with multiple slots. The Can. J. Chem. Eng. **91**, 1768–1775 (2013)

Chen, D., Liu, X., Zhong, W., Shao, Y., Jin, B.: Interactions of spout jets in a multiple spouted Bed. Can. J. Chem. Eng. **92**, 1150–1159 (2014)

Foong, S.K., Barton, R.K., Ratcliffe, J.S.: Characteristics of multiple spouted beds. Mech. Chem. Eng. Trans. Instn. Engrs. Australia. **11**, 7–12 (1975)

Hu, G., Gong, X., Wei, B., Li, Y.: Flow patterns and transitions of a novel annular spouted bed with multiple air nozzles. Ind. Eng. Chem. Res. **47**, 9759–9766 (2008)

Huang, C.C., Chyang, C.S.: Multiple spouts with two dimensional bed with a perforated plate distributor. J. Chem. Eng. Japan. **26**, 607–614 (1993)

Ishikura, T., Nagashima, H., Ide, M.: Hydrodynamics of a spouted bed with a porous draft tube containing a small amount of finer particles. Powder Technol. **131**, 56–65 (2003)

Mathur, K.B., Gishler, P.E.: A technique for contacting gases with coarse solid particles, A.I.Ch.E. J. **1**(2), 157–164 (1955)

Murthy, D.V.R.: Hydrodynamic studies in gas–solid and liquid–solid multiple spouted beds. Ph.D Thesis, Mangalore university, Karnataka, India (1991)

Murthy, D.V.R., Singh, P.N.: Minimum spouting velocity in multiple spouted beds. Can. J. Chem. Eng. **72**, 235–239 (1994)

Murthy, D.V.R., Singh, P.N.: Minimum spouting velocity in multiple spouted beds, Chapter 32. In: Cheremisinoff, N.P. (ed.) Mixed-Flow Hydrodynamics: Advances in Engineering Fluid Mechanics Series, pp. 741–758. Gulf publishing co., USA (1996)

Olazar, M., Lopez, G., Altzibar, H., Bilbao, J.: Modelling batch drying of sand in a draft-tube conical spouted bed. Chem. Eng. Res. Des. **89**(10), 2054–2062 (2011)

Peterson, W.: US 3242586, Nat. Res. Council, invs. (1966)

Rajashekhara, S., Murthy, D.V.R.: Grain drying in a multiple draft tube spouted bed—A preliminary study. In: Proceedings of the 8th Asia-Pacific Drying Conference, Kuala Lumpur, Malaysia, 10–12 August, 2015, pp. 80–84 (2015). ISBN 978-967-11279-2-6

Studies on Drying Characteristics of Rhizomes of Ginger (*Zingiber officinale Rosc*)

J. Prakash Maran, K. Swathi, J. Jayalakshmi, A. Sangeetha, J. Jony Blessing Manoj, N. Sivarajasekar, B. Priya, Naif Abdullah Al-Dhabi and K. Ponmurugan

1 Introduction

India is one of the leading producer of ginger (*Zingiber officinale Rosc*) (Family: *Zingiberaceae*) in the world, with an estimated production of 7,56,000 tons in 2012–13 against the global production of about 1.6 million tons. About 1,55,000 ha of land is under ginger cultivation in India with an average productivity of 1,514 kg per hectare. India accounts for about 21 % of the total ginger production, followed by China (20.5 %), Indonesia (12.7 %), Nepal (11.5 %) and Thailand (10 %). After harvesting, ginger rhizomes are cured and then dried to stabilize rhizomes to avoid degradation during storage and transportation. For the preparation of dried ginger, the produce (harvested after 8 months) is soaked in water for 6–7 h and then dried using sun drying method. Drying is the oldest preservation technique used for processing the special food and ingredients to enhance its shelf-life. There are numerous methods of drying among which microwave drying is an emerging alternative method and it retains the product quality (Bal et al. 2010). In recent

J. Prakash Maran (✉) · K. Swathi · J. Jayalakshmi · A. Sangeetha · J. Jony Blessing Manoj
Department of Food Technology, Kongu Engineering College, Perundurai, Erode 638052, Tamil Nadu, India
e-mail: prakashmaran@gmail.com

N. Sivarajasekar
Department of Chemical Engineering, Kongu Engineering College, Erode 638052, TN, India

B. Priya
Department of Food Process Engineering, SRM University, Chennai 603203, TN, India

N.A. Al-Dhabi · K. Ponmurugan
Department of Botany and Microbiology, Addiriyah Chair for Environmental Studies, College of Science, King Saud University, Riyadh 11451, Kingdom of Saudi Arabia

I. Regupathi et al. (eds.), *Recent Advances in Chemical Engineering*,
DOI 10.1007/978-981-10-1633-2_7

times, applications of microwave drying (MW) were increased abundantly because of its advantages such as elevated drying rate, lesser drying time, fast and consistent heating, reduced energy consumption and improved quality of the dried products (Ozkan et al. 2007; Paengkanya et al. 2015; Fennell and Boldor 2014; Pickles et al. 2014). The literature analysis revealed that, most of the researchers studied about the drying characteristics of ginger using conventional drying techniques such as hot air, solar, rotary and thin layer drying methods. The microwave drying of ginger has not been investigated till the date. Hence, the study has been aimed to evaluate the efficacy of microwave drying characteristics of ginger (soaked and unsoaked).

2 Materials and Methods

2.1 Materials

Fresh ginger (*Zingiber officinale Rosc*) was procured from local vendors near Perundurai, Tamil Nadu, India and washed in running tap water in order to make samples free from foreign materials. Surface moisture of the samples was eliminated by blotting with hankie paper and was stored in a refrigerator at 4–7 °C prior to the drying. Gravimetric method was used to estimate the moisture content (initial) of ginger.

2.2 Microwave Drying of Ginger

Domestic microwave oven (LG MH-3948 WB) was used to study the drying characteristics of ginger. Both microwave power and emission time could be controlled by the control terminal in the microwave oven. The dimensions of the microwave cavity are 327 × 370 × 207 mm. The rotating disc of oven was used to reduce the level of the reflected microwaves from the magnetrons in order to obtain uniform heating. The experiment was performed using fresh and soaked rhizomes of known weight of ginger at different microwave power levels (160, 320, 480, 640 and 800 W).A set of samples were pretreated by soaking overnight (12 h) in normal water and increase in weight was noted. Another set of samples were placed as a distinct layer on the centre of the rotating glass plate of the oven. Weight loss of the samples was examined at the interval of 10 s using a weighing balance until no evident in weight change was observed. Equilibrium moisture content was considered while the sample attained the constant weight. Each experiment was repeated thrice at different microwave power level (160–800 W at the interval of 160 W).

2.3 *Drying Characteristics of Ginger*

The moisture content and drying rate of ginger were determined by the method described by Darvishi et al. (2013). The effective moisture diffusivity, energy consumption and microwave drying efficiency were determined based on the method illustrated by Pillai (2013).

2.4 *Mathematical Modelling and Statistical Analysis*

Experimental data obtained from microwave drying at various power levels were fitted to various models (Table 1) and the constant in the models were derived by non-linear regression analysis using MATLAB software.

The integrity of fit for each model was assessed based on coefficient of determination (R^2), root mean square error (RMSE) and chi square (χ^2) by evaluating the predicted values with observed values.

3 Results and Discussion

3.1 *Drying Characteristics of Ginger*

The value of moisture ratio reduces quickly and then gradually decreases with increase in drying time because of the initial moisture content of the sample during the initial stage of drying process (Fig. 1). Due to higher moisture diffusion, the absorption of MW output power was better at higher power level (Barba et al. 2013). As the process proceeds, the absorption of MW output power decreased due to the loss of moisture in the product resulting in a falling rate (Dadali et al. 2007). The value of moisture ratio of soaked sample is high compared to the sample without soaked as soaking enhances the movement of moisture from internal to external region of ginger at higher rate (Lee and Kim 2013).

The drying rate increased simultaneously as the power increases due to higher moisture movement from inner region and evaporation at the surface of product due to higher difference in partial vapour pressure between the sample and the

Table 1 Different models for thin layer drying

Model name	Equation
Page model	$MR = \exp(-kt^n)$
Modified page—I	$MR = \exp(-(kt)^y)$
Wang and Singh	$MR = 1 + at + bt^2$
Simplified Fick's diffusion	$MR = a\exp(-c(t/L^2))$
Parabolic	$MR = c + bt + at^2$

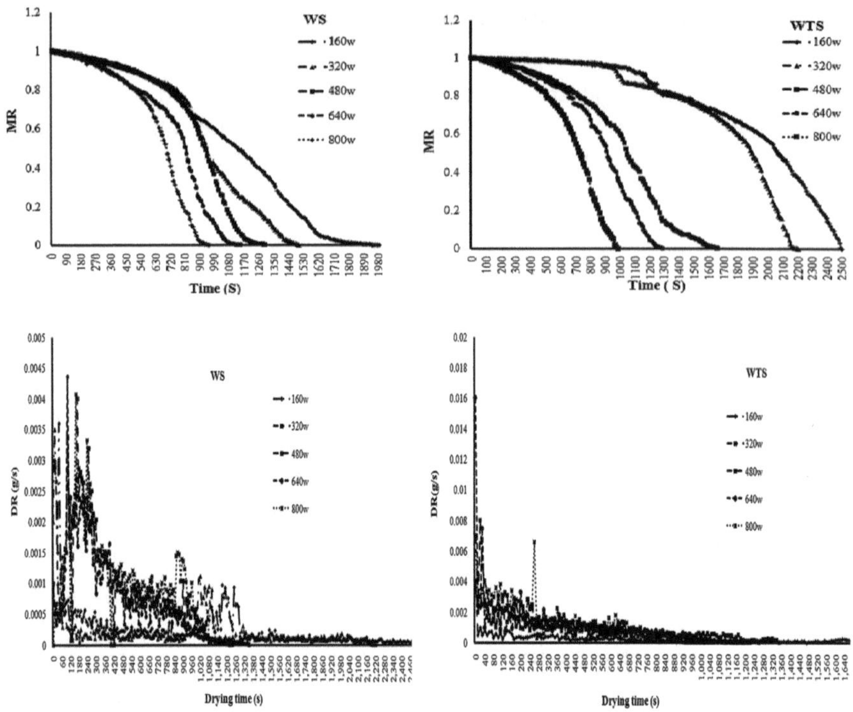

Fig. 1 Influence of MW power levels on moisture ratio and drying rate

surrounding (Botha et al. 2012a, b).So drying time reduces by 60 % for higher power levels. The drying rate value is more for soaked sample because during soaking there is increase in the moisture content by 2–3 % which is free water and hence upon drying the amount of moisture evaporated is fast which had direct influence on drying rate and so drying time reduces by 20 % upon soaking.

The moisture diffusivity values (D_{eff}) were calculated from the slope of the linear regression of ln (MR) and time. Decrease in power increases the diffusion of water from the inner regions of the Ginger was confirmed by the moisture diffusivity values (Botha et al. 2012a). The cause for the increase in diffusion of water with increased power level was the partial vapour pressure difference between the surroundings and sample (Walde et al. 2002). Pre-treated ginger also had higher D_{eff} values.

The energy consumption is low at lower power level and as power level increases from 180 to 800 W the energy consumption value rose from 3.154 to 8.845 MJ/kg (Fig. 2) and this occurs due to high power level and in case of soaking sample show less energy as high initial moisture content results in higher heat absorption of microwave energy (Sarimeseli 2011). The energy consumption values were plotted against Microwave power level.

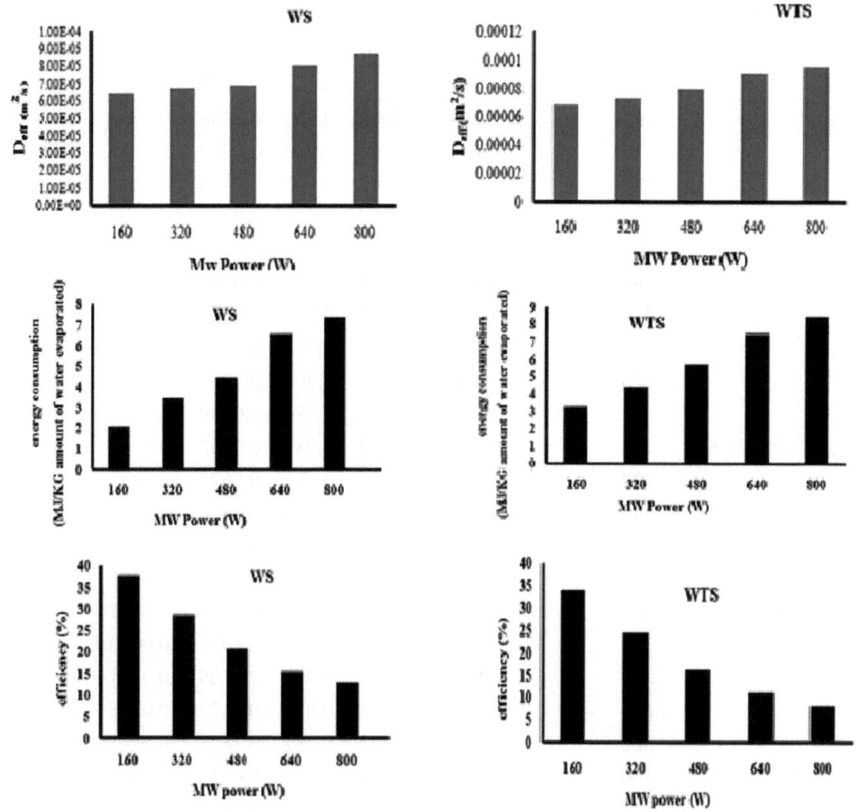

Fig. 2 Effect of process variables on effective moisture diffusivity, energy consumption and drying efficiency

Since drying efficiency is based on energy consumption value, the efficiency is more at lower power and as power increases drying efficiency value decreases. The soaking of sample in water increases the efficiency of drying and also enhanced the drying rate.

3.2 Mathematical Modelling of Drying Process

The obtained moisture content for both soaked and without soaked ginger at the different microwave power levels were used to calculate moisture ratio and the same were fitted to the thin layer drying models as shown in Table 1. The empirical drying constant for dried ginger at each microwave power level is given in Table 2. Although the R^2 value for the models was greater than 0.90 for all microwave

Table 2 Fitting of various different thin layer models and its statistical analysis

Model name	Microwave power	WS			WTS		
		RMSE	χ^2	R^2	RMSE	χ^2	R^2
Page	160	0.0358	0.0002	0.920	0.0358	0.0002	0.994
	320	0.0388	0.0003	0.979	0.0860	0.0075	0.979
	480	0.6508	0.0003	0.962	0.6508	0.0005	0.962
	640	0.0397	0.0008	0.953	0.0397	0.0008	0.993
	800	0.0715	0.0052	0.947	0.0715	0.0052	0.957
Modified page	**160**	**0.0258**	**0.0001**	**0.999**	**0.0218**	**0.000102**	**0.999**
	320	0.0231	0.0002	0.998	0.0631	0.0040	0.993
	480	0.5396	0.0004	0.977	0.5396	0.0004	0.977
	640	0.0269	0.0007	0.994	0.0269	0.0007	0.994
	800	0.0439	0.0019	0.986	0.0439	0.0019	0.986
Wang and Singh	160	0.0362	0.0003	0.989	0.0362	0.0013	0.989
	320	0.0675	0.0046	0.977	0.0675	0.0046	0.977
	480	0.0464	0.0003	0.979	0.0464	0.0043	0.979
	640	0.0539	0.0029	0.975	0.0539	0.0029	0.975
	800	0.0507	0.0026	0.977	0.0507	0.0026	0.977
Simplified ficks diffusion	160	0.1327	0.0005	0.856	0.1200	0.0007	0.884
	320	0.1896	0.0054	0.749	0.1916	0.0010	0.749
	480	2.1456	0.0167	0.764	2.1305	0.0174	0.844
	640	0.1545	0.0278	0.874	0.1522	0.0236	0.812
	800	0.1845	0.0345	0.787	0.1828	0.0343	0.702
Parabolic	160	0.0356	0.0008	0.995	0.0310	0.0009	0.991
	320	0.0654	0.0045	0.953	0.0669	0.0045	0.977
	480	0.2788	0.0027	0.967	0.2894	0.0023	0.939
	640	0.0534	0.0025	0.956	0.0.053	0.0028	0.976
	800	0.0467	0.0027	0.989	0.0455	0.0021	0.981

power levels, but modified page model is found to be have highest R^2 values with lowest χ^2 and RMSE. Therefore modified page model have the ability to predict the MR for microwave dried Indian ginger at various microwave power levels.

4 Conclusion

The drying characteristics of ginger such as moisture ratio, drying rate, effective moisture diffusivity, energy consumption and drying efficiency were investigated in this study using different microwave power levels for both raw and soaked ginger samples. The value of moisture ratio reduces quickly and then gradually decreases with increase in drying time. The drying rate was also increased concurrently as the power increases due to higher moisture migration from inner region and evaporation at product surface because of greater difference in partial vapour pressure between ginger and surrounding. Decrease in power increases the diffusion of water from the inner region of the Ginger. The energy consumption is low at lower power level and as power level increases from 180 to 800 W, the energy consumption values also raised. The drying efficiency is more at lower power and as power increases drying efficiency value decreases. The modified page model was fitted very well with the experimental data with highest R_2 and lowest χ^2 and RMSE values for both soaked and unsoaked ginger samples.

References

Bal, L.M., Kar, A., Satya, S., Naik, S.N.: Drying kinetics and effective moisture diffusivity of bamboo shoot slices undergoing microwave drying. Int. J. Food Sci. Technol. **45**, 2321–2328 (2010)

Barba, A.A., Dalmoro, A., D'Amore, M.: Microwave assisted drying of cellulose derivative (HPMC) granular solids. Powder Technol. **237**, 581–585 (2013)

Botha, G.E., Oliveira, J.C., Ahrné, L.: Quality optimisation of combined osmotic dehydration and microwave assisted air drying of pineapple using constant power emission. Food Bioprod. Process. **90**, 171–179 (2012a)

Botha, G.E., Oliveira, J.C., Ahrné, L.: Microwave assisted air drying of osmotically treated pineapple with variable power programmes. J. Food Eng. **108**, 304–311 (2012b)

Dadali, G., Apar, D.K., Ozbek, B.: Microwave drying kinetics of okra. Drying Technol. **25**, 917–924 (2007)

Darvishi, H., Azadbakht, M., Rezaeiasl, A., Farhang, A.: Drying characteristics of sardine fish dried with microwave heating. J. Saudi Soc. Agric. Sci. **12**, 121–127 (2013)

Fennell, L.P., Boldor, D.: Continuous microwave drying of sweet sorghum bagasse biomass. Biomass Bioenergy **70**, 542–552 (2014)

Lee, J.-Y., Kim, J.-H.: Microwave-assisted drying of paclitaxel for removal of residual solvents. Process Biochem. **48**, 545–550 (2013)

Ozkan, A.I., Akbudak, B., Akbudak, N.: Microwave drying characteristics of spinach. J. Food Eng. **78**, 577–583 (2007)

Paengkanya, S., Soponronnarit, S., Nathakaranakule, A.: Application of microwaves for drying of durian chips. Food Bioprod. Process. **96**, 1–11 (2015)

Pickles, C.A., Gao, F., Kelebek, S.: Microwave drying of a low-rank sub-bituminous coal. Miner. Eng. **62**, 31–42 (2014)

Pillai, M.G.: Thin layer drying kinetics, characteristics and modeling of plaster of paris. J. Chem. Eng. Res. Des. **91**, 1018–1027 (2013)

Sarimeseli, A.: Microwave drying characteristics of coriander (Coriandrum sativum L.) leaves. Energy Convers. Manag. **52**, 1449–1453 (2011)

Walde, S.G., Balaswamy, K., Velu, V., Rao, D.G.: Microwave Drying and grinding characteristics of wheat (*Triticumaestivum*). J. Food Eng. **55**, 271–276 (2002)

Cationic Dyes Sequestration from Aqueous Phase Using Biosurfactant Based Reverse Micelles

N. Sivarajasekar, S. Ramasubbu, J. Prakash Maran and B. Priya

1 Introduction

Dyes are one of the organic pollutant and well known for their adverse effects on the aquatic and human life (Sivarajasekar and Baskar 2014a). Therefore, the dye bearing effluent must be carefully treated in order to ensure the environmental health (Sivarajasekar and Baskar 2014b, c). Economical efficient and easily scaled-up methods those can sequestrate, concentrate, and refine dye contaminated wastewater continuously are seeking attention now (Sivarajasekar et al. 2016; Sivarajasekar and Baskar 2015a, b). One such novel separation method is reverse micelles extraction due to its skill to extract a targeted pollutant selectively (Jarudilokkul et al. 1999). Reverse micellar extraction is governed by factors, such as, dye concentration, surfactant concentration and carrier concentration. Reverse micellar extraction is more effective if the chemical surfactants used are replaced by biosurfactants. Surfactants produced via microbial route require only trivial production conditions. Yet, their surface activity is higher and compatible to environment comparatively than chemical surfactants (Hasmann et al. 2007).

N. Sivarajasekar (✉) · S. Ramasubbu
Department of Chemical Engineering, Kongu Engineering College,
Erode 638052, TN, India
e-mail: sivarajasekar@gmail.com

J. Prakash Maran
Department of Food Technology, Kongu Engineering College,
Erode 638052, TN, India

B. Priya
Department of Food Process Engineering, SRM University,
Chennai 603203, TN, India

I. Regupathi et al. (eds.), *Recent Advances in Chemical Engineering*,
DOI 10.1007/978-981-10-1633-2_8

Response Surface Methodology (RSM) is an efficient device for optimizing any process where numerous factors and their interactions considerably influence the response. RSM is a semi-empirical statistical approach that avails quantifiable data from an appropriate experimental design. Box-Behnken Design (BBD) is a widely used RSM which solves a second order multivariate equation by a least squares procedure (Sivarajasekar and Baskar 2014b; Montgomery 2001).

Objectives of this work were to assess *Pseudozyma Antarctica* produced bio-surfactant for reverse micellar extraction of cationic dyes. To model the combined effects of the cationic dye concentration, bio-surfactant concentration, and sulphuric acid concentration over the extraction efficiency of dyes by using response surface methodology.

2 Materials and Methods

2.1 Reagents and Biosurfactant Preparation

Basic red 9 (BR9, $C_{19}H_{18}ClN_3$, 323.82 mol, CI: 42500, λ_{max}: 550 nm), basic violet 3 ($C_{25}H_{30}ClN_3$, 407.98 mol, CI: 42555, λ_{max}: 591 nm), sulfuric acid, amyl alcohol, sodium nitrate, magnesium sulphate, mono sodium phosphate and other chemicals used are purchased from Merk, India Ltd., and of analytical grade. Deionised water was used for solution preparations.

Pseudozyma antarctica was grown in the soybean oil medium (carbon source: soybean oil 80 g l^{-1}, $NaNO_3$ 2.5 g l^{-1}, $MgSO_4.7H_2O$ 0.2 g l^{-1}, KH_2PO_4 0.2 g l^{-1}, yeast extract 1 g l^{-1}) as per the procedure given by the Adamczak and Bednarski (2000) and the glycolipid (biosurfactant) was extracted from foam phase. Properties of the biosurfactant such as surface tension and critical micelle concentration (Desai and Banat 1997) were determined. Those are found to be 37 mN m^{-1} and 2.7×10^{-6} M respectively.

2.2 Extraction Experiments

The organic phase was prepared in a 100 ml beaker by adding amyl alcohol and biosurfactant. The mixture was homogenized for 5 min and H_2SO_4 drops were injected into the mixture with help of a syringe. The resultant milky white suspension was transferred into another 250 ml beaker holding dye solution. The entire mix was stirred carefully using a magnetic stirrer. The samples were drawn on a consistent time interval and the aqueous phase was separated by using gravity settling. The obtained aqueous dye solution was filtered and analysed using

Table 1 BBD matrix for extraction efficiency of BR9 and BV3

X_1	X_2	X_3	%E BR9		%E BV3	
			Expt.	Pred.	Expt.	Pred.
1.00 (−1)	0.10(−1)	10.00(0)	77.51	76.27	74.82	73.61
3.00(+1)	0.10(−1)	10.00(0)	65.52	64.47	62.79	61.77
1.00(−1)	1.00(+1)	10.00(0)	79.31	78.04	76.53	75.29
3.00(+1)	1.00(+1)	10.00(0)	70.48	69.35	67.76	66.66
1.00(−1)	0.55(0)	5.00(−1)	54.82	53.94	52.11	51.27
3.00(+1)	0.55(0)	5.00(−1)	49.74	48.94	47.03	46.27
1.00(−1)	0.55(0)	15.00(+1)	87.29	85.89	84.58	83.21
3.00(+1)	0.55(0)	15.00(+1)	74.86	73.66	72.15	70.98
2.00(0)	0.10(−1)	5.00(−1)	48.52	47.74	45.81	45.07
2.00(0)	1.00(+1)	5.00(−1)	52.72	51.88	50.01	49.20
2.00(0)	0.10(−1)	15.00(+1)	77.49	76.25	74.78	73.57
2.00(0)	1.00(+1)	15.00(1)	82.12	80.81	79.41	78.12
2.00(0)	0.55(0)	10.00(0)	67.48	66.40	64.77	63.72
2.00(0)	0.55(0)	10.00(0)	66.46	65.40	63.75	62.72
2.00(0)	0.55(0)	10.00(0)	67.47	66.39	64.76	63.71
2.00(0)	0.55(0)	10.00(0)	67.48	66.40	64.77	63.72
2.00(0)	0.55(0)	10.00(0)	67.47	66.39	64.76	63.71

UV-Visible spectrophotometer. The extraction efficiency is taken as response and was measured by Eq. (1):

$$\text{Extraction efficiency} = \frac{C_0 - C_i}{C_0} \times 100 \qquad (1)$$

where C_0 is the initial aqueous phase dye concentration and C_i is the final aqueous phase dye concentration.

Design-Expert 7.1.3 software was utilised for statistical analysis and modelling of experimental data. Three factors namely, concentration of surfactant (1–3 %, X_1), concentration of sulphuric acid (0.1–1.0 N, X_2) and concentration of dye (5–15 mg l^{-1}, X_3) were considered to analyse the extraction efficiency of selected dye from aqueous solution. Response function coefficients were estimated using orthogonal 2^3 BBD of five replicates at the centre point. The results for this design are shown in Table 1. Experiments were conducted in random order to reduce error. The regression analysis was executed to evaluate the response function. The following quadratic model (Eq. (2)) can predict the extraction efficiency (%E):

$$\%E = \beta_0 + \sum_{i=1}^{k} \beta_i X_i + \sum_{j=1}^{k} \beta_i X_i^2 + \sum_{i<j}^{k} \beta_{ij} X_{ij} + \cdots \qquad (2)$$

where %E is the predicted response (Extraction efficiency), β_i, β_j, and β_{ij} are the coefficients from regression analysis, k is the number of factors, X_j and X_j represent the uncoded value of the ith (linear coefficient), and jth (quadratic coefficient) parameter respectively (Sivarajasekar and Baskar 2014b).

2.3 *Results and Discussions*

2.3.1 Statistical Analysis

BBD design matrix for each dye is given in Table 2 along with the experimental and the predicted extraction efficiency. The batch extraction experiments were conducted in order to visualize the combined effects of selected independent factors on the response. Multivariate regression analysis of the empirical data generated the following quadratic equations:

Table 2 ANOVA for response quadratic model

Source	Sum of squares		Degree of freedom		F Value		p Value	
	BR9	BV3	BR9	BV3	BR9	BV3	BR9	BV3
Model	2059.98	2058.81	9.00	9.00	571.10	565.60	<0.0001	<0.0001
X_1	183.65	183.46	1.00	1.00	458.23	453.60	<0.0001	<0.0001
X_2	30.38	30.07	1.00	1.00	75.80	74.35	<0.0001	<0.0001
X_3	1680.84	1680.84	1.00	1.00	4193.93	4155.86	<0.0001	<0.0001
X_1X_2	2.50	2.66	1.00	1.00	6.23	6.57	0.0412	0.0374
X_1X_3	13.51	13.51	1.00	1.00	33.70	33.39	0.0007	0.0007
X_2X_3	0.05	0.05	1.00	1.00	0.12	0.11	0.7441	0.7452
X_1^2	57.61	57.30	1.00	1.00	143.75	141.67	<0.0001	<0.0001
X_2^2	21.01	20.83	1.00	1.00	52.43	51.49	0.0002	0.0002
X_3^2	77.62	77.26	1.00	1.00	193.67	191.02	<0.0001	<0.0001
Residual	2.81	2.83	7.00	7.00				
Lack of Fit	1.98	2.01	3.00	3.00	3.20	3.25	0.1450	0.1426
Pure error	0.82	0.82	4.00	4.00				
Cor total	2062.79	2061.64	16.00	16.00				
R^2	0.9986	0.9942						
R^2_{adj}	0.9969	0.9958						
R^2_{pre}	0.9840	0.981						
Adequate precision	80.17	79.7						

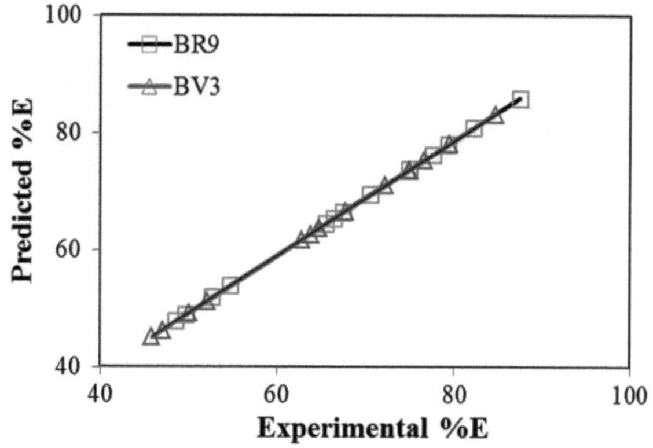

Fig. 1 Predicted extraction efficiency versus actual values

$$\%E(BR9) = 67.27 - 4.79 \times X_1 + 1.95 \times X_2 + 14.50^{*}C + 0.79 \times X_1 \times X_2 - 1.84 \times X_1 \times X_3 + 0.11 \times X_2 \times X_3 + 3.70 \times X_1^2 + 2.23 \times X_2^2 - 4.29 \times X_3^2 \quad (3)$$

$$\%E(BV3) = 64.56 - 4.79 \times X_1 + 1.94 \times X_2 + 14.50 \times X_3 + 0.81 \times X_1 \times X_2 - 1.84 \times X_1 \times X_3 + 0.11 \times X_2 \times X_3 + 3.69 \times X_1^2 + 2.22 \times X_2^2 - 4.28 \times X_3^2 \quad (4)$$

The extraction efficiency predicted from the Eqs. (3) and (4) were well correlated with the experimental data as shown in Fig. 1. Therefore, these quadratic models are quite fit for this experimental set up. Nevertheless, the connotation and the aptness of these models were confirmed by using (Analysis of Variance) ANOVA test. ANOVA results are presented in Table 2.

From Table 2, values of p less than 0.05 imply that the model coefficients are noteworthy. In the case of BR9, parameters (X_1, X_2, X_3) and their interactions (X_1X_2, X_1X_3, X_1^2, X_2^2, X_3^2) were significant. Whereas, X_1, X_2, X_3, X_1X_2, X_1X_3, X_1^2, X_2^2 and X_3^2 were significant terms in BV3 extraction. The model F value (BR9: 571.10, BV3: 565.60) and values of p less than 0.0001 indicate that the predicted quadratic models are significant for these dyes. The regression coefficient (BR9: 0.9986, BV3: 0.9946) is nearer to one shows that these relationships are reasonable in predicting the model coefficients for the current extraction systems. The predicted R^2 (BR9: 0.9840, BV3: 0.9821) and adjusted R^2 (BR9: 0.9969, BV3: 0.9958) are in sensible adjustment with each other. Greater than 4 of adequate precision values specify an adequate signal. Hence, the predicted quadratic models could navigate the design space to understand the factor interactions.

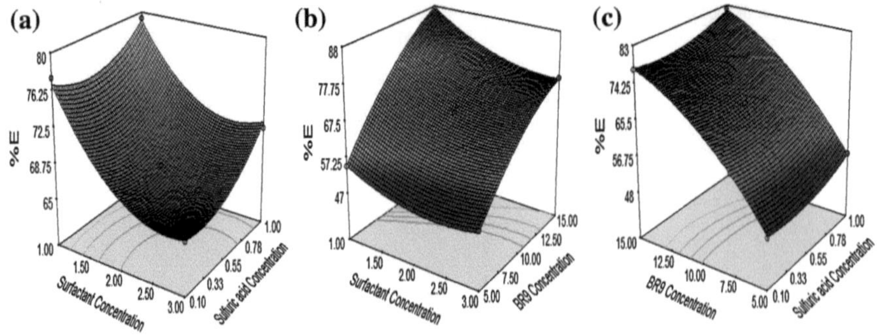

Fig. 2 Effect of process parameters of BR9 extraction

2.4 Significant Interactions of Parameters

Contour and 3D plots were produced to get the clear picture about the interactions of the independent parameters and optimum conditions for the dyes extraction. The nonlinear nature of the contours suggests that the interactions between the parameters are significant and the optimum parameters cannot be found simply. Each 3D plot characterizes the combinations of the two selected parameters and the shaded surface indicates the maximum extraction efficiency.

Figure 2. illustrates the response surface plots of BR9 extraction by amyl alcohol. The nonlinear contours evidenced significant parameters interactions. Extraction efficiency of BR9 increases when the concentration of surfactant increased gradually up to a level (Fig. 2a, b). After that value, the extraction efficiency reduces slowly due to reduction in the driving force (i.e., the concentration gradient of BR9 between organic phase and aqueous phase). Increase in sulphuric acid concentration from lower level to upper level gradually increases the extraction efficiency (Fig. 2a, c). With the increase in BR9 concentration in the aqueous phase, the driving force for dye transport to the organic phase increases (Fig. 2b, c). This leads to the sharp decline of BR9 concentration of the aqueous phase. Similar trends (Fig. 3) were noticed in the BV3 extraction as in the BR9 extraction. However, the extraction efficiency of BV3 is less than the BR9 comparatively. The molecular size is proportional to the molecular weight, hence, the heavy and bigger BV3 molecules were transported slowly than the small BR9 molecules. Thus for the same parameter ranges BV3 extraction efficiency were smaller than that of the BV3 extraction efficiency.

2.5 Optimization Using Derringer's Desirability Function

Process parameters were optimized using Derringer's desirability function (Sivarajasekar and Baskar 2014b). This function optimizes the response at selected

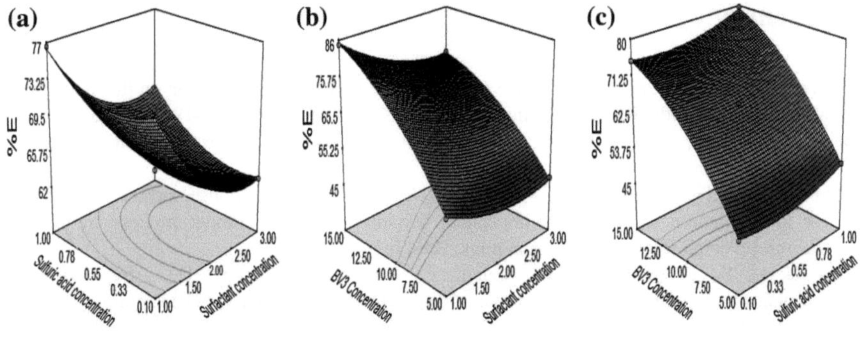

Fig. 3 Effect of process parameters of BR9 extraction

parameter level by satisfying the necessities in the BBD design. A minimum surfactant concentration, maximum dye concentration, and minimum sulfuric acid concentration were fixed to optimize maximum extraction efficiency. By seeking from 30 starting points in the optimum results, the best local maximum for BR9 extraction was found to be at surfactant concentration 1 %, BR9 concentration 15 mg l^{-1}, sulfuric acid concentration 0.42 mg l^{-1}, extraction efficiency 87.62 % at 0.998 desirability. In the case of BV3, the optimum values were found to be at surfactant concentration 1 %, BV3 concentration 15 mg l^{-1}, sulfuric acid concentration 0.84 mg l^{-1}, extraction efficiency 87.23 % at 0.996 desirability. Validating confirmatory experiments were done at the optimized parameters and the extraction efficiency was found to be 87.54 % for BR9 and 87.12 % for BV3 respectively. The obtained values were associated to the calculated data, indicating BBD joined with desirability function could be efficiently employed to optimize the batch extraction process parameters of the selected dyes.

3 Conclusions

Reverse micellar extraction technique was employed to extract cationic dyes from its aqueous phase. Glycolipids produced naturally using *Pseudozyma antarctica* from soybean oil medium were used as surfactant. Batch experiments were carried out using two dyes, namely Basic red 9 (BR9) and basic violet 3 (BV3). Extraction efficiency of single component system of each dye was investigated by using amyl alcohol as solvent and sulphuric acid. The process parameters such as dye concentration, sulphuric acid concentration and bio-surfactant concentration were optimized using BBD. The effects of parameters were studied by using 3D response surface plots. The optimum conditions were calculated by using Derringer's desirably.

References

Adamczak, M., Bednarski, W.: Influence of medium composition and aeration on the synthesis of biosurfactants produced by Candida Antarctica. Biotechnol. Lett. **22**, 313–316 (2000)

Desai, J.D., Banat, I.M.: Microbial production of surfactants and their commercial potential. Microbiol. Mol. Biol. Rev. **61**, 47–64 (1997)

Hasmann, F.A., Gurpilhares, D.B., Roberto, I.C., Pessoa Jr., A.: Response surface methodology for the evaluation of glucose-6-phosphate dehydrogenase enrichment process by soybean lecithin reversed micelles. J. Chromatogr. B **847**, 262–266 (2007)

Jarudilokkul, S., Poppenborg, L.H., Valetti, F., Gilardi, G., Stuckey, D.C.: Separation and purification of periplasmic cytochrome c553 using reversed micelles. Biotechnol. Tech. **13**, 159–163 (1999)

Maran, J.P.: Statistical optimization of aqueous extraction of pectin from waste durian rinds. Int. J. Biol. Macromol. **73**, 92–98 (2015)

Montgomery, D.C.: Design and analysis of experiments, 5th edn. Wiley, New York (2001)

Sivarajasekar, N., Baskar, R.: Adsorption of basic magenta II onto H_2SO_4 activated immature Gossypium hirsutum seeds: kinetics, isotherms, mass transfer, thermodynamics and process design. Arab. J. Chem. accepted manuscript, (2014a). doi:10.1016/j.arabjc.2014.10.040

Sivarajasekar, N., Baskar, R.: Adsorption of basic red 9 on activated waste Gossypium hirsutum seeds: process modeling, analysis and optimization using statistical design. J. Ind. Eng. Chem. **20**, 2699–2709 (2014b)

Sivarajasekar, N., Baskar, R.: Adsorption of basic red 9 onto activated carbon derived from immature cotton seeds: isotherm studies and error analysis. Desalin Water Treat **52**, 7743–7765 (2014c)

Sivarajasekar, N., Baskar, R.: Biosorption of basic violet 10 onto activated Gossypium hirsutum seeds: Batch and fixed-bed column studies. Chinese J. Chem. Eng. **23**, 1610–1619 (2015a)

Sivarajasekar, N., Baskar, R.: Agriculture waste biomass valorisation for cationic dyes sequestration: a concise review. J. Chem. Pharm. Res. **7**(9), 737–748 (2015b)

Sivarajasekar, N., Baskar, R., Ragu, T., Sarika.K., Preethi. N., Rathika. T.: Appl. Water Sci. accepted manuscript, (2016). doi:10.1007/s13201-016-0379-2

Screening of Solvents for Deacidification of Sardine Oil

Chandrasekar Vaisali, Belur D. Prasanna and I. Regupathi

1 Introduction

Marine oils are considered to have an effective role in human health and nutrition, due to the presence of n-3 polyunsaturated fatty acids (n-3 PUFA) (Rubio-Rodríguez et al. 2010). Hence, replacing vegetable oils with fish oils has been given increasing importance. However, crude fish oil contains many impurities like phospholipids, free fatty acids, metal ions, pigments and oxidation products (Vaisali et al. 2015) that reduce oil quality. Hence, in order to meet the safety standards refining treatments have to be employed. The various stages in edible oil refining and the different methods under each stage has been critically reviewed by Vaisali et al. (2015). Of all the stages, removal of free fatty acids (FFA) is a crucial step due to the role of FFA in further oxidation and its economic impact on oil production. As fish oils are said to have higher FFA content than vegetable oils, this step is crucial in sardine oil refining. Deacidification of oils is usually performed by chemical, physical and miscella methods. Though physical refining combines deacidification with deodorisation, it cannot be applied to fish oils, as it leads to formation of side products such as polymers and trans isomers (Rodrigues et al. 2014). Neutralization of free fatty acid is usually done in batch or continuous process using sodium hydroxide. However, increased loss of neutral oil has been documented in conventional alkali neutralisation process (Wang and Johnson 2001), leading to the search of novel processes for effective deacidification in fish oil.

The different rates at which solvents solubilise free fatty acids and triglycerides can be considered, to develop a relatively simpler deacidification process with reduced oil loss. Furthermore, liquid-liquid extraction has the advantage of avoiding waste product formation and the recovery of solvent is relatively easier (Rodrigues et al. 2007). Though many patents are available on the solvent

C. Vaisali (✉) · B.D. Prasanna · I. Regupathi
Department of Chemical Engineering, NITK, Surathkal, India
e-mail: vaisali31@gmail.com

I. Regupathi et al. (eds.), *Recent Advances in Chemical Engineering*,
DOI 10.1007/978-981-10-1633-2_9

extraction of free fatty acids from vegetable oils (Bollmann 1921; Swaboda 1984), the deacidification of sardine oil involving mild process conditions are megre. The current study focuses on such softer processing of sardine oil, involving the application of solvent extraction of FFA from sardine oil. In order to have a better understanding of the factors affecting liquid-liquid extraction, different solvents were tested for their ability to separate FFA from oil. The lack of literature on the refining of sardine oil without the loss of its nutritional properties provides much scope for the current work, thus providing a novel approach to the problems in conventional deacidification.

2 Materials and Methods

2.1 Materials

Crude sardine oil was purchased from a local seafood industry. All chemicals and reagents were purchased from Merck, India and were of analytical grade.

2.2 Methods

The FFA content was analysed by determining the acid value as described in (AOCS 2009). The acid value was calculated by the following equation.

$$\text{Acid value} = ((\text{Sample titre} - \text{Blank titre}) \times \text{Molarity of KOH solution} \times 56.1)/\text{sample weight} \tag{1}$$

$$\%\,\text{FFA(as oleic acid)} = \text{Acid value}/1.99 \tag{2}$$

Moisture content of the samples was determined by the method of AOCS. The moisture content was estimated as in below.

$$\text{Moisture}(\%) = ((\text{Initial weight} - \text{final weight})/\text{Initial weight}) \times 100 \tag{3}$$

3 Results and Discussion

The effect of different solvents on the reduction of free fatty acids from sardine oil was studied (Fig. 1). The efficiency of extraction increased with increase in the solvent content in the mixture. Methanol showed maximum reduction in FFA, followed by ethanol, propanol and acetonitrile. At a concentration of 50 % oil/solvent (w/v), methanol, ethanol, propanol and aetonitrile showed 45, 33, 24,

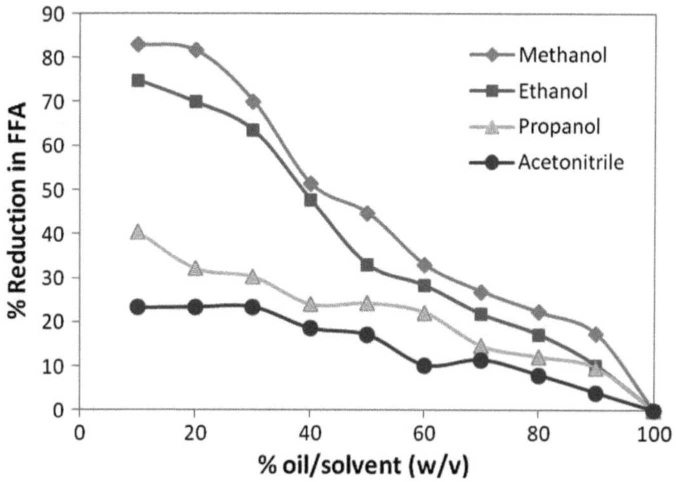

Fig. 1 Effect of different solvents on reduction of free fatty acids from oil

and 17 % reduction respectively (Fig. 1). Since many fatty acids have mutual solubility effects in a particular solvent, maximum efficiency can be achieved by increasing the solvent quantity in the mixture (Kolb and Brown 1955).

The extraction of FFA using solvents was already reported feasible based on several liquid-liquid equilibration data (Calvo and Cepeda 2008; Calvo et al. 2009). The varying degree of extraction by different solvents can be attributed to its increase in the polarity with increase in the chain length of the alcohol. This could probably explain the reason for the complete solubility of oil with butanol (data not shown). However, acetonitrile has a similar polarity index (6.3) to that of methanol (6.6) but failed to show efficiency similar to that of methanol. The probable reason could be the participation of the hydroxyl group of alcohol in the formation of hydrogen bonds, unlike acetonitrile. Also, as the chain length of the alcohol increases, the mutual solubility increases with decrease in the two-phase region (Rodrigues et al. 2007). This could explain the high efficiency of methanol and ethanol relative to propanol. In case of acetonitrile and propanol, two phase coexistence is very small, due to which very low efficiencies in FFA reduction were noted (Rodrigues et al. 2007). This data shows the feasibility of solvent extraction in deacidification of sardine oil. However, experimental equilibrium data and thermodynamic modelling for the estimation of activity coefficients should be performed to design industrial scale equipments for the deacidification of sardine oil.

The effect of solvents on oil loss was also calculated to determine the efficiency of the process. As the major disadvantage of the available conventional deacidification processes is related to the loss of neutral oil (Bhosle and Subramanian 2005), proper quantification of the same ensures a better approach for development of a process. The percentage of oil loss was found to be in correlation with the percentage reduction of FFA as well. As the FFA reduction increased, there was an

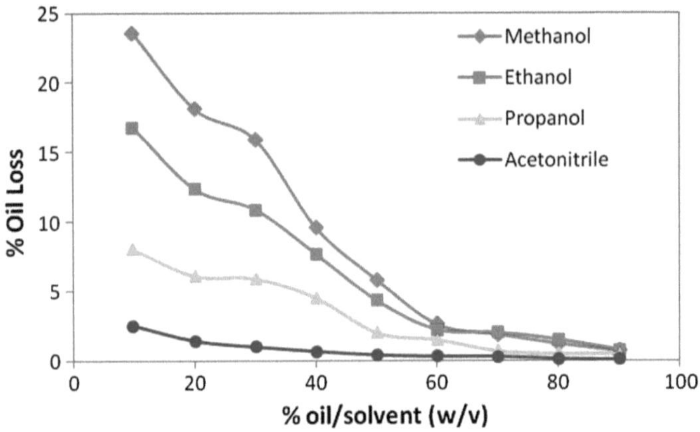

Fig. 2 Effect of solvents on oil loss

increase in the percentage oil loss (Fig. 2). However, this can be correlated to the removal of FFA rather than the neutral oil loss, which is evident from the consistent iodine value (Table 1). As the solvent content in the oil increase from 50 to 90 %, the percentage oil loss was more evident for ethanol and methanol. Acetonitrile showed the lowest amount of oil loss (2.5 %), which could be attributed to its inability to form hydrogen bonds unlike alcohols. In order to reduce the oil loss at higher concentrations of solvents, two stage extraction was proposed (Kale et al. 1999). Though the oil loss was significantly reduced by this approach, there are several disadvantages viz., large solvent requirements, increased solvent traces in oil and the energy required to remove the solvent traces which make the process economically insignificant at industrial level. Hence, acetonitrile can be considered for the industrial deacidification of oils with low FFA content, as it reduces oil loss.

The chemical characterisation of the oil after deacidification using different solvents is shown in Table 1. Preliminary experiments were carried out for deacidification without degumming process. However, fresh crude sardine oil with initial FFA content of 3.7 % contained higher quantities of phospholipids, which hindered with the subsequent deacidification process (Vaisali et al. 2015). Hence, a laboratory defined degumming process was developed prior to deacidification stage. Since 50 % oil/solvent (w/v) was found to be ideal for both effective FFA reduction

Table 1 Characterisation of the sardine oil after solvent extraction with 50 % (oil/solvent w/v)

Solvent extracted oil				
Parameters analysed	Methanol	Ethanol	Propanol	Acetonitrile
Acid value	3.75	4.01	5.78	6.1
% FFA	1.88	2.01	2.90	3.1
Iodine value	154.5	155.3	155.1	158.9
% oil loss	5.80	4.30	2.00	0.39
% moisture	7.66	8.87	8.45	5.67

and reduced oil loss, chemical characterisation of deacidified samples was performed at 1:1 oil-solvent ratio. It was evident from Table 1 that, regardless of the nature of solvents used, the iodine value of the samples remained almost similar. This could be attributed to the retainment of unsaturated fatty acids that are usually present in the oil as triglycerides (De and Bhattacharyya 1999). This further proves the efficiency of the process, by maintaining the initial iodine value, thus protecting the nutritionally rich compounds of interest. The residual solvent traces in the oil after extraction should also be considered to understand the efficiency of the process. The solvent traces in the oil after extraction determine the energy requirement of the process. The moisture content of oil after extraction was analysed to indicate the amount of solvent traces in the oil. It was found that the moisture content was relatively high for all the alcohol extracted samples when compared to acetonitrile (5.67 %). This could probably be attributed to the hydrogen bond formation between the hydroxyl group of the alcohol, which leaves higher traces of solvents in oil. Though acetonitrile was found to be ideal for the deacidification of sardine oil in terms of percentage oil loss, solvent traces and iodine value, it showed relatively low efficiency in reducing the acid value (Table 1) in comparison with other alcohols.

4 Conclusions

The use of solvent extraction for the deacidification of sardine oil gave promising results, hence it could be considered for further applications in sardine oil refining. Of all tested solvents, ethanol and methanol showed a maximum efficiency in terms of FFA reduction. However, oil loss and moisture content analysis indicated that acetonitrile could be better suited for deacidification application. Though reasonable amount of literatures are available on the solvent extraction of vegetable oil, its applications in the refining of fish oil are yet to be explored, providing further scope for the current study.

References

AOCS: Official Methods and Recommended Practices of the AOCS. AOCS Methods For Biodiesel Feedstock Quality Analytical Guidelines for Assessing Feedstock to Ensure Biodiesel Quality (2009)

Bhosle, B.M., Subramanian, R.: New approaches in deacidification of edible oils—a review. J. Food Eng. **69**, 481–494 (2005)

Bollmann, H.: Process for the removal of free fatty acids, resins, bitter and mucilaginous substances from fats and oils. US Patent US 1371342 (1921)

Calvo, B., Cepeda, E.A.: Solubilities of stearic acid in organic solvents and in azeotropic solvent mixtures. J. Chem. Eng. Data **53**, 628–633 (2008)

Calvo, B., Collado, I., Cepeda, E.A.: Solubilities of palmitic acid in pure solvents and its mixtures. J. Chem. Eng. Data **54**, 64–68 (2009)
De, B.K., Bhattacharyya, D.K.: Deacidification of high acid rice bran oil by re esterification with monoglycerides. J. Am. Oil Chem. Soc. **76**, 1243–1246 (1999)
Kale, V., Katikaneni, S.P.R., Cheryan, M.: Deacidifying rice bran oil by solvent extraction and membrnae technology. J. Am. Oil Chem. Soc. **76**, 723–727 (1999)
Kolb, D.K., Brown, J.B.: Low temperature solubilities of fatty acids in selected organic media. J. Am. Oil Chem. Soc. **32**, 357–362 (1955)
Rodrigues, C.E.C., Goncalves, C.B., Batsta, E., Meirelles, A.J.A.: Deacidification of vegetable oils by solvent extraction. Recent Pat. Eng. **1**, 95–102 (2007)
Rodrigues, C.E.C., Gonclaves, C.B., Marcon, E.C., Batista, E.A.C.: Deacidification of rice bran oil by liquid-liquid extraction using a renewable solvent. Sep. Purif. Technol. **132**, 84–92 (2014)
Rubio-Rodríguez, N., Beltran, S., Jaime, I., Diego, S.M., Sanz, M.T., Carballido, J.R.: Production of omega-3 polyunsaturated fatty acid concentrates: a review. Innovative Food Sci. Emerg. Technol. **11**, 1–12 (2010)
Swaboda, P.A.T.: Refining of palm oils. United Kingdom Patent GB 2144143 (1984)
Vaisali, C., Charanyaa, S., Belur, P.D., Regupati, I.: Refining of edible oils: a critical appraisal of current and potential technologies. Int. J. Food Sci. Technol. **50**, 13–23 (2015)
Wang, T., Johnson, L.A.: Refining high-free fatty acid wheat germ oil. J. Am. Oil Chem. Soc. **78**, 71–76 (2001)

Forward Osmosis Membrane Concentration of Raw Sugarcane Juice

H.N. Shalini and Chetan A. Nayak

1 Introduction

Sugarcane is scientifically known as *Saccharum officinarum,* largely produced in tropical and sub-tropical regions of India and relatively low cost agricultural resource (Nene et. al. 2002). Sugarcane juice consists of organic non-sugars which are in the form of soluble solids such as starch, protein, waxes and gums; it varies in 0.85–1.45 %. In addition 3–5 % of soluble solids constitute other components such as colour precursors, colorants etc. (Saha et al. 2006). In industries, the sugarcane is processed and converted to syrup; extracted juice after potential sugarcane coarse filtration consists of tiny particles such as proteins, salts of acids, polysaccharides and tiny particles of bagasse. Dark brown color of the juice is mainly due to the presence of polyphenolics, organic acids and flavonoids in small amounts (Laksameethanasan et al. 2012). The impurities are not precipitated completely in conventional liming process; moreover the calcium concentration is enhanced in the treatment (Saha et al. 2009).

Sugar processing is one of the most energy intensive processes in food industry, which challenges membrane separation processes like microfiltration (MF), ultra-filtration (UF), nanofiltration (NF) or reverse osmosis (RO) (Hinkova et al. 2000). Vastly employed method for food preservation, shelf-life extension and for concentration is thermal processing. However, negative impacts may be encountered on sensory parameters (colour, aroma, flavor), nutritious components and technological properties among others in industrial thermal treatments (Sant'Anna et al. 2012). Concentrated sucrose solutions are enormously used in bakery products, candies and in jams and marmalades during the formulation and have many other applications in food process engineering. Sucrose solutions can also be used as

H.N. Shalini · C.A. Nayak (✉)
Department of Chemical Engineering, BMS College of Engineering,
Bull Temple Road, P B No. 1908, Bengaluru 560019, Karnataka, India
e-mail: canayak.che@bmsce.ac.in

I. Regupathi et al. (eds.), *Recent Advances in Chemical Engineering*,
DOI 10.1007/978-981-10-1633-2_10

preservatives and are especially used to preserve fruits and vegetables (sometimes in combination with salt). These sugar concentrates have a prominent role in fermentation processes. For instance, sugar should be added to grape juice in wine making, if the grape does not have sufficient sugar content to reach the required level of alcohol during fermentation (Garcia-Castello et al. 2009).

A new membrane separation technology i.e, Forward Osmosis (FO) which stands next in cutting edge technology has gained wide attention in many applications such as desalination (Williams et al. 2015; Qasim et al. 2015; Akther et al. 2015; Zhao et al. 2016), power generation (Zhu et al. 2015), food processing (Rastogi 2016; Nayak et al. 2011), wastewater treatment (Ansari et al. 2015; Jamil et al. 2015; Qin et al. 2012) and pharmaceuticals (Rastogi and Nayak 2011). In forward osmosis process, water is transferred from low osmotic pressure feed solution (FS) to a high osmotic pressure draw solution (DS); concentration of the FS and dilution of the DS takes place mainly because of osmotic pressure difference which acts as driving force between the two aqueous solutions that are on two different sides of semi-permeable membrane (Nayak and Rastogi 2010a, b). The objective of this present work is to extensively study an alternative process for concentration of raw sugarcane juice.

2 Materials and Methods

2.1 Chemicals

The sugarcane juice was obtained from local cane juice centre, Bengaluru. The chemicals procured were of analytical grade (SRL Chemicals, Mumbai).

2.2 Membranes

Osmotek, Inc., Corvallis, OR, USA develops forward osmosis asymmetric membrane and was employed in the present study. The membrane consists of nylon mesh (a porous support layer) to provide strength and cellulose triacetate (active layer) which is made up of a very thin semi-permeable non porous skin layer, embedded in nylon mesh. The forward osmosis membrane is highly hydrophilic in nature and the thickness of membrane is less than 50 μm (Nayak and Rastogi 2010a, b; McCutcheon et al. 2005).

2.3 Forward Osmosis Membrane Module and System

FO experiments were carried out on a bench scale laboratory system. Experiments were performed using a flat sheet membrane module which has a membrane area of 0.0144 m^2, as shown in Fig. 1. The membrane module consists of forward osmosis membrane, which was placed and supported in between two frames made of acrylic plastic. Similar system was described in the previous investigation (Nayak and Rastogi 2010a, b). Feed solution (raw sugarcane juice) and osmotic agent solution (NaCl, 6M) were circulated on either side of the membrane in co-current mode using peristaltic pumps (Make: Masterflex, Model:77200-62). A ratio of 1:5 was maintained in all the experiments for feed solution to draw solution. Difference in volume of osmotic agent solution was measured every hour to estimate the trans-membrane flux. Average values of the flux were reported by conducting experiments for a period of 3 h. Experiments were conducted in triplicates and the room temperature in range of 25 ± 2 °C.

2.4 Feed and Draw Solutions

The suspended solids, fibres, coarse pulp and other impurities were removed from sugarcane raw juice (feed solution) by filtering the juice using filter mesh. Osmotic agent solutions for FO experiments were prepared; sodium chloride was dissolved in distilled water in varied proportions (1.0, 2.0, 3.0, 4.0, 5.0 and 6.0M) with aid of magnetic stirrer (Make: Hot plate magnetic stirrer, MS 200). To make sure complete dissolution of sodium chloride before use, the solutions were kept overnight at normal temperature. The flowrate was fixed to 100 mL/min for both FS and DS

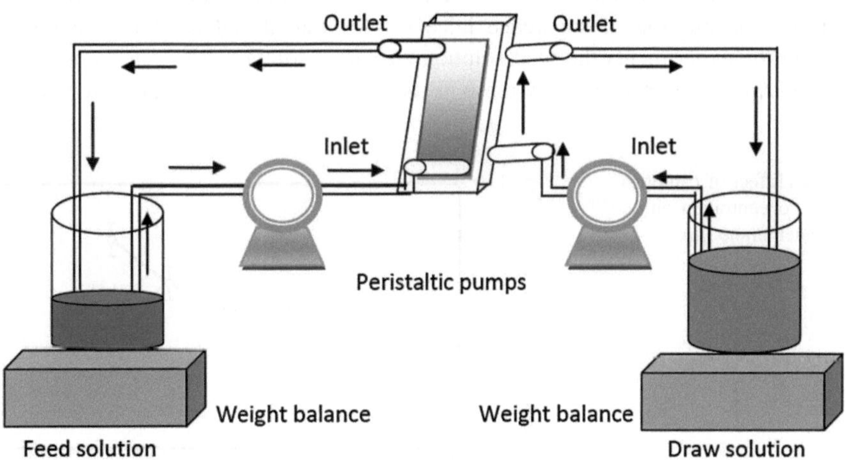

Fig. 1 Schematic diagram of forward osmosis membrane process system

side. Negligible amount of NaCl was diffused across the membrane from DS to FS. The concentration of Na^+ ions transferred across the membrane is tested by using flame photometer (Make: Systronic, Model: 128, India). However, the salt rejection of the membrane used is relatively high, in most cases it is over 98 % as reported by McCutcheon and Elimelech (2006).

Digital Refractometer (Make: Hanna) was used to estimate the total soluble solids of Sugarcane juice at 25 ± 2 °C.

Rotary evaporator (Make: Buchi Rotavapor R-210) was used to concentrate sugarcane juice thermally. The forward osmosis membrane process was compared with thermal concentration at 40 °C.

3 Results and Discussion

Sugarcane raw juice as feed material and sodium chloride as draw solution was used to perform FO experiments. The experiments at different concentrations were carried out to study the effect of draw solution on transmembrane flux. The effect of flow rate of draw solution on transmembrane flux was evaluated by varying the draw solution flow rate from 25 to 125 mL/min and feed flow rate was set constant. The effect of feed flow rate on transmembrane flux was evaluated by varying the feed from 25 to 125 mL/min and by setting osmotic agent flow rate constant.

3.1 Influence of Osmotic Agent Concentration on Transmembrane Flux

The effect of draw solution (sodium chloride) concentration on transmembrane flux during concentration of sugarcane juice was evaluated and as shown in Fig. 2. During the experiments the feed solution and draw solution flow rates were set at 100 mL/min. In FO, with an increase in sodium chloride concentration from 1.0 to

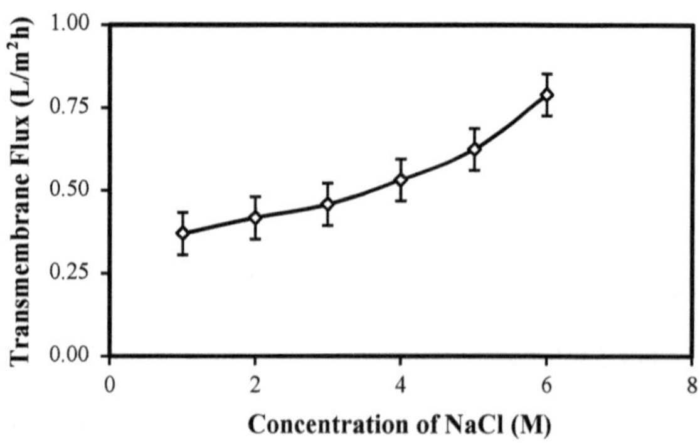

Fig. 2 Effect of osmotic agent concentration on transmembrane flux

6.0M, the transmembrane flux was increased from 0.37 to 0.79 L/m^2 h. This increase in the transmembrane flux is attributed due to the osmotic pressure difference across the membrane with increase in sodium chloride concentration. Similar results were reported for water and sodium chloride solution wherein the lower transmembrane flux was attributed to excessive concentration polarization on the draw solution side because of the diffusion of osmotic solute into the nylon support layer of CTA membrane (Chanukya et al. 2013). The increase in flux during forward osmosis membrane process is attributed to an increase in osmotic pressure difference across the membrane due to increase in the concentration of osmotic agent solution, which resulted in an increased driving force for water transport through the membrane (Nayak 2014).

3.2 *Influence of Feed and Osmotic Agent Flow Rate on Transmembrane Flux*

The effect of feed flow rate and osmotic agent flow rate on transmembrane flux for concentration of sugarcane juice is shown in Fig. 3. The transmembrane flux was increased from 25 mL/min (0.836 L/m^2 h) to 100 mL/min (0.98 L/m^2 h) and it was found to decrease at 125 mL/min (0.609 L/m^2 h). There was sudden decrease in flux at high flowrate because there was no sufficient contact time between the feed solution and osmotic agent solution. The concentration and flow rate of osmotic agent was maintained at 6.0M and 100 mL/min, respectively during the experiments.

The effect of osmotic agent flow rate on transmembrane flux for concentration of sugarcane juice is shown in Fig. 3. The transmembrane flux was increased from 25 mL/min (0.566 L/m^2 h) to 100 mL/min (0.778 L/m^2 h) and it was found to decrease at 125 mL/min (0.653 L/m^2 h). There was sudden decrease in flux at high flowrate because there was no sufficient contact time between the feed solution and osmotic agent or draw solution. The concentration of osmotic agent and feed flow rate was maintained at 6.0M and 100 mL/min, respectively during the experiments.

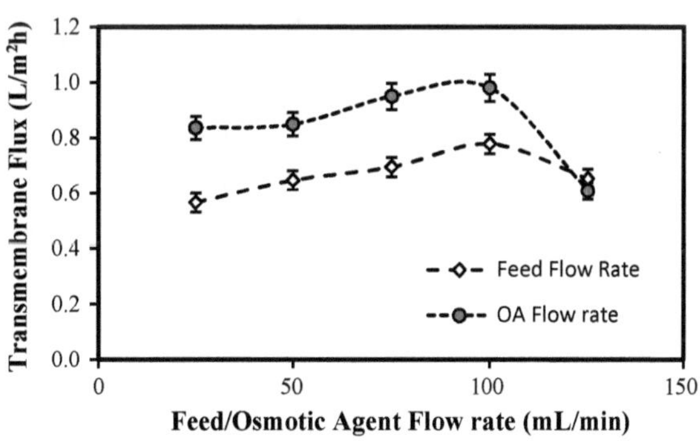

Fig. 3 Influence of feed and osmotic agent flow rate on transmembrane flux

The concentration build up in the solution at the vicinity of the membrane surface can be prevented by increasing the flow rate of feed solution, thus transmembrane flux can be increased by reducing the accumulated solute on the surface of the membrane. Similar results were found in the literature for different concentrates of beetroot, pineapple, grape juices (Nayak et al. 2011) and kokum extract (Nayak and Rastogi 2010a, b) using varied concentration of NaCl as osmotic agent solution.

3.3 Performance of Forward Osmosis in Large Scale and Comparison with the Thermal Concentration

Feed solution and draw solution flow rate was set at 100 mL/min and sodium chloride (6.0M) as osmotic agent conditions to concentrate sugarcane juice in a large scale experiment. Sugarcane juice (250 mL, 17.6 °Brix) was concentrated for 17 h. During the concentration of sugarcane juice the transmembrane flux and total dissolved solids (TSS) were measured for time interval of 1 h which is depicted in Fig. 4. The final concentration after 18 h, the total dissolved solids (TSS) was 31.7 °Brix. Due to increased feed concentration the driving force across the membrane was reduced therefore the transmembrane flux was found to decrease with time. The final concentrate was analysed for transfer of Na^+ ions using flame photometer, the amount of Na^+ ions transferred is found to be 179.3 ppm.

The Thermal evaporation was carried out by using rotovap to concentrate raw sugarcane juice. The evaporation was carried out for duration of 5 h at 50 °C and 370 mbar pressure, the final total dissolved solids (TSS) was 57.7 °Brix from initial TSS of 17.6 °Brix. When compared to FO process, thermal concentration is energy intensive and there was loss of color and flavour. The loss of colour and flavour may be attributed to Maillard reaction of reducing sugars due to high temperature.

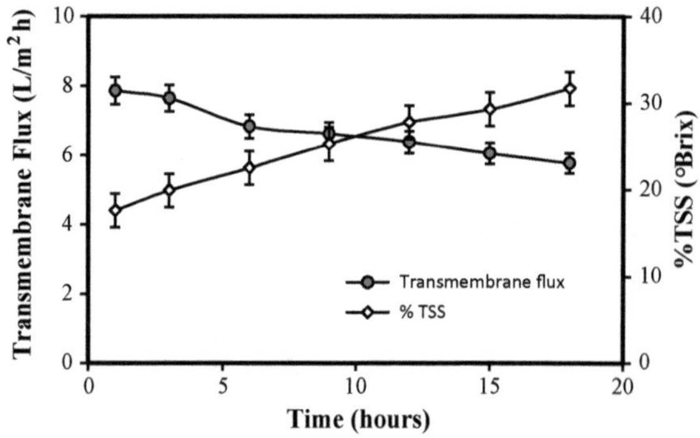

Fig. 4 Variation of Flux and Brix with time

4 Conclusion

Forward osmosis processes play a vital role to concentrate liquid foods especially when it contains various sugars, nutrients and heat sensitive compounds. During the concentration of sugarcane raw juice by forward osmosis process the effect of various process parameters such as osmotic agent concentration, flow rate of draw solution and feed solution on transmembrane flux were studied. The sugarcane juice was concentrated from 250 to 99.6 mL for 18 h. Higher transmembrane flux was resulted for 6M concentration in forward osmosis membrane process. However, the feed solution was coupled with 179.3 ppm of sodium ions that are being transferred from draw solution to feed solution across the membrane.

Acknowledgments Authors thank VGST and TEQIP-II for the support. Technical Education Quality Improvement Programme [TEQIP-II] of the MHRD, Government of India, supported the work reported in this paper. This project is continuation of the project granted by Vision Group of Science and Technology (VGST), Karnataka State to Dr. Chetan A Nayak, Project Grant No GRD-204 for 2013–14.

References

Akther, N., Sodiq, A., Giwa, A., Daer, S., Arafat, H.A., Hasan, S.W.: Recent advancements in forward osmosis desalination: a review. Chem. Eng. J. **281**, 502–522 (2015)

Ansari, A.J., Hai, F.I., Guo, W., Ngo, H.H., Price, W.E., Nghiem, L.D.: Selection of forward osmosis draw solutes for subsequent integration with anaerobic treatment to facilitate resource recovery from wastewater. Bioresour. Technol. **191**, 30–36 (2015)

Chanukya, B.S., Patil, S., Rastogi, N.K.: Influence of concentration polarization on flux behavior in forward osmosis during desalination using ammonium bicarbonate. Desalination **312**, 39–44 (2013)

Garcia-Castello, E.M., McCutcheon, J.R., Elimelech, M.: Performance evaluation of sucrose concentration using forward osmosis. J. Membr. Sci. **338**, 61–66 (2009)

Hinkova, A., Bubnik, Z., Kadlec, P., Pour, V., Starhova, H.: Membrane filtration in the sugar industry. Chem. Papers **54**(6a), 375–382 (2000)

Jamil, S., Loganathan, P., Kazner, C., Vigneswaran, S.: Forward osmosis treatment for volume minimisation of reverse osmosis concentrate from a water reclamation plant and removal of organic micropollutants. Desalination **372**, 32–38 (2015)

Laksameethanasana, P., Somla, N., Janprem, S., Phochuen, N.: Clarification of sugarcane juice for syrup production. Procedia Eng. **32**, 141–147 (2012)

McCutcheon, J.R., Elimelech, M.: Influence of concentrative and dilutive internal concentration polarization on flux behavior in forward osmosis. J. Membr. Sci. **284**, 237–247 (2006)

McCutcheon, J.R., McGinnis, R.L., Elimelech, M.: A novel ammonia–carbon dioxide forward (direct) osmosis desalination process. Desalination **174**, 1–11 (2005)

Nayak, C.A.: Forward osmosis membrane concentration process of anthocyanins from pomegranate fruit. Res. Rev.: J. Biotechnol. **4**(1), 2231–3826 (2014)

Nayak, C.A., Valluri, S.S., Rastogi, N.K.: Effect of high or low molecular weight of components of feed on transmembrane flux during forward osmosis. J. Food Eng. **106**, 48–52 (2011)

Nayak, C.A., Rastogi, N.K.: Comparison of osmotic membrane distillation and forward osmosis membrane processes for concentration of anthocyanin. Desalin. Water Treat. **16**, 134–145 (2010a)

Nayak, C.A., Rastogi, N.K.: Forward osmosis for the concentration of anthocyanin from GarciniaindicaChoisy. Sep. Purif. Technol. **71**, 144–151 (2010b)

Nene, S., Kaur, S., Sumod, K., Joshi, B., Raghavarao, K.S.M.S.: Membrane distillation for the concentration of raw cane sugar syrup and membrane clarified sugarcane juice. Desalination **147**, 157–160 (2002)

Qasim, M., Darwish, N.A., Sarp, S., Hilal, N.: Water desalination by forward (direct) osmosis phenomenon: A comprehensive review. Desalination **374**, 47–69 (2015)

Qin, J.J., Lay, W.C.L., Kekre, K.A.: Recent developments and future challenges of forward osmosis for desalination: a review. Desalin. Water Treat. **39**, 123–136 (2012)

Rastogi, N.K.: Opportunities and challenges in application of forward osmosis in food processing. Crit. Rev. Food Sci. Nutr. **56**(2), 266–291 (2016). doi:10.1080/10408398.2012.724734

Rastogi, N.K., Nayak, C.A.: Membranes for forward osmosis in industrial applications. In: Basile, A., Nunes, S.P. (eds.) Advanced membrane Science and Technology for sustainable energy and environmental applications, pp. 680–717. Woodhead Publishers, Cambridge UK (2011). doi:10.1533/9780857093790.5.680

Saha, N.K., Balakrishnan, M., Ulbricht, M.: Polymeric membrane fouling in sugarcane juice ultrafiltration: role of juice polysaccharides. Desalination **189**, 59–70 (2006)

Saha, N.K., Balakrishnan, M., Ulbricht, M.: Fouling control in sugarcane juice ultrafiltration with surface modified polysulfone and polyethersulfone membranes. Desalination **249**, 1124–1131 (2009)

Sant'Anna, V., Marczak, L.D.F., Tessaro, I.C.: Membrane concentration of liquid foods by forward osmosis: process and quality view. J. Food Eng. **111**, 483–489 (2012)

Williams, P.M., Ahmad, M., Connolly, B.S., Oatley-Radcliffe, D.L.: Technology for freeze concentration in the desalination industry. Desalination **356**, 314–327 (2015)

Zhao, D., Chen, S., Guo, C., Zhao, Q., Lu, X.: Multi-functional forward osmosis draw solutes for seawater desalination. Chin. J. Chem. Eng. **24**(1), 23–30 (2016)

Zhu, X.Z., Zhang, F., Li, W.W., Liu, H.Q., Wang, Y.K., Huang, M.S.: Forward osmosis membrane favors an improved proton flux and electricity generation in microbial fuel cells. Desalination **372**, 26–31 (2015)

Microencapsulation of *C-Phycocyanin* by Microfludics

H.N. Pradeep and Chetan A. Nayak

1 Introduction

C phycocyanin (CPC) is a natural blue colorant phycobiliprotein widely extracted from *Spirulina plantesis*. It is highly fluorescent (Alexander 1994), anti-inflamtory, antioxidant (Romey et al. 2003) and has high therapeutic value (Patil and Raghavarao 2007). It has many application in dietary supplements, colorants in cosmetics (Chethana et al. 2015), health drink and beverages (Yoshida et al. 1996), anticancer medicines (Nagaraj et al. 2011). *C-Phycocyanin* is unstable to heat and light in aqueous solution (Sarada et al. 1999). *Phycocyanin* is insoluble in acidic solution (pH 3.0) and denatures above 45 °C. Exposure to light of 3×10^5 Flux causes color change and degrades nearly 80 % of color (Lone et al. 2005; Ratana and Bung 2012). Due to its limited stability of CPC, it has not explored for various applications. Thus there is a huge requirement to improve its stability.

Microencapsulation technology has been used to encapsulate various materials including agrochemicals, biomolecules, pharmaceutical drugs for various reasons like protect the materials from degradation, controlled and targeted drug delivery systems, improve the stability in final desired product. In this microencapsulation technology, an active material is entrapped within a coating or wall material (Dubey et al. 2009; Nayak and Rastogi 2010). With respect to the advantages of microencapsulation over conventional encapsulation techniques i.e., spray drying, extrusion, fluidized beds, coacervation or emulsification may not be suitable for highly sensitive biomolecules.

H.N. Pradeep
Department of Chemical Engineering, DayanandaSagar College of Engineering, ShavigeMalleshwara Hills, Kumaraswamy Layout, Bengaluru 560078, Karnataka, India

H.N. Pradeep · C.A. Nayak (✉)
Department of Chemical Engineering, BMS College of Engineering, P B NO. 1908 Bull Temple Road, Bengaluru 560019, Karnataka, India
e-mail: canayak.che@bmsce.ac.in

I. Regupathi et al. (eds.), *Recent Advances in Chemical Engineering*,
DOI 10.1007/978-981-10-1633-2_11

This may be due to many parameters such as incompatible operating conditions i.e., high pressure, high temperature, as well as use of organic solvent.

Microfluidic devices are used for continuous production, control, and characterization of high-quality droplet particles (Tan et al. 2006, Kand et al. 2014; Skurtys and Aguilera 2008). The size of droplet formation depends up on various factors such as properties of fluid, surface tension, interfacial stress the channel dimensions and geometry (Huang et al. 2006; Garstecki et al. 2006). Coating materials polysaccharides such as chitosan, alginate are widely used for encapsulation (Park et al. 2005; Wan 2012). Sodium alginate anionic natural biodegradable polymer is broadly used for encapsulation of protein, nutrients, drugs, cells because of its bio-compatibility, inert in nature, non toxic, mild gelation process and controlled release of active material (Tønnesen and Karlsen 2002, Gombotz and Wee 1998).

The objective of this present work is to encapsulate food grade C-phycocynin within alginate matrix by using cross junction microfluidic chip.

2 Materials and Methods

2.1 Materials

Food grade C-Phycocyanin (Purity > 1.00) was been used in the experiment gifted by M/s. Delhi Neutraceuticals Private Limited., New Delhi. The grade was estimated by ratio of absorbance at 620 and 280 nm (A_{620}/A_{280}). AR Grade Cetrimoniumbromide (CTAB), Sodium alginate and anhydrous calcium chloride used were purchased from SRL Chemicals, Mumbai. Sunflower oil was used as continuous phase was purchased from local supermarket.

2.2 Microfludic Chip

Microfludic chip was fabricated in Poly Methyl Methacrylate (PMMA) substrate. Design was designed by Autocad software and fabricated at M/s. Achira Lab, Bengaluru. The chip consist of two inlet and one outlet. The details of the dimensions of the channels are depicted in the Fig. 1a and the photograph of the chip is shown in Fig. 1b.

2.3 Experimental Setup

The experiment setup for production and observation of the formation of CPC droplets is shown in Fig. 2. Sodium alginate solution 1.5 % (w/w) was prepared and a ratio of 1:1 of CPC powder was added and the solution was mixed thoroughly

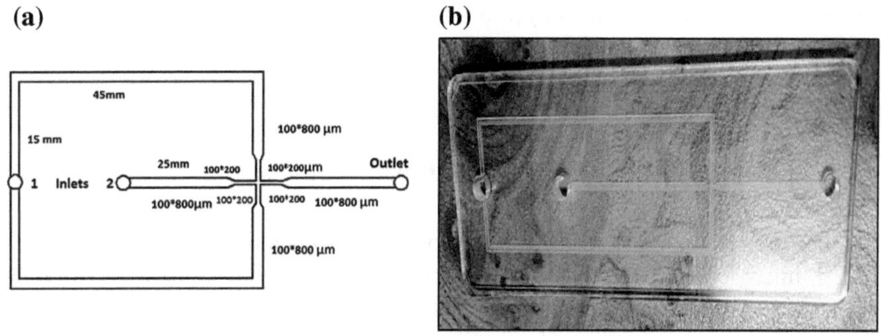

Fig. 1 **a** Chip design **b** fabricated chip

Fig. 2 Experimental setup

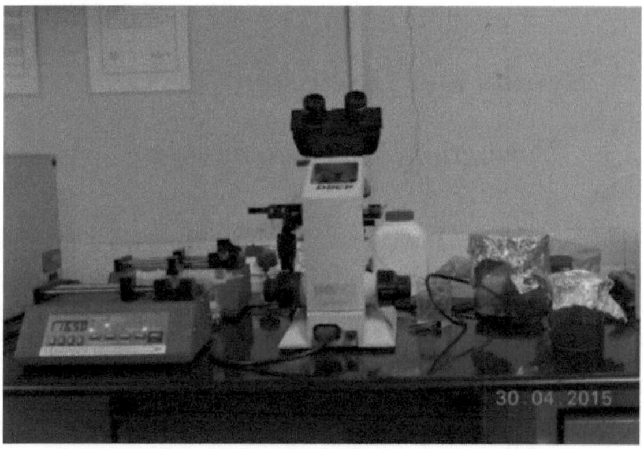

by magnetic stirrer. Further, an amount of 0.25 % of CTAB surfactant was added to the solution. This solution is considered as a dispersed phase for experiment. The dispersed phase was pumped through microfluidic channel using a syringe pump (Make: New Era Model: 300 Infusion). For the other inlet, Sunflower oil as a continuous phase was pumped for facilitating the formation of droplets. The flow rate of the syringe pump manually controlled. The flow rates for dispersed phase and the continuous phase were maintained at 1.5 µL/min was 16.5 µL/min, respectively. The droplet starts forming inside the chip channel and can be viewed using an optical microscope. The droplet formed was led into the beaker consists of 3 % calcium chloride. The encapsulated spherical beads are collected and analysed.

2.4 Determination of C-Phycocyanin Purity

The Purity of C-Phycocyanin was determined using a UV-vis spectrophotometer (Model: Cary 8454, Agilent Technologies, Inc.) by measuring the absorbance at

620 and 280 nm. The C-Phycocyanin purity was estimated by ratio of absorbance at 620 and 280 nm (A_{620}/A_{280}).

2.5 Morphology of Microencapsulate

The diameter of the encapsulated C phycocyanin droplet size was measured using optical microscope (Motic Microscope) consists of digital camera. A total 50 droplet was collected and measured for its size and average value is been reported. The morphology was confirmed using scanning electron microscope (SEM Make: HITACHI). The scanning electron microscope (SEM) is an electron microscope that images the sample surface by scanning using a high-energy beam of electrons.

3 Results and Discussion

3.1 Formation Microencapsulate

Encapsulated C-Phycocyanin of micrometre size was prepared by calcium alginate coating material. The sodium alginate react with calcium chloride to form calcium alginate. The purity of the CPC was found to be 1.009 (food grade). The continuous phase and dispersed phase is pumped using syringe pump to input 1 and 2, respectively. Several combinations of flow rates was attempted to get a droplet formation inside the chip channel. The droplet forms at a combination of 16.5 μL/min flow rate of continuous phase and 1.5 μL/min flow rate of dispersed phase. The formation of the droplet is shown in Fig. 3a, b. The sunflower oil which is been used as a continuous phase in the experiment. The sunflower oil which generate shear force and pressure gradient to form the droplet. This shear force and pressure gradient cause the dispersed phase to thin and squeeze until (Fig. 3a) it pinch off into formation of a droplet. Thus the droplet formed which is formed and flows through the output channel. The flow of the spherical droplets is shown in the Fig. 3b. Further, the droplet is made to react with 3 % calcium chloride solution for 20 min. The sodium alginate reacts with 3 % calcium chloride solution to form matrix of calcium alginate with encapsulated CPC. After 20 min, the encapsulated material is removed using filter and washed using distilled water and stored in a refrigerator at 4 °C for further analysis.

Similar work is reported by Yeh et al. 2009 for encapsulation of silver nanoparticles (AgNPs) and gold nanoparticles (AuNPs). The simulation of flow velocity distribution and the process of formation of monodisperse calcium alginate microparticles by using microfluidic T-junction geometries are studied and reported (Yeh et al. 2009). Ca-alginate microspheres, using a microfluidic chip, for the

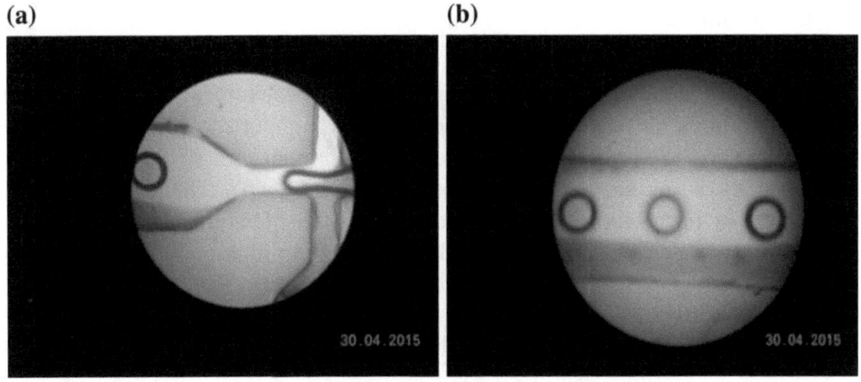

Fig. 3 **a** Squeezing of the dispersed phase **b** Flow of the spherical droplet formed

encapsulation of Sertoli Cells a biomolecule is reported by Capretto et al. 2010. This research paper reports on encapsulation using Y junction and explains the squeezing mechanisms for droplets and its morphology.

3.2 Morphology of the Encapsulate

The diameter of the encapsulated CPC droplet size was measured using optical microscope (Motic) consists of digital camera. A total of 50 droplets was collected and measured size and averaged, the average size was found 95–100 μm. This was confirmed by Scanning electro microscope and the SEM image of one of the encapsulated C Phycocyaninis shown in Fig. 4. It was found that the droplet formed by microfluidicchip is of spherical in shape.

Fig. 4 SEM image of encapsulate CPC

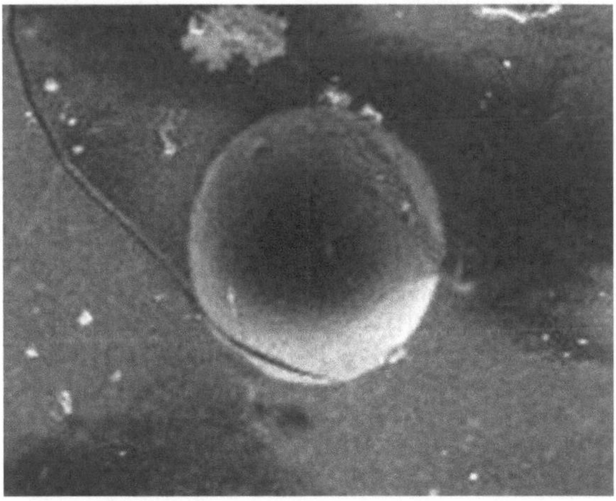

4 Conclusion

C-phycocyanin biomolecule is encapsulated within calcium alginate gel matrix using microfluidic device. Monodispersed high throughput encapsulated *C-phycocyanin* was produced with a microencapsulated droplet size of 95–100 μm and confirmed by scanning electron microscope for its shape. The drop formation mechanism is been studied and compared with the literature. The microfluidic method of microencapsulation is capable of generating high-quality gel matrix of uniform size.

Acknowledgments The work reported in this paper is supported by the college through the TECHNICAL EDUCATION QUALITY IMPROVEMENT PROGRAMME [TEQIP-II] of the MHRD, Government of India. The authors thank Dr. Samita Maitra, Head, Department of Chemical Engineering, BMSCE and Dr. Ravishankar R., Head, Department of Chemical Engineering, DSCE for constant encouragement and support.

References

Alexander, N.G.: Phycobiliproteins–a family of valuable, widely used fluorophores. J. Appl. Phycol. **6**, 105–112 (1994)

Capretto, L., Mazzitelli, S., Luca, G., Nastruzzi, C.: Preparation and characterization of polysaccharidicmicrobeads by a microfluidic technique: Application to the encapsulation of Sertoli cells. Acta Biomater. **6**(2), 429–435 (2010)

Chethana, S., Nayak, C.A., Madhusudhan, M.C., Raghavarao, K.S.M.S.: Single step aqueous two-phase extraction for downstream processing of C-phycocyanin from Spirulinaplatensis. J. Food Sci. Technol. **52**(4), 2415–2421 (2015)

Dubey, R., Shami, T.C., Bhasker Rao, K.U.: Microencapsulation technology and applications. Defence Sci. J. **59**(1), 82–95 (2009)

Garstecki, P., Fuerstman, M.J., Stone, H.A., Whitesided, G.M.: Formation of and bubbles in a microfludic T Junction-scaling and mechanisim of break-up. Lab Chip **6**, 437–446 (2006)

Gombotz, W.R., Wee, S.: Protein release from alginate matrices. Adv. Drug Deliv. Rev. **31**(3), 267–285 (1998)

Huang, K., Lai, T., Lin, Y.C.: Manipulating the generation of Ca-alginate microfludic channels as carrier gold nanoparticles. Lab Chip **6**(7), 954–957 (2006)

Kand, A., Park, J., Ju, J., Jeong, G., Lee, S.: Cell encapsulation via microtechnologies. Biomaterials **35**, 2651–2663 (2014)

Lone, J., Strømdahl, L.D., Olsen, K., Skibste, L.H.: Heat and light stability of three natural blue colorants for use in confectionery and beverages. Eur. Food Res. Technol. **220**(3), 261–266 (2005)

Nayak, C.A., Rastogi, N.K.: Effect of selected additives on microencapsulation of anthocyanin by spray drying. Drying Technol. **28**(12), 1396–1404 (2010)

Nagaraj, S., Arulmurgan, P., Karuppasamy, K., Jayappriyan, K.R., Sundararaj, R., Vijayanand, N., Rengasamy, R.: Hepatoprotective and antioxidative effects of C–phycocyanin in CCl_4 induced heptic damag rats. Acad. J. Cancer Res. **4**(2), 29–34 (2011)

Park, J., Ye, M., Park, K.: Biodegradable polymers for microencapsulation of drugs. Molecules **10**(1), 146–161 (2005)

Patil, G., Raghavarao, K.S.M.S.: Aqueous two phase extraction for purification of C-phycocyanin. Biochem. Eng. J. **34**(2), 156–164 (2007)

Ratana, C.N., Bung, B.: Stability of phycocyanin extracted from Spirulina sp.: influence of temperature, pHand preservatives. Process Biochem. **47**, 659–664 (2012)

Romey, C., Gonzalez, R., Ledon, N., Remirez, D., Rimbau, V.: C Phycocyanin: Abiliprotein with Antioxidant, Anti-Inflammatory and Neuroprotective Effects. Curr. Protein Peptide Sci. **4**, 207–216 (2003)

Sarada, R., Manoj, G.P., Ravishankar, G.A.: Phycocyanin from Spirulina sp: influence of processing of biomass on phycocyanin yield ,analysis of efficacy of extraction methods and stability studies on phycocyanin. Process Biochem. **34**, 795–801 (1999)

Skurtys, O., Aguilera, J.M.: Applications of microfluidic devices in food engineering. Food Biophys. **3**, 1–15 (2008)

Tan, Y.C., Cristini, V., Lee, C.A.P.: Monodispersd Micro droplet generation by shear focusing Microfludic device. Sens. Actuat. B-Chem. **114**, 350–356 (2006)

Tønnesen, H.H., Karlsen, J.: Alginate in drug delivery system. Drug Dev. Ind. Pharm. **28**(6), 621–630 (2002)

Wan, J.: Microfluidic–based synthesis of hydrogel particle for cell microencapsulation and cell–based drug delivery. Polymers **4**(2), 1084–1108 (2012)

Yeh, C., Zhoa, Q., Lee, S., Lin, Y.: Using a T-junction microfluidic chip for monodisperse calcium alginate microparticles and encapsulation of nanoparticles. Sens. Actuators A **151**(2), 231–236 (2009)

Yoshida, A., Takagaki, Y., Nishimune, T.: Enzyme immunoassay for Phycocyanin as the main components of spirulina color in food. Biosci. Biotechnol. Biochem. **60**, 57–60 (1996)

Part III
Adsorption

Demineralization Study of Low Grade Indian Coal by Aqueous Caustic Leaching

Sushanta Kumar Behera, S. Chakraborty and B.C. Meikap

1 Introduction

Large reserves of low grade coal in the various parts of world. The usability and energy efficiency of these coal quality is very low in comparison to high grade coal. Usually the low grade coal features with low carbon, high mineral matter (impurity) and moisture. By virtue of the formation of Indian coal, the characteristics of these coals different from other countries. Due to drift origin formation of Indian coal which signifies more chance of variations in property and high amount of inorganic inclusions resulting in formation of high ash content. The inorganic inclusions in coal is the major amount of silicates (kaolinite or dolomite), quartz, pyrite and carbonates. The coal ash formed by decomposition inorganic mineral matter (Bolat et al. 1998). The ash content of low grade coal varies from 35 to 45 % and the coals have high ash fusion temperature about 1500 °C in spite of this these coal contains low sulfur content about (0.5 %). With growing problems of energy supply, attention is turning increasingly to low grade coals. The reduction of ash content to a certain limit from the coal is very difficult by conventional physical beneficiation techniques due to low quality of coal (Dash et al. 2013). So to develop more efficient technologies for production of clean coal. In the physical method the coal

S.K. Behera (✉) · S. Chakraborty · B.C. Meikap
Department of Chemical Engineering, IIT Kharagpur,
Kharagpur 721302, West Bengal, India
e-mail: skbehera@che.iitkgp.ernet.in

S. Chakraborty
e-mail: sc@che.iitkgp.ernet.in

B.C. Meikap
e-mail: bcmeikap@che.iitkgp.ernet.in

B.C. Meikap
Chemical Engineering, School of Engineering, Howard College,
University of Kwazulu-Natal, Durban, South Africa

I. Regupathi et al. (eds.), *Recent Advances in Chemical Engineering*,
DOI 10.1007/978-981-10-1633-2_12

is crushed into fine particles, and the removal of mineral matter based on their physical properties like surface phenomena (wettability) and specific gravity (Meshram et al. 2015). It is only possible in chemical method for removal of associated mineral and bound to coal matrix or structure. In the chemical method, the mineral species reacts with the mineral matter and dissolute in water acid or alkali solution (Choudhury and Bhaktavatsalam 1997; Mukherjee and Borthakur 2001). The method used in chemical cleaning of coals include acid leaching or alkali leaching and or followed both acid and alkali at varying physical parameters. The mineral rich component were reduced in chemical demineralization process without modifying the organic structure of coal matrix (Gulen et al. 2005). The extent of demineralization is vary place to place because it depends upon the type of mineral structure and composition during coalification (Sharma and Gihar. 1991; Kara and Ceylan. 1988). Chemical cleaning of coal is using simple equipment and aqueous solution of common industrial chemicals and operation is easier when it operate under mild condition but the cost of chemicals needed (Waugh and Bowling 1984). In chemical cleaning the mineral acid are effective reducing the minerals and sulfur content of the coal to different extent. However the combined method of alkali leaching followed by acid washing is the best treatment method for effective demineralization because the alkali are extremely liberation the most of minerals from coal (Wang et al. 1996; Mukherjee et al. 2001).

A review of literature signifies the coal demineralization is improved by leaching effect. The aim of this work is to examine leaching effect of temperature, time, concentration and particle size on the demineralization of coal. The coal characterized and coal minerals at different stages during leaching was analyzed by XRF spectrometer and SEM-EDS analysis.

2 Materials and Methods

2.1 Materials

In the present investigation the coal sample were collected from Bhubaneswari OCP (Open Cast Project) of Talcher Coal field, Orissa. The coals were crushed and sieved to −16+100 British standard sieves (BSS) mesh size particle, was used for the experiments. Before the actual experiment starts the coal sample was analysed for its ash content using standard method proximate and ultimate analyses and summarized in Table 1.

Table 1 Chemical analysis of coal samples

Coal seam	Proximate analysis (wt%, dry)			Ultimate analysis (wt%, daf)					Calorific value (Kcal/kg)
	Ash	Volatile matter	Fixed Carbon	C	H	N	S	O (diff.)	
Bhubaneswari	26.95	38.71	34.34	52.9	4.53	1.5	1.28	39.79	4631.5

2.2 Experimental Methods

The leaching experiments were conducted in a bench scale unit consist of a batch reactor of capacity 2.5 l with heating facilities, condenser provision for reflux and temperature measured by thermometer. The batch reactor made up stainless steel reactor (SS-316) which consist of 11 cm diameter and 22 cm height with Teflon coating inside. The reactor is heated with four heating coils of 1 KW capacity each, which fitted outside of the circumference of the reactor. A stainless steel stirrer, is firmed in motor and whose propeller placed at 6 cm above from the bottom of reactor. A motor moves the stirrer and its speed measured by the sensor which mounted in front of motor. The stirrer usually operate at 600–900 rpm. The layout of bench scale experimental setup is shown in Fig. 1.

The experiment were carried out in stainless steel reactor with 50 gm of coal sample for each test was added to 500 ml NaOH solution of known concentration (50–150 g L^{-I}) for each test at temperature 100 °C for 60 min. The coal slurry was heated from a room temperature to required leaching temperature and maintained constant for period of time. In second set of experiments, 50 gm of coal was treated with 50 g L^{-I} NaOH concentration of 500 ml solution for 60–120 min at reaction temperature 65–150 °C. In both stages, after leaching the hot coal slurry is cooled, filtered and washed with distilled water for neutralize the alkali concentration. The sample is than dried at 75–85 °C and weighed. The slurry after washed with distilled water, filtered and dried. The ash content was determined for before and after treated coal by using standard methods (Bureau of Indian Standards 1969).

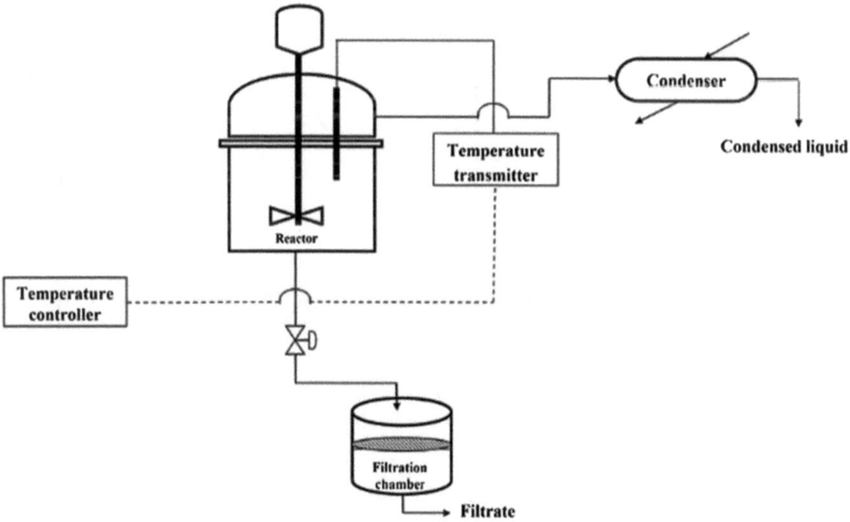

Fig. 1 Experimental set up for leaching process

3 Results and Discussion

3.1 Effect of Concentration on Demineralization of Coal

The effect of NaOH concentration for the demineralization of coal has been studied in the range of 50–150 g/L of NaOH solution was treated at 100 °C for 60 min. The caustic concentration has direct effect on degree of demineralization and the results depicted in Fig. 2. As the concentration of NaOH with which treated coal is increased, the percent demineralize of the coal increases up to 100 g/L NaOH then it decreases at higher concentration. The percent demineralization reduce at higher caustic concentration (above 100 g/L NaOH) may be due to adsorption the sodium complex compound on the reactive site of coal surface and forms film coating on the coal.

In this case maximum percent demineralization was 27 % obtained at 100 g/L NaOH concentration than it decreases further increases in NaOH concentration. The decrease in percent demineralization with increase in alkali concentration is revealed that formation of insoluble acid silicate and aluminate. Several authors also observed similar case as decrease demineralization with increases alkali concentration. During alkali leaching, the coal containing silicate and aluminate may dissolute and formed sodium compound silicate and aluminate that leached to the solution but the dissolution decreases at higher concentration of alkali solution as from reaction (1), (2) and (3).

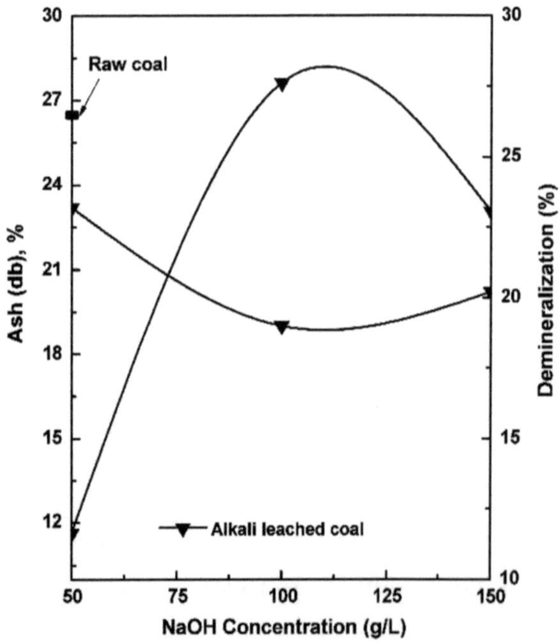

Fig. 2 Effect of concentration on demineralization of coal

$$2NaOH + Al_2O_3 \rightarrow 2NaAlO_2 + H_2O \quad (1)$$

$$2NaOH + SiO_2 \rightarrow Na_2SiO_3 + H_2O \quad (2)$$

At higher concentration of alkali solution the following reaction

$$\begin{gathered} NaOH(aq) + NaAlO_2(aq) + Na_2SiO_3(aq) \\ \rightarrow Na_X(AlO_2)_Y(SiO_2)_Z NaOH \cdot H_2O \quad (3) \\ \text{(Sodium-aluminosilicate complex compound)} \end{gathered}$$

At time of leaching in stirred batch reactor, the film layer limited formation and disappear by stirring at desired concentration (up to 100 g/L in this investigation) but at higher concentration the thickness of film appear was more prominent which inhibit the diffusion of leachant to the reactive site of coal and resist the demineralization process. The formation of Na compound of silicate and aluminate dissolve up to 100 g/L NaOH and after decreases its solubility and forms complex gel. The gel disperse to the coal matrix through pores and block the reactive site and create resistant to diffusion. It may also possible during diffusion of leachant up to 100 g/L, then after some mineral drifted from the pores and become trapped where resist to diffusion of leachant.

3.2 *Effect of Contact Time and Temperature on Demineralization*

The experiment were carried out at three different temperature specially 65, 100 and 150 °C and each temperature operated three different leaching time such as 60, 90 and 120 min. The plot of percent demineralization versus time for leaching −16 +100 mesh size particle with sodium hydroxide solution. The effect of time and temperature is one of the most important parameter for leaching process. The degree of demineralization of coal increases with time and temperature as shown on Fig. 3. The graph reveals that elimination of mineral matter from Bhubaneswari coal increases accordingly with increase of temperature up to studied range. In the investigation, the maximum demineralization was obtained 23 % at 150 °C and less demineralization achieved at lower temperature. This is due to with rise of temperature which enhance the rate of reaction of leaching process and increases the activation energy favored easier to reaction between the ash bearing mineral particle and leachant and enhance the degree of demineralization. At higher temperature, greater magnitude of demineralization while less at lower temperature at similar contact time. This means the higher temperature less time required to dissolution of mineral matter and leached out to solution.

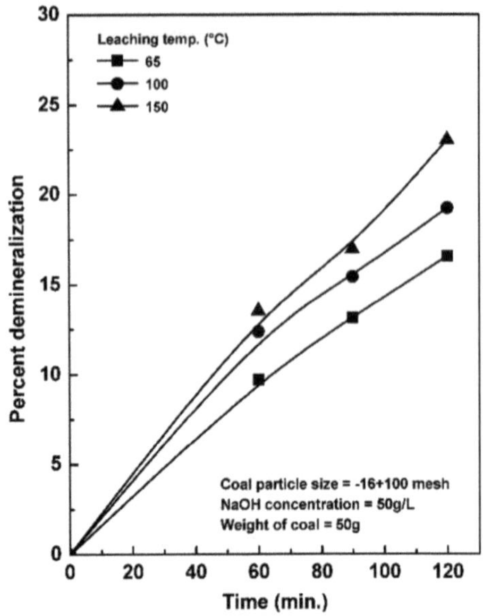

Fig. 3 Effect of time and temperature on demineralization of coal

3.3 Effect of Particle Size on Demineralization

The particle size is also affect degree of demineralization of coal. The ash content of raw coal varies with particle size and the results were shown in Fig. 4. The ash content of raw coal decreases with increase particle size while it increases for leached coal because the exposed surface area towards the leachants for larger particle size is less in comparison to finer particles. Therefore, the ash forming mineral matter less gap to diffuse in the leachant for smaller size particle while the exposed rich mineral of finer raw coal particle directly formed greater amount of ash after combustion.

It is evident from Fig. 4, the extent of demineralization increases with decrease particle size. In this investigation, −72 mesh particle size giving maximum demineralization and throughout the process, −60 and −72 mesh particle size is best demineralization of coal. However very fine ground particle (over −100 mesh particle size) were inhibit to demineralization. This is because, the very finer particle stick to each other and greater is cohesive force in between the particle (above 100 mesh) and also difficult to handling. It reduces the rate of diffusion of leachant to the reactive part in coal and very less yield and also more time to take for filtration.

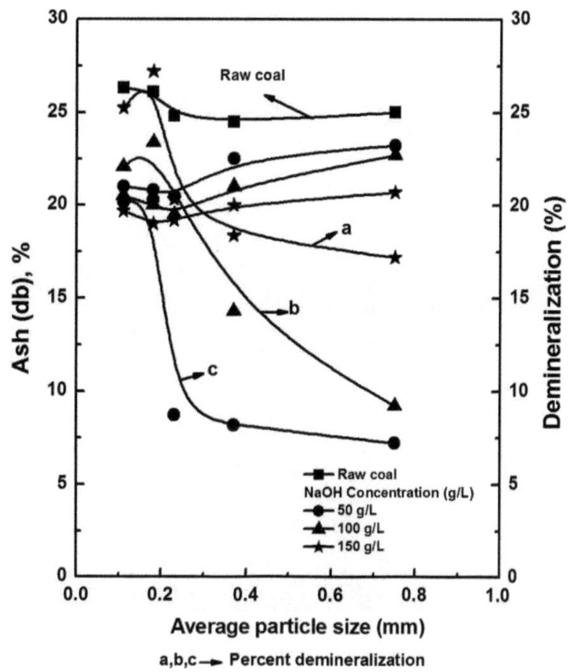

Fig. 4 Effect of particle size on demineralization of coal

3.4 *SEM-EDS Analysis of Coal*

The SEM has been carried out for the raw and chemical leached coals by using scanning electron microscope with an energy dispersive X-ray analyzer (SEM-EDS model JSM5800 from JEOL Company in Japan). In this investigation the surface morphology and elementary composition of mineral group in the raw coal and leached coal was studied. The SEM micrograph of raw and leached sample were provided in Fig. 5. In the chemical leaching most of the minerals are removed and also enrich the carbon content. From Table 2 and Fig. 5a represents the SEM—EDS image of the raw coal sample contains large amount quartz and aluminate and less percentage of carbon content as well as small proportion of elements such as calcium, potassium, sulfur etc. The SEM study Fig. 5b–d with corresponding EDS analysis from Table 2 revealed that the carbon content in the coal is increased by demineralization of coal and there is no loss of carbon content. The carbon content increases from 48 to 68 % during the alkali leaching. It also noticed that the liberation of mineral content drastically changes and the 'Fe' content increases with alkali concentration solution. The liberation of 'Fe' increases from the pyrite and hematite group of minerals during leaching. The increase of porosity was indication of significant amount of elimination of inorganic elements. Maximum amount of

Fig. 5 SEM image of **a** Raw coal and **b** 50 g/L NaOH **c** 100 g/L NaOH **d** 150 g/L NaOH treated coal

Table 2 EDS analysis of coal samples

Elements (wt%)	Raw coal	50 g/L NaOH	100g/L NaOH	150 g/L NaOH
C	48.95	65.23	68.38	65.85
O	36.48	24.24	22.69	16.51
Mg	0.13	0.38	0.42	0.51
Al	4.29	1.88	1.22	3.15
Si	9.16	5.78	3.35	8.51
S	0.22	0.27	0.62	0.4
K	0.2	0.12	0.2	0.96
Ca	0.08	0.32	0.08	0.69
Ti	0.3	0.37	0.12	0.99
Mn	0.04	0.06	0.09	0.03
Fe	0.15	1.25	2.26	2.4

silica (Quartz), alumina and clay containing mineral were leached out during alkali leaching due to high affinity of OH^- (hydroxyl group ions) towards the silicate and aluminates. However, the surface layer in the SEM image still showing bright and luminous, that indicating maybe presence of mineral phases.

Table 3 Ash composition of coals

Coal seams	Coal sample	Ash composition (wt%)							
		SiO_2	Al_2Oj	Fe_2O_3	CaO	MgO	Na_2O	K_2O	SO_3
Bhubaneswari	Raw coal	44.51	20.12	3.67	1.45	0.98	0.15	0.97	0.741
	Alkali leached coal	22.70	7.67	3.93	1.24	0.52	17.96	0.756	5.94

3.5 Ash Composition

The ash composition has been analyzed by XRF spectrometer and the analysis results were summarized in Table 3. Most of silica (SiO_2) and alumina (Al_2O_3) and other clay minerals soluble in caustic leaching while Fe_2O_3, Na_2O, SO_3 content of ash were increased during caustic treatment. From the results, It reduced to above 50 % minerals with caustic treatment by dissolution forming sodium silicate (Na_2SiO_3) and sodium aluminate ($NaAlO_2$).

3.6 Calorific Value

The calorific value (C.V) or heating value of coal is one of the most important indexes in assessing the quality of coal. High pressure oxygen Bomb calorimeter (LECO AC 500) was used for measuring the calorific value of coal. The calorific value of coal increases with extent of demineralization of coal. The calorific value of treated coal increases from 4631.5 to 5241.7 kcal/kg. The increase of heating value may be the removal of non-carbonaceous materials or mineral matter with the alkali treatments.

4 Conclusions

From the investigation, The maximum demineralization has been achieved 27.6 % by alkali treated coal from 27 % ash in Bhubaneswari coal seam. The size of particles also effects the extent of demineralization; in the present work −72 mesh particle size has been found to the most effective demineralization. The calorific value of coal increases with the demineralization due to the liberation of non heating valuable minerals from the coal. The efficiently extent of degree of demineralization depends on the increasing the solvent concentration, temperature and leaching time and decreasing the size of particle. The percent of demineralization improves with alkali concentration and at higher concentration reduce the demineralization due to the formation sodium complex gel like liquor and coat to reactive part of coal. Also, It showed from SEM-EDS analysis, there is no loss of carbon content during the leaching effect and most of the clay bearing minerals were leached out.

References

Bolat, E., Sağlam, S., Pişkin, S.: Chemical demineralization of a Turkish high ash bituminous coal. Fuel Process. Technol. **57**, 93–99 (1998)

Choudhury, R., Bhaktavatsalam, A.K.: Beneficiation of Indian coal by chemical techniques. Energy Convers. Manag. **38**, 173–178 (1997)

Dash, P.S., Kumar, S.S., Banerjee, P.K., Ganguly, S.: Chemical leaching of high ash Indian coals for production of low ash clean coal. Miner. Process. Extr. Metall. Rev. **34**, 223–239 (2013)

Gulen, J., Doymaz, I., Piskin, S., Toprak, S.: Removal of mineral matter from Silopi-Harput asphaltite by acid treatment. Energy Sources **27**, 1457–1464 (2005)

Kara, H., Ceylan, R.: Removal of sulfur from four Central Anatolian lignites by NaOH. Fuel **67**, 170–172 (1988)

Meshram, P., Purohit, B.K., Sinha, M.K., Sahu, S.K., Pandey, B.D.: Demineralization of low grade coal—a review. Renew. Sustain. Energy Rev. **41**, 745–761 (2015)

Mukherjee, S., Borthakur, P.C.: Chemical demineralization/desulphurization of high sulphur coal using sodium hydroxide and acid solutions. Fuel **80**, 2037–2040 (2001)

Mukherjee, S., Mahiuddin, S., Borthakur, P.C.: Demineralization and desulfurization of subbituminous coal with hydrogen peroxide. Energy Fuels **15**, 1418–1424 (2001)

Sharma, D.K., Gihar, S.: Chemical cleaning of low grade coals through alkali-acid leaching employing mild conditions under ambient pressure. Fuel **70**, 663–665 (1991)

Wang, J., Zhang, Z., Kobayashi, Y., Tomita, A.: Chemistry of $Ca(OH)_2$ leaching on mineral matter removal from coal. Energy and Fuel **10**, 386–391 (1996)

Waugh, A.B., Bowling, K.M.: Removal of mineral matter from bituminous coals by aqueous chemical leaching. Fuel Process. Technol. **9**, 217–233 (1984)

Isotherm, Thermodynamic and Kinetics Studies for Removal of Disperse Orange 25 Dye from Contaminated Water Using Adsorption on Acid-Treated Red Mud

G.M. Ratnamala, K. Vidya Shetty and G. Srinikethan

1 Introduction

Textile industries are one of the largest industries that produce colorful wastewater. Among many dyes used in these industries, quantities of azo dyes are widely used in a various products, such as food stuffs, textiles, paper and leather (Perez-Urquiza et al. 2000; Suzuki et al. 2001). The release of disperse dyes into the environment is a problem for life and a threat to the environment (Sahin et al. 2007). Adsorption is a physicochemical method, which is widely used for dye adsorption also has wide applications in wastewater treatment (Bansal and Goyal 2005, Danis et al. 1998, Imamura et al. 2002). Adsorbents, which are generally used for dye wastewater treatment are alumina, silica gel, activated carbon, zeolite etc. Studies have shown that activated carbons, silica gel, zeolites and alumina are good adsorbents for the adsorption of different types of dyes, but their use is sometimes restricted in view of their high cost. This has resulted in attempts by various researchers to prepare alternative adsorbents which are of low cost (Gupta and Suhas 2009) which may replace the above said commercial adsorbents in pollution control through adsorption process. Red mud is an industrial by product obtained in production of alumina from bauxite. Red mud presents a promising application in waste water

G.M. Ratnamala (✉)
Department of Chemical Engineering, KLE Dr M S Sheshagiri
College of Engineering and Technology, Udyambag, Belgaum 590008,
Karnataka, India
e-mail: ratna_chem@yahoo.com

G.M. Ratnamala · K. Vidya Shetty · G. Srinikethan
Department of Chemical Engineering, National Institute of Technology
Karnataka Srinivasnagar Post, Surathkal 575025, Karnataka, India

I. Regupathi et al. (eds.), *Recent Advances in Chemical Engineering*,
DOI 10.1007/978-981-10-1633-2_13

treatment for adsorption of toxic heavy metal and metalloid ions, inorganic anions, as well as organics compounds and phenolic compounds (Rai et al. 2012). A few studies have reported on the use of red mud for adsorbing dyes such as Acid Blue 92 (Norouzi et al. 2010), Congo red (Namasivayam and Arasi 1997), Rhodamine B, fast green, and Methylene Blue (Gupta et al. 2004,Wang et al. 2005), and Acid Violet (Namasivayam et al. 2001), Acid blue 113 and Reactive black 5 (Shirzad-Siboni et al. 2014). However the red mud as an adsorbent has not been tested for its efficacy as an adsorbent for dispersive dyes. The present work focuses on the use of red mud, a solid waste from aluminium industry as an adsorbent for the adsorption of Disperse Orange-25, a disperse azo dye. Experimental investigations were conducted on studies of various factors affecting adsorption.

2 Materials and Methods

2.1 Red Mud and Dye

Red mud was procured from HINDALCO LTD, Belgaum, India, Aluminium industry. The chemical composition of the red mud used in the present study was Fe_2O_3 (35–42 %), Al_2O_3 (18–22 %), TiO_2 (7–10 %), SiO_2 (9–12 %), Na_2O (6–8 %), CaO (1.5–3.5 %), P_2O_5 (0.3–0.4 %), V_2O_5 (0.3–0.35 %) and Loss on ignition (10.51 %) on dry weight basis as reported by HINDALCO LTD, Belgaum, India. 10 g of raw red mud was treated with 20 mL sulphuric acid solution (98 %) for a contact time of 5 h. Then the red mud was separated from the solution by decantation and centrifugation, washed with distilled water several times and dried in an oven at 110 °C overnight. The activated red mud thus prepared was sieved and the red mud of average size 120 μm was obtained from sieve analysis. The red mud pre-treated with H_2SO_4 is here in after referred as sulphuric acid treated red mud (SARM). Disperse Orange 25 dye (DO25) was gifted by Campbell Knitwear Ltd, Belgaum, India.

The dye concentrations in aqueous solution was analyzed by measurement of the absorbance of the solutions at 475 nm for using a pre-calibrated UV–VIS bio-spectrophotometer (Elico BL-198). The batch adsorption experiments were carried out in 250 mL conical flasks containing 100 mL aqueous dye solution of known concentration, with the red mud added at the desired quantity. Initial pH of dye solution was adjusted to the desired pH value with 1 N HCl or 1 N NaOH solutions. The solution was agitated with constant speed of 145 rpm in an incubator shaker. The dye concentration in the supernatant was determined from the measurement of absorbance using UV–VIS bio-spectrophotometer.

3 Results and Discussions

3.1 Red Mud Characterization

The surface areas of original red mud and acid-treated red mud were determined by BET analysis using ASAP 2020 V3.04 H (Micromeritics, USA) surface area analyzer. The raw red mud and SARM have specific surface area of 20.2 and 32.28 m^2/g respectively. The surface area and pore volume of acid-treated red mud is higher than the raw red mud.The acid treatment leads to removal of sodium ions and compounds (Shing 1997; Pratt and Christoverson 1982; Snigdha et al. 2012), further it causes dispersion of dissolved metal oxides as hydroxides, which leads to formation of pores there by increasing the total surface area of red mud.

3.2 Effect of Initial PH

The initial pH of dye solution affects the chemistry between the dye and the red mud surface. Effect of pH on percentage adsorption was observed during the adsorption of DO25 on SARM.

Figure 1 shows the effect of initial pH on percentage adsorption of DO25 at equilibrium, with initial dye concentration of 130 mg/L and SARM dosage of 1.2 g/L. The percentage adsorption of dye by adsorption has increased from 82 to 98 % with the increase in pH from 1 to 4. The maximum adsorption of dye was obtained at a pH of 4. As observed in Fig. 1, at the pH values above 4, the percentage adsorption of dye decreased from 98 to 77 % as the pH was increased from 4 to 12. Such a decrease in dye adsorption may be due to repulsion between anionic dye molecules and negatively charged red mud surface.

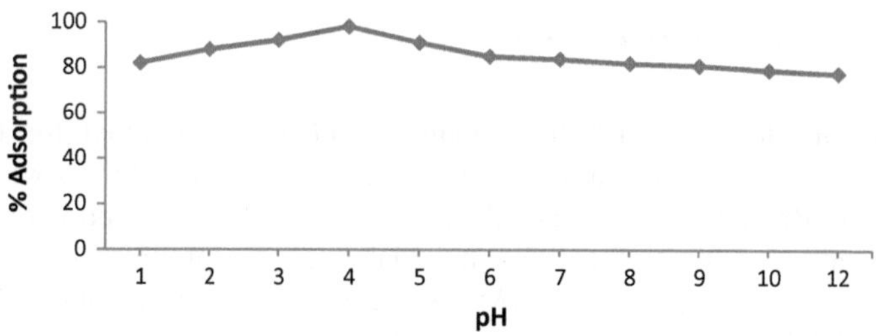

Fig. 1 Effect of initial pH on percentage adsorption of DO25 on SARM. (Temperature 30 °C Initial dye concentration 150 mg/L, and SARM dosage 1.2 g/L)

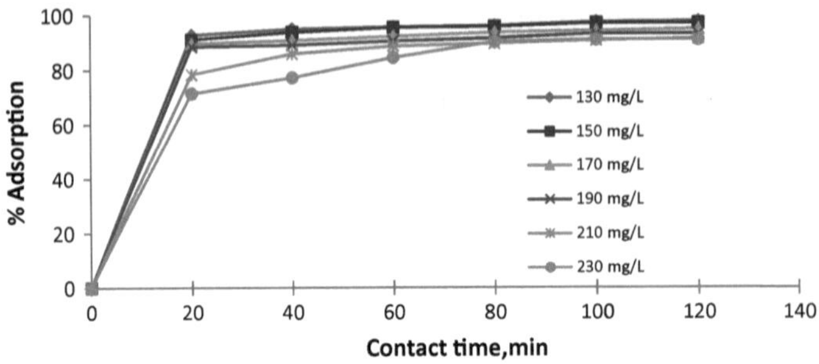

Fig. 2 Effect of contact time on percentage adsorption of DO25 using SARM at different initial dye concentration. Conditions: pH 4 temperature 30 °C and SARM dosage 1.2 g/L

3.3 *Effect of Contact Time and Initial Concentration of Dye*

The contact time between the adsorbate and the adsorbent plays an important role in adsorption. The effect of contact time on percentage adsorption of DO25 on SARM, with different initial dye concentrations ranging from 130–230 mg/L is shown in Fig. 2. It can be observed that, the adsorption is fast within 20 min of initial time, but the adsorption proceeded at a lower rate for the remaining time. It can also be observed from Fig. 2 that, around 97 % of 130 mg/L and around 91 % of 230 mg/L DO25 dye could be removed from the contaminated water in 120 min of contact time. Adsorption equilibrium was found to be achieved within 120 min at all the initial dye concentrations studied.

The effects of initial dye concentration on percentage adsorption of dye at equilibrium are presented in Fig. 2. It can be observed from Fig. 2 that, the adsorption of DO25 at equilibrium has decreased from 97.85 to 91 % with increase in DO25 concentration from 130 to 230 mg/L.

3.4 *Effect of SARM Dosage*

It is necessary to determine the amount of adsorbent required for the given adsorption process. The variation in percentage adsorption of DO25 with varying amount of SARM dosage was studied by conducting batch adsorption experiments at a temperature of 30 °C for different initial dye concentrations in the range of 130–230 mg/L for DO25. The SARM dosage was varied in the range of 0.15 and 1.8 g/L for DO25 adsorption. Initial pH of 4 was maintained for experiments. It is observed from Fig. 3 that, as SARM dosage was increased from 0.15 to 1.5 g/L, percentage adsorption of DO25 on SARM was found to increase. With further increase in dosage to 1.8 g/L, the adsorption of DO25 has remained constant. Thus, 1.5 g/L of adsorbent was considered to be quite appropriate dosage, as further increase in dosage did not show much increase in percentage adsorption of DO25.

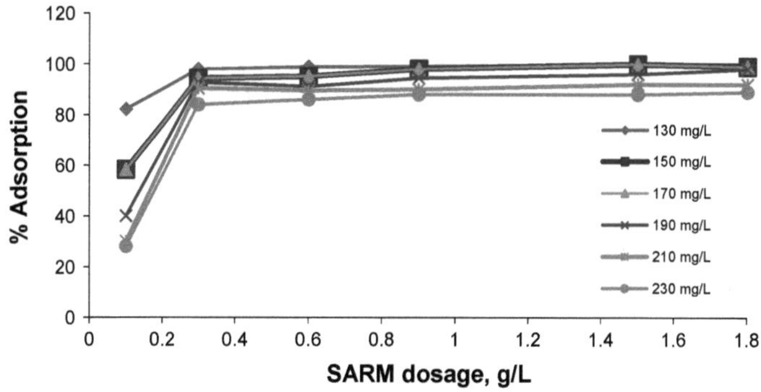

Fig. 3 Effect of SARM dosage on percentage adsorption for adsorption of DO25 with different initial dye concentrations. (pH = 4 and temperature = 30 °C)

3.5 *Adsorption Isotherm Analysis for Dye-SARM System*

Adsorption isotherms provide information about adsorption capacity of the adsorbent and its affinity towards the dye molecules. The two most common isotherms for studying the adsorption system are the Langmuir and the Freundlich isotherms. The model of monolayer adsorption is the Langmuir isotherm, the linearized form of which is shown in Eq. (1)

$$\frac{C_e}{q_e} = \frac{1}{Q_o b} + \frac{1}{C_e Q_o} \tag{1}$$

Where C_e is equilibrium dye solution concentration of DO25 (mg/L), q_e is the amount of adsorbed dye per unit mass of adsorbent (mg/g), Q_o the monolayer capacity of the adsorbent (mg/g) and b is the Langmuir adsorption constant indicating the adsorption energy (L/mg). The Freundlich model in the linear form is given as in Eq. (2)

$$lnq_e = lnk + \frac{lnC_e}{n} \tag{2}$$

Where, k is the adsorption capacity (mg/g) and n is an adsorption intensity. The equilibrium data obtained from batch experiments were fitted into Langmuir and Freundlich isotherm models to test the validity of these models for the dye adsorbent system. The values of parameters at three different temperatures of 20, 30 and 40 °C, along with the corresponding values of R^2 are presented in Table 1. The values of the parameters for Freundlich isotherm were obtained from the slope and intercept of lnq_e versus lnC_e. The values of parameters for Langmuir isotherms were obtained from the slope and intercept of $\frac{1}{q_e}$ versus $\frac{1}{C_e}$. The R^2 values indicated that, the DO25-SARM adsorption system can be represented by Langmuir isotherm

Table 1 Adsorption isotherm parameters for adsorption of DO25 at different temperature

Temperature °C	Langmuir model	Freundlich model
20	Q_o = 108 mg/g b = 1.11 L/mg R^2 = 0.98	n = 2.19 k = 34 mg/g, R^2 = 0.925
30	Q_o = 116 mg/g b = 1.27 L/mg R^2 = 0.99	n = 3.54 k = 52 mg/g R^2 = 0.88
40	Q_o = 127 mg/g b = 1.33 L/mg R^2 = 0.989	n = 6 k = 65.3 mg/g R^2 = 0.92

model under the conditions of the study owing to very small dye concentrations involved in the study.

3.6 *Thermodynamic Analysis for the Adsorption of DO25 on SARM*

Thermodynamic analysis of adsorption process for adsorption of DO25 dye on SARM was carried out to test for the feasibility and spontaneity of the process. Thermodynamic parameters, i.e. enthalpy change ΔH^o and entropy change ΔS^o are related as in Van't Hoff's equation given by Eq. (3). The values ΔH^o and ΔS^o were calculated for adsorption of DO25 on SARM, from the slopes and intercepts of the linear plot of *ln b* versus 1/*T* (calculated using the data in Table 1). ΔH^o The free energy ΔG^o of adsorption was ΔG^o calculated using Eq. (4).The values of thermodynamic parameters for DO25-SARM adsorption systems are presented in Table 2.

$$lnb = \frac{\Delta S^o}{R} - \frac{\Delta H^o}{RT} \tag{3}$$

Where b is the Langmuir adsorption equilibrium constant, R is the universal gas constant (8.314 J/mol. K), T is the absolute temperature. ΔH^o and ΔS^o are the standard enthalpy and entropy changes of adsorption respectively and the values ΔH^o and ΔS^o are calculated from the slopes and intercepts of the linear plot of *lnb* versus 1/*T*. The free energy ΔG^o of specific adsorption can be calculated using Eq. (4).

Table 2 Thermodynamic parameters for the DO25 (pH 4, Dosage 1.5 g/L) on SARM

Dye	ΔGo (kJ/mol)			ΔH^o (kJ/mol)	ΔS^o (J/mol)
	20 °C	30 °C	40 °C		
DO25	−0.23	−0.602	−0.702	67	23

$$\Delta G^o = -RTlnb \tag{4}$$

The adsorption equilibrium constant can be obtained from Langmuir isotherms at different temperatures. Negative values of ΔG^o indicate the spontaneous nature and feasibility of the adsorption process.

The negative values of ΔG^o as shown in Table 2 indicate the feasibility and spontaneous nature of DO25 adsorption on SARM. The change in enthalpy $\Delta H°$ for DO25-SARM adsorption process were found to be positive and it confirms the endothermic nature of adsorption (Wang et al. 2005). In the present study, the $\Delta H°$ value for DO25 adsorption on SARM is 67 kJ/mol.

3.7 Adsorption Kinetics for Adsorption of DO25 on SARM

Adsorption kinetic studies help in understanding the mechanism and performance of adsorption process. For kinetic studies, adsorption experiments were carried out at different initial DO25 concentrations in the range of 130–230 mg/L, with dosage of 1.5 g/L, pH 4 and temperature of 30 °C. In order to evaluate the kinetics of adsorption process of DO25 on the SARM from single dye aqueous solution, linearized form of pseudo-first-order and pseudo-second-order kinetic models (Eq. 5 and 6) were tested with the experimental data.

$$\log(q_e - q) = \log(q_e) - \left(\frac{K_1}{2.303}\right)t \tag{5}$$

$$\frac{t}{q} = \frac{1}{q_e^2 K_2} + \frac{t}{q_e} \tag{6}$$

Where q is the amount of adsorbate adsorbed at time t (mg g^{-1}), q_e is the adsorption capacity in equilibrium (mg g^{-1}), K_1 and K_2 are the pseudo-first and second order model rate constant respectively. The constants were obtained from the slope and intercept of the plot. The values of first-order rate constants obtained with analysis of experimental data with different initial dye concentrations and the corresponding correlation coefficient (R^2 value) are given in Tables 3. The R^2 value from the fit of experimental data on first order kinetic models are much lesser than ‘1’, indicating that first order kinetic model fails to describe the kinetics of DO25 on SARM accurately. To test the validity of second order kinetics, $\frac{t}{q}$ versus t were plotted for different initial concentrations of dye. The constants were obtained from the slope and intercept. R^2 value being closure to ‘1’ in the case of second order

Table 3 Rate constants and q_e values of first and second order kinetic models obtained at different initial DO25 concentrations with R^2 values

First order				
Initial concentration (mg/L)	q_e experimental (mg/g)	K_1 (min^{-1})	$q_{e'}$ calculated (mg/g)	R^2
130	54.53	0.046	30.2	0.842
150	90.66	0.048	41.40	0.812
170	102.533	0.047	40.74	0.792
190	168	0.043	60.26	0.81
Second order				
Initial concentration (mg/L)	q_e experimental (mg/g)	K_2 (g/mg min)	$q_{e'}$ calculated (mg/g)	R^2
130	54.53	0.0103	55.6	0.999
150	90.66	0.0057	91.74	0.99
170	102.53	0.0045	104.2	0.99
190	168	0.0027	169.5	0.99
		Mean = 0.00102		

kinetic model describes accurately the kinetics of adsorption of DO25 on SARM. The values of second-order rate constants data with different initial dye concentrations and the corresponding correlation coefficient (R^2 value) indicating the goodness of fit are given in Table 3. The rate constant decreased with the increase in initial dye concentration.

4 Conclusions

Batch adsorption experimental investigations were carried out to evaluate the efficacy of concentrated sulfuric acid-treated red mud as an adsorbent for the removal of Disperse orange 25 dye from aqueous solution. Acid treatment was found to increase the surface area of red mud. Acidic pH value of 4 was found to favour the adsorption of DO25. Adsorption rate was found to be faster initially, but decreased with the increase in contact time during the batch adsorption process. Red mud dosage of 1.5 g/L has been found to be adequate for dye solutions with initial dye concentrations ranging from 130 to 230 mg/L. Langmuir isotherm model has been found to represent the equilibrium data for DO25-SARM adsorption system. Thermodynamic analysis confirmed that DO25-SARM adsorption is an endothermic process. Adsorption kinetics may be represented by second-order kinetic models for the adsorption of DO25 onto SARM and the kinetic constant was estimate to be 0.00102 g/mg min.

References

Bansal, R.C., Goyal, M.: Activated Carbon Adsorption. Taylor & Francis Group, Boca Raton (2005)

Danis, T.G., Albanis, T.A., Petrakis, D.E., Pomonis, P.J.: Removal of chlorinated phenols from aqueous solutions by adsorption on alumina pillared clays and mesoporous alumina aluminum phosphates. Water Res. **32**(2), 295–302 (1998)

Gupta, V.K., Suhas, Ali, I., Saini, V.K.: Removal of rhodamine B, fast green, and methylene blue from wastewater using red mud, an aluminum industry waste. Ind. Eng. Chem. Res. **43**(7), 1740–1747 (2004)

Gupta, V.K., Suhas.: Application of low-cost adsorbents for dye removal—a review. J. Environ. Manage. **90**, 2313–2342 (2009)

Imamura, K., Ikeda, E., Nagayasu, T., Sakiyama, T., Nakanishi, K.: Adsorption behavior of methylene blue and its congeners on a stainless steel surface. J. Colloid Interface Sci. **245**(1), 50–57 (2002)

Namasivayam, C., Arasi, J.S.E.: Removal of congo red from wastewater by adsorption onto waste red mud. Chemosphere **34**, 401–417 (1997)

Namasivayam, C., Yamuna, R., Arasi, D.J.S.E.: Removal of acid violet from wastewater by adsorption on waste red mud. Environ. Geol. **41**(3–4), 269–273 (2001)

Norouzi, S., Badii, K., DoulatiArdejani, F.: Activated bauxite waste as an adsorbent for removal of Acid Blue 92 from aqueous solutions. Water Sci. Technol. **62**, 2491–2500 (2010)

Perez-Urquiza, M., Ferrer, R., Beltran, J.L.: Determination of sulfonatedazo dyes in river water samples by capillary zone electrophoresis. J. Chromatogr. A **883**(1), 277–283 (2000)

Pratt, K.C., Christoverson, V.: Hydrogenation of a model hydrogen-donor system using activated red mud catalyst. Fuel **61**(5), 460–462 (1982)

Rai, S., Wasewar, K.L., Mukhopadhyay, J., Yoo, C.K., Uslu, H.: Neutralization and utilization of red mud for its better waste management. Arch. Environ. Sci. **6**, 13–33 (2012)

Suzuki, T., Timofei, S., Kurunczi, L., Dietze, U., Schüürmann, G.: Correlation of aerobic biodegradability of sulfonated azo dyes with the chemical structure. Chemosphere **45**(1), 1–9 (2001)

Şahin, S., Demir, C., Guçer, S.: Simultaneous UV–vis spectrophotometric determination of disperse dyes in textile wastewater by partial least squares and principal component regression. Dyes Pigm. **73**(3), 368–376 (2007)

Shing, J. S. Method of activation of Red Mud, US Patent 4017425, (1997)

Shirzad-Siboni, M., Jafari, S.J., Giahi, O., Kim, I., Lee, S.M., Yang, J.K.: Removal of acid blue 113 and reactive black 5 dye from aqueous solutions by activated red mud. J. Ind. Eng. Chem. **20**(4), 1432–1437 (2014)

Snigdha, S., Vidya, S. and Batra. Modification of red mud by acid treatment and its application for CO removal. J. Hazard. Mater. 264–273 (2012)

Wang, S., Boyjoo, Y., Choueib, A., Zhu, Z.H.: Removal of dyes from aqueous solution using fly ash and red mud. Water Res. **39**(1), 129–138 (2005)

Removal of Cd(II) Ions from Aqueous Solution by Adsorption on $ZnCl_2$-Activated Carbon: Equilibrium and Kinetic Study

P.D. Meshram and S.S. Bhagwat

1 Introduction

The natural surface and ground water sources are greatly contaminated by heavy metal ions released from various sources including industrial processes and mining activities. This has posed a serious environmental threat because of the potential toxicity and irreparable effects of the heavy metal ions on the environment. Cadmium, one of the potentially carcinogenic heavy metal, can enter into the ecosystem from effluents of electroplating, production of batteries, mining, smelting and alloy industries (Farooq et al. 2010; Purkayastha et al. 2014). Bioaccumulation of cadmium in liver causes the severe liver and reproductive damage (Hrudey et al. 1996). Moreover, higher doses of cadmium build up in kidney is also responsible for the malfunction in spilling proteins in urine and disrupting protein metabolism (Waalkes 2000). A number of physio-chemical methods have been used over the past few decades for the removal of cadmium from wastewater. These include chemical precipitation, coagulation–flocculation, flotation, membrane filtration, ion exchange and electro-dialysis (O'Connell et al. 2008; Sud et al. 2008). However, most of these methods have their own intrinsic limitations of high capital and operational cost, inefficiency to treat effluents of dilute metal concentrations and generation of toxic sludge which requires further treatment (Moussous et al. 2012; Purkayastha et al. 2014). In this concern, an efficient and economically viable process is needed to be developed for effective removal of cadmium from wastewater to environmentally safe levels.

P.D. Meshram
University Institute of Chemical Technology, North Maharashtra University, Jalgaon, Maharashtra, India
e-mail: pawan.dm@gmail.com

P.D. Meshram · S.S. Bhagwat (✉)
Institute of Chemical Technology, Mumbai, Maharashtra, India
e-mail: ss.bhagwat@ictmumbai.edu.in

I. Regupathi et al. (eds.), *Recent Advances in Chemical Engineering*,
DOI 10.1007/978-981-10-1633-2_14

Adsorption process using the low-cost and abundantly available renewable sources offers an attractive potential alternative to the conventional methods. An extensive literature is available on the utilization of different types of biomass as adsorbent for removal of heavy metal ions from wastewater. However, untreated biomass suffers with low adsorption efficiency, system contamination due to release of soluble organic compounds contained in the biomass that further requires the higher chemical and biological oxygen demand (Wan Ngah and Hanafiah 2008). To resolve this, biomass has to be either chemically treated and/or transformed to carbonized mass using thermal activation for the effective surface area and texture to achieve desired adsorbent reactivity. Comprehensive efforts have been made to develop the activated adsorbent material for efficient metal ion removal (Leyva-Ramos et al. 2005; Kula et al. 2008; Wang et al. 2010; Machida et al. 2012; Bohli et al. 2015). Nevertheless, there is still having scope to find the cost-effective alternative adsorbent for heavy metal ion removal from contaminated streams. India being an agrarian country, sweet lime peels are readily available as inexpensive waste from fruit processing industry. An appropriate treatment and activation of this waste biomass may result into the suitable adsorbent for metal ion removal from contaminated water resources.

Hence, aim of the present study is to utilize sweet lime peels as a cheap renewable material for preparation of activated carbon by chemical treatment with $ZnCl_2$, followed by carbonization. The adsorption experiments in batch mode were performed to evaluate the potentiality of the prepared activated carbon for the adsorption of Cd (II) ions from aqueous solution. The effects of pH, temperature and contact time were investigated. Furthermore, the equilibrium experimental data were assessed using isotherm and kinetic sorption models to elucidate adsorption mechanism.

2 Experimental

2.1 *Metal Solution and Reagents*

All the reagents including the metal salt ($3CdSO_4 \cdot 8H_2O$) were of AR grade and procured from Merck Scientific-India. Experimental stock solution of Cd(II) ion (1000 mg L^{-1}) was prepared using double distilled water. Fresh dilutions of Cd(II) ion stock solution were used to prepare the synthetic water of different concentrations for experimental adsorption studies.

2.2 *Preparation of Activated Carbon*

The precursor (sweet lime peels) was obtained from M/S Mapro Foods Pvt. Ltd., Kangra (Himachal Pradesh, India). The preparation of activated carbon from the peels using $ZnCl_2$ as activating agent was carried out according to the procedure described by Subha and Namasivayam (2010). 200 g of dried peels were impregnated (70 °C, 4 h) with concentrated solution of $ZnCl_2$ (1:1, w/w). An impregnated whole mass was then set in a cotton filter to remove the excess solution, and oven-dried at 80 °C for 24 h. This impregnated dried mass was filled in a purpose made stainless still tubular reactor (*L*, 120 mm × Ø, 80 mm) and heated to carbonization temperature of 500 °C under N_2 flow at the heating rate of 10 °C min^{-1} for 2 h. The carbonized mass thus prepared was cooled in N_2 atmosphere and immersed in 200 ml of 2 M HCl solution at 80 °C for 1 h to leach out the excess $ZnCl_2$. Acid treated carbonized mass was repeatedly washed with warm distilled water to remove the traces of $ZnCl_2$ and other residue. An effective removal of $ZnCl_2$ was monitored through silver nitrate addition. Obtained material was dried at 100 °C for 24 h and sieved through particle size $\leq$0.425 mm.

2.3 Characterization of Activated Carbon

The proximate analysis viz. moisture content, volatile matter and ash content of the activated carbon was determined using ASTM D 2867-99, ASTM D 5832-98, ASTM D 2866-94 (reapproved 1999) methods respectively, and fixed carbon content was obtained as difference. The apparent density was determined according to ASTM D 2854-96 (reapproved 2000). The point of zero charge (pHpzc) determination was carried out as per the method described by Cerovic et al. (2007). Specific surface area and pore volume were determined using multipoint surface area analyzer NOVA–1000 (Quantachrome Instruments, USA). Temperature of the system (−195.6 °C) was maintained using liquid nitrogen. Brunaur, Emmet and Teller (BET) method was followed to determine the specific surface area (S_{BET}) over a relative pressure range of 0.01–0.3. Pore volume (V_t) was estimated from the liquid volume of adsorbate (N_2) at a relative pressure, $p/p_0 = 0.99$. The average pore diameter (D_{ap}) was obtained using the relation of $D_{ap} = 4V_t/S_{BET}$ (Tsai et al. 2001). The surface morphology of the activated carbon was observed by Field Emission-Scanning Electron Microscope (FE-SEM) S4800 Type-II (Hitachi High-Technologies Corporation, Tokyo, Japan) coupled to Energy dispersive x-ray (EDX) elemental mapping facility. A thin layer of gold was sputter-coated on the carbon for charge dissipation during SEM imaging.

2.4 Batch Adsorption Study

Batch mode adsorption studies were performed in a series of 100 mL Erlenmeyer flasks for the parameters of specified set conditions. Rotation speed of the shaker was kept at 200 rpm (REMI CIS-24BL orbital shaker). An investigated pH range of Cd(II) ion aqueous solution was 2–10. The pH of each solution was adjusted using 0.1 M HCl or 0.1 M NaOH solutions. A fixed mass of adsorbent carbon (2.0 g L^{-1})

was added to 25 mL Cd(II) ion solution (50 mg L^{-1}) and shaken to achieve equilibrium condition. An equilibrated mass was then filtered through Whatman filter paper (no. 42) and extent of Cd(II) ion adsorbed was determined as difference between residual and initial Cd(II) ion concentrations obtained using S-series atomic absorption spectrometer (AAS) equipped with SOLAAR software (Thermo Fisher Scientific Inc., India).

The equilibrium adsorption isotherms were generated for varying concentrations (5 mg L^{-1} to 300 mg L^{-1}) of Cd(II) ion solutions at optimized pH condition and at the temperatures 303, 313 and 323 K. Mass of the carbon was kept constant at 2.0 g L^{-1}. All the experiments were replicated twice and the average results are reported.

Percent Cd(II) ion adsorption and adsorption capacity (q_e, mg g^{-1}) at equilibrium were calculated from the experimental results of the batch study using the Eqs. (1) and (2), respectively, as given below:

$$\text{Cd(II) Adsorption (\%)} = \left(\frac{C_0 - C_t}{C_0}\right) \times 100 \tag{1}$$

$$q_e = \left(\frac{C_0 - C_e}{M}\right) \times V \tag{2}$$

where C_0—initial Cd(II) ion concentrations (mg L^{-1}), C_t and C_e—residual concentrations of Cd(II) ion (mg L^{-1}) after adsorption at time t and at equilibrium time respectively, V—volume of solution treated (L) and M—mass of the adsorbent (g).

3 Result and Discussion

The preparation of $ZnCl_2$—activated carbon has produced the yield of 27.1 and 40.8 % on the basis of dried untreated peel mass and $ZnCl_2$ impregnated mass respectively. The yield is the relative measure of the vaporization of volatile products from the hydroaromatic structure of the precursor due to intensified dehydration and elimination reactions. The physio-chemical characteristics of the prepared activated carbon are presented in Table 1.

3.1 Effect of PH

The pH of the solution medium to be treated for remediation of metal ion contaminated resources plays a worthy role in adsorption process. The variation in pH influences the surface charge on the binding sites of the adsorbent. Also, the ionization and solubility of metal species in the solution is affected (Remenarova et al. 2012). The Cd(II) ion removal efficiency as a function of varying pH in the range of

Table 1 Characteristics of $ZnCl_2$-activated carbon prepared from sweet lime peels

Properties	Value
Moisture content (%)	14.1
Volatile matter (%)	11.9
Fixed carbon (%)	67.9
Ash content (%)	6.1
Apparent density (g cm^{-3})	0.627
Point of zero charge (pH_{PZC})	5.8
BET specific surface area (S_{BET}, m^2 g^{-1})	1068.7
Total pore volume (V_t, cm^3 g^{-1})	0.6101
Average pore diameter (D_{ap}, Å)	22.83

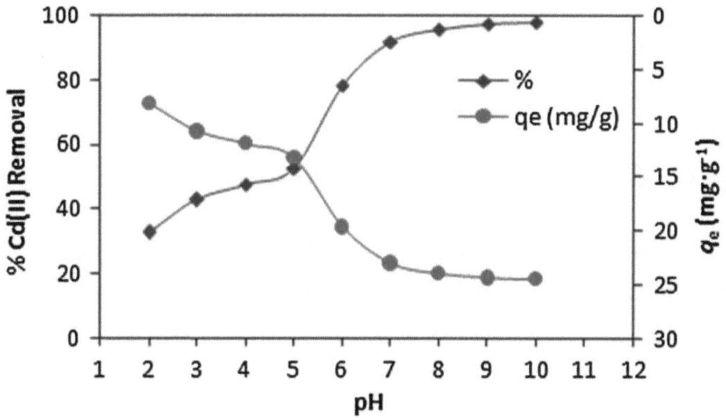

Fig. 1 Effect of pH on the removal of Cd(II) ions using $ZnCl_2$-activated carbon (*conditions* Initial Cd(II) ion concentration = 50 mg L^{-1}; adsorbent dose = 2 g L^{-1}; contact time = 1080 min; and temperature = 303 K)

2–10 was investigated and the results are shown in Fig. 1. A gradual increase in Cd (II) ion adsorption from 32.7 to 52.5 % is evident with increasing pH from 2 to 5. Further increase in pH value above 5 has shown a steep rise in the Cd(II) ion removal till pH, 8 (from 52.5 to 95.9 %); thereafter almost a plateau regime with no significant increase was observed. These changes in the Cd(II) ion adsorption is related to surface charge properties of the activated carbon in solutions of different pH conditions. This can be elucidated well by point of zero charge (pHpzc) theory, which defines the pH at which the adsorbent surface gains a neutral charge in absence of specific sorption. The experimentally determined pHpzc of activated carbon is 5.8. At solution pH > pHpzc the active sites on the carbon surface became negatively charge favoring the better adsorption of cationic Cd(II) ions due to increased electrostatic attraction (Borai et al. 2007). Therefore, changing the solution pH from 5 to 8 resulted in higher Cd(II) uptake rate from 13.1 to 23.9 mg g^{-1}. On the contrary, when the solution pH is maintained below pHpzc the surface charge on the carbon is positive and thus, would repulse more significantly

Cd(II) ions in the solution (Srivastava et al. 2006). Therefore, the adsorptive uptake of Cd(II) ion decreased from 19.6 to 8.1 mg g^{-1} with decline in the pH from 6 to 2.

Besides, the adsorption process is also dependent upon the species distribution of the metal ions in the aqueous medium, which is strongly influenced by pH. At pH < 6, cadmium solely exists in the form of Cd^{2+} ionic species in the solution, while at pH below 8 the prevailing species are Cd^{2+} and $Cd(OH)^+$ (Leyva-Ramos et al. 1997). The shift in the solution pH beyond 8 resulted into decreased Cd^{2+} concentration in solution, and concurrently cadmium could start precipitating at pH, 8.3 due to formation of insoluble Cd-hydroxy species, $Cd(OH)_{2(s)}$ (Ajmal et al. 2003). From Fig. 1, the majority of Cd(II) ion adsorption was seen at pH, 7 corresponding to adsorption efficiency of 92.7 % and the equilibrium capacity was 23.0 mg g^{-1}. Therefore, the solution pH of 7 was chosen to be optimum, at which adsorption efficiency was greater than 90.0 % and is well below the pH levels of cadmium precipitation.

3.2 *Effect of Temperature and Contact Time*

The rate of Cd(II) ion adsorption on activated carbon was investigated at different temperatures in the range of 303–323 K and the results are shown in Fig. 2.

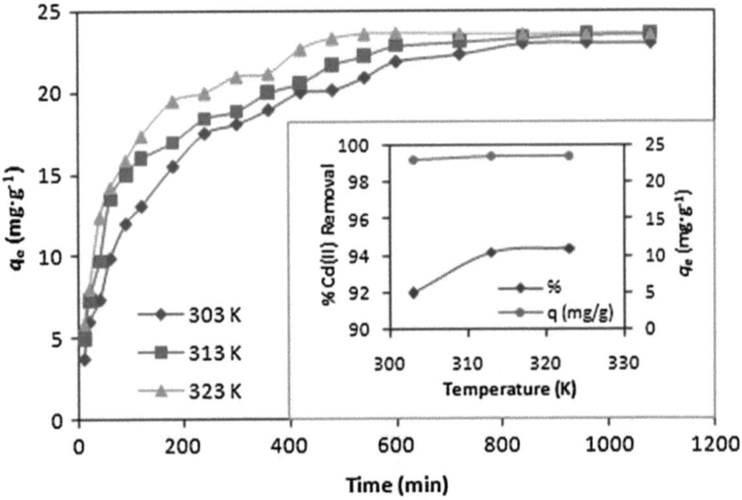

Fig. 2 Effect of temperature and contact time on the removal of Cd(II) ions using $ZnCl_2$-activated carbon (*conditions* pH = 7.0; Initial Cd(II) ion concentration = 50 mg L^{-1}; adsorbent dose = 2 g L^{-1})

It has been observed that percent Cd(II) ion removal and adsorption capacity (q_e) increased for temperature rise from 303 to 313 K, but further increase in temperature showed no significant contribution in sorption process. A rapid Cd(II) ion adsorption stage was seen initially within 180 min of contact time at different temperatures, and then subsequently much slower stage of adsorption before equilibrium was established. It can also be noted that the time required to reach sorption equilibrium was not affected by increasing the temperature from 303 to 323 K. In general, increase in temperature provides sufficient energy to overcome the resisting hydration (adsorbate solvation) energy acting on the diffusing metal ions, thereby increasing the mobility of metal ions towards adsorbent surface (Wang et al. 2012). However, a very slight change in q_e values (23.0–23.6 mg g^{-1}) was observed for activated carbon in the tested temperature range suggesting non-significant influence of temperature rise on adsorptive removal of Cd(II) ions.

3.3 Adsorption Isotherm

The fitment of experimental equilibrium data to one or more standard isotherm models is of immense importance to understand the interaction of adsorbate (metal ions) with solid adsorbent (activated carbon) and to elucidate the governing mechanism in sorption process (Arshadi et al. 2014). In this work, the two typical adsorption isotherm models viz. Freundlich isotherm and Langmuir isotherm were used to fit the equilibrium data. The non-linear form of the Freundlich and Langmuir isotherm models are given by Eqs. (3) and (4) respectively, as follows:

$$q_e = K_F C_e^{1/n} \tag{3}$$

$$q_e = \frac{q_m K_L C_e}{1 + K_L C_e} \tag{4}$$

where q_e—adsorption capacity of adsorbent at equilibrium (mg g^{-1}), C_e—equilibrium concentration of the adsorbate in solution (mg L^{-1}), K_F—Freundlich constant related to sorption capacity ($mg^{1-1/n}$ g^{-1} $L^{1/n}$), ($1/n$)—dimensionless constant related to sorption intensity, q_m—saturated monolayer adsorption capacity of the sorbent (mg g^{-1}) and K_L—Langmuir affinity constant (L mg^{-1}).

The goodness of fit of these applied models was assessed using coefficient of determination (r^2) and non-linear chi-square (χ^2) error function. An optimization approach was used for model parameter estimation and fitting of the isotherm curves. In the method, non-linear regression using repetitive attempts to succeed at maximizing the r^2 values between the experimental data and the predicted isotherms was performed using Solver add-in function facility in Microsoft excel. The optimized non-linear Freundlich and Langmuir isotherm fits for adsorption of Cd (II) ions on activated carbon at different temperatures are shown in Fig. 3.

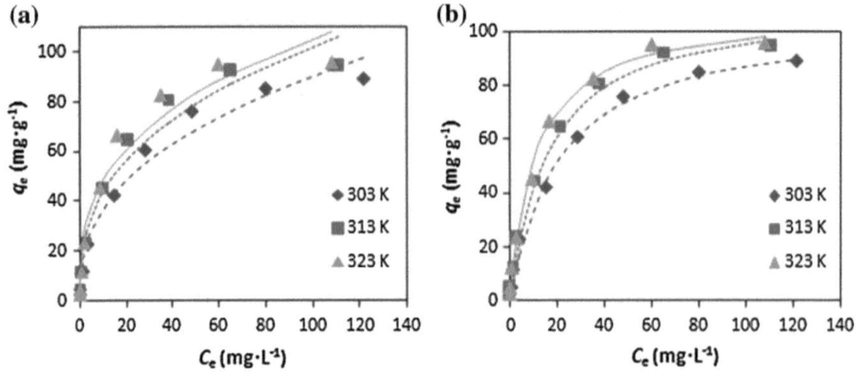

Fig. 3 Adsorption isotherms for Cd(II) ion adsorption on $ZnCl_2$-activated carbon at different temperatures; **a** Freundlich model fit and **b** Langmuir model fit

The resultant Freundlich and Langmuir parameters and error function (χ^2) values are listed in Table 2. The r^2 values of Langmuir isotherm obtained for activated carbon were extremely high (≥ 0.99) in comparison to that of Freundlich isotherm. In addition, the relatively very low χ^2 values of Langmuir isotherm suggest that the Langmuir model is able to provide the perfect fit to the equilibrium data, confirming the monolayer adsorption of Cd(II) ions on activated carbon. The Langmuir constant, K_L is related to adsorption energy. The rise in temperature from 303 to 323 K of sorption process resulted to give increasing K_L values, signifying higher heat of adsorption associated with increasing temperature. Moreover, increase in q_m values with temperature was observed. This indicated that higher temperatures favor the adsorption of Cd(II) ions on activated carbon, suggesting the sorption process to be endothermic.

3.4 Adsorption Kinetics

The study of rate kinetics of sorption process is of great significance in determination of the performance parameters, which subsequently forms a basis for effective designing and scaling of adsorption process (Elkady et al. 2011). The experimental kinetic data were fitted to reaction rate models viz. Pseudo-first-order and Pseudo-second-order kinetic models, the linear forms of which are given by the Eqs. (5) and (6) respectively, as follows:

$$\log(q_e - q_t) = \log q_e - \frac{k_1}{2.303}t \tag{5}$$

$$\frac{t}{q_t} = \frac{1}{k_2 q_e^2} + \frac{1}{q_e}t \tag{6}$$

Table 2 Isotherm parameters for Cd(II) ion adsorption on $ZnCl_2$-activated carbon at different temperatures

Temperature (K)	Freundlich				Langmuir			
	K_F ($mg^{1-1/n}$ g^{-1} $L^{1/n}$)	n (−)	r^2	χ^2	q_m ($mg\ g^{-1}$)	K_L ($L\ mg^{-1}$)	r^2	χ^2
303	14.6605	2.5335	0.9767	8.4137	105.1312	0.0485	0.9906	0.2298
313	18.9531	2.7402	0.9650	13.0650	107.6748	0.0622	0.9889	0.3589
323	21.5903	2.9065	0.9587	13.9306	110.9858	0.0937	0.9925	0.6484

Table 3 Kinetic parameters for Cd(II) ion adsorption onto $ZnCl_2$-activated carbon at different temperatures

Temperature (K)	$q_{e,exp}$ ($mg\ g^{-1}$)	Pseudo-first-order model			Pseudo-second-order model		
		$q_{e,cal}$ ($mg\ g^{-1}$)	$k_1 \times 10^{-3}$ (min^{-1})	R^2	$q_{e,cal}$ ($mg\ g^{-1}$)	$k_2 \times 10^{-3}$ ($g\ mg^{-1}\ min^{-1}$)	R^2
303	23.00	23.74	5.7575	0.9092	25.00	0.4186	0.9974
313	23.55	19.09	5.5272	0.9578	24.94	0.5989	0.9977
323	23.61	18.40	8.5211	0.9293	24.81	0.9038	0.9991

where k_1—first-order rate constant (min^{-1}), k_2—second-order rate constant ($g\ mg^{-1}\ min^{-1}$), q_e—equilibrium adsorption uptake at time $t = \infty$($mg\ g^{-1}$) and q_t—adsorption uptake at time t ($mg\ g^{-1}$).

The experimental data at different temperatures for the adsorption of Cd(II) ions on activated carbon was fitted to kinetics models using linear regression analysis of Eqs. (5) and (6). The calculated values of rate parameters (k_1, k_2), adsorption capacity ($q_{e,cal}$) and the corresponding regression coefficients (R^2) are summarized in Table 3.

Pseudo-first-order model showed poor fit to the experimental data as R^2 values at tested temperatures were ≤ 0.95 and the calculated equilibrium sorption capacity values ($q_{e,cal}$) were far lower than the experimentally determined values ($q_{e,exp}$). As against, a highly linear relationship over the whole range of contact time with R^2 values close to unity (>0.99) was obtained for Pseudo-second-order kinetic plot at each temperature. Additionally, $q_{e,cal}$ from the second-order model at different temperature approximated well to $q_{e,exp}$ values. Therefore, the sorption kinetics of Cd(II) ion adsorption on activated carbon is more precisely described by the mechanism of surface site-adsorbate reaction of Pseudo-second-order model. The model is based on the assumption that the rate-limiting step is chemisorption, and thus may involving valence forces through sharing or exchange of electrons between the adsorbate, Cd(II) ions and adsorbent, activated carbon (Mohan et al. 2006). Similar findings were reported for the uptake of Cd(II) ions by multi-walled carbon nanotubes (Salam et al. 2012) and activated carbon prepared chemically from olive stones (Bohli et al. 2015).

These results further allowed determination of the activation energy (E_a) using the Arrhenius equation in logarithmic form as follows:

$$\ln k_2 = \ln A - \frac{E_a}{RT} \tag{7}$$

where A—constant called the frequency factor ($g\ mg^{-1}\ min^{-1}$), R—gas constant ($J\ mol^{-1}\ K^{-1}$) and T—solution temperature (K).

Linear regression was applied to the plot of $\ln(k_2)$ versus $(1/T)$ to determine the activation energy (E_a) from the slope, ($-E_a/R$). Besides, magnitude of E_a is commonly taken as a measure to identify the underlying phenomenon of physical and chemical sorption in the process. Physical adsorption involves weaker electrostatic forces yielding usually less than 4.2 kJ mol^{-1} of E_a. While chemical adsorption is specific, involving stronger forces and thus, requires larger E_a from 8.4 to 83.7 kJ mol^{-1} (Saha and Chowdhury 2011). For the sorption of Cd(II) ions on activated carbon the E_a was obtained to be 31.27 kJ mol^{-1}. Positive value of E_a signifies the energy that must be overcome for adsorption to occur. Also, it can be confirmed from E_a value that the chemisorption is the primary governing mechanism in sorption of Cd(II) ions on activated carbon.

3.5 SEM-EDX Analysis

SEM micrographs and EDX spectra of activated carbon before and after Cd(II) ion adsorption are presented in Fig. 4. The carbon surface is shown to possess the porous structure with highly heterogeneous and rough texture. After Cd(II) ion adsorption, distinct alterations in surface morphology of carbon can be clearly visualized. The carbon surface shows filling of pores covered with precipitous layers formed upon complexation with Cd(II) ions. Furthermore, the carbon-metal interaction has shown appearance of a distinctive peak signal of Cd(II) ion in the

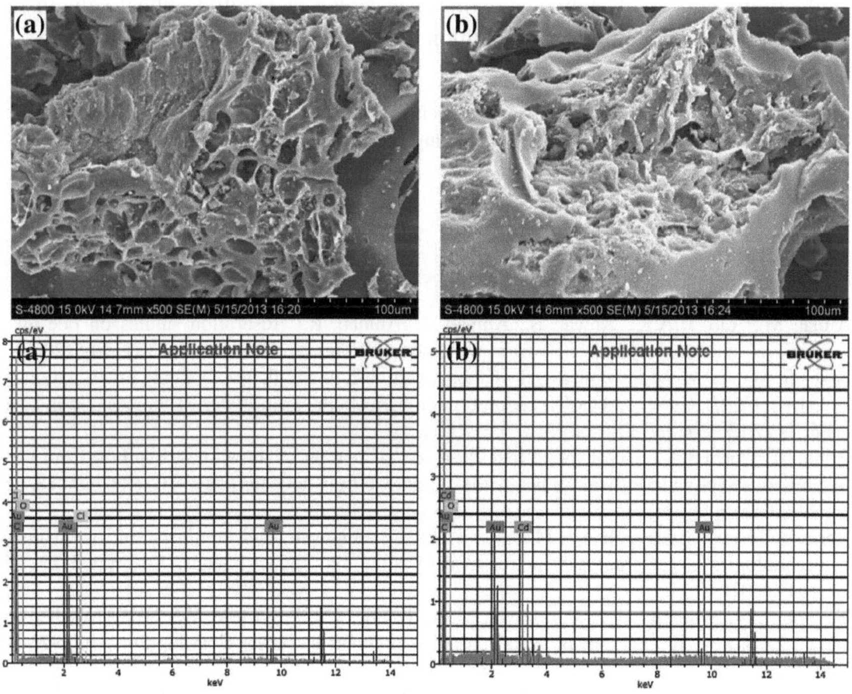

Fig. 4 FE-SEM and EDX analysis of $ZnCl_2$-activated carbon **a** before and **b** after Cd(II) ion adsorption

EDX spectra (Fig. 4b), which confirms the ability of the activated carbon for Cd(II) ion binding. The elemental composition as determined by EDX analysis consist of C = 69.92 %, O = 8.16 % and Cd = 3.82 % on normalized wt percent basis.

4 Conclusion

The following important conclusions were drawn based on the results of the present study:

(i) The pH of the aqueous medium is the controlling parameter for adsorption of Cd(II) ions on $ZnCl_2$-activated carbon. The Cd(II) ion removal efficiency increased with increasing pH, and the optimized pH was found to be 7.
(ii) Increasing the temperature, neither affected the time to reach equilibrium nor showed any substantial influence on Cd(II) ion adsorption capacity.
(iii) Adsorption of Cd(II) ion is better described by Langmuir isotherm model ($r^2 \geq 0.99$) suggesting the dominance of monolayer coverage of Cd(II) ions on activated carbon surface.
(iv) Kinetics of Cd(II) ion adsorption obeyed the Pseudo-second-order model indicating the rate limiting step may be chemical sorption.
(v) In overall, $ZnCl_2$-activated carbon has shown its ability to effectively adsorb Cd(II) ions from aqueous solution.

Acknowledgments The authors are very much thankful to the TEQIP-II, funded by MHRD, Government of India, Delhi (India) for providing financial support to carry out this work.

References

Ajmal, M., Rao, R.A.K., Anwar, S., Ahmad, J., Ahmad, R.: Adsorption studies on rice husk: removal and recovery of Cd(II) from wastewater. Bioresour. Technol. **86**, 147–149 (2003)

Arshadi, M., Amiri, M.J., Mousavi, S.: Kinetic, equilibrium and thermodynamic investigations of Ni(II), Cd(II), Cu(II) and Co(II) adsorption on barley straw ash. Water Resources and Industry **6**, 1–17 (2014)

Bohli, T., Ouederni, A., Fiol, N., Villaescusa, I.: Evaluation of an activated carbon from olive stones used as an adsorbent for heavy metal removal from aqueous phases. C. R. Chim. **18**, 88–99 (2015)

Borai, E.H., El-Sofany, E.A., Morcos, T.N.: Development and optimization of magnetic technologies based processes for removal of some toxic heavy metals. Adsorption **13**, 95–104 (2007)

Cerovic, L.S., Milonjic, S.K., Todorovic, M.B., Trtanj, M.I., Pogozhev, Y.S., Blagoveschenskii, Y., Levashov, E.A.: Point of zero charge of different carbides. Colloids Surf. A **297**, 1–6 (2007)

Elkady, M.F., Abu-Saied, M.A., Rahman, A.A., Soliman, E.A., Elzatahry, A.A., Yossef, M.E., Eldin, M.M.: Nano-sulphonated poly (glycidyl methacrylate) cations exchanger for cadmium ions removal: effects of operating parameters. Desalination **279**, 152–162 (2011)

Farooq, U., Kozinski, J.A., Khan, M.A., Athar, M.: Biosorption of heavy metal ions using wheat based biosorbents—a review of the recent literature. Bioresour. Technol. **101**, 5043–5053 (2010)

Hrudey, S.E, Chen, W., Rousseaux, C.G.: Bioavailability in environment risk assessment, pp. 91–100. CRC Press, Boca Raton, Florida (1996)

Kula, I., Ugurlu, M., Karaoglu, H., Celik, A.: Adsorption of Cd(II) ions from aqueous solutions using activated carbon prepared from olive stone by $ZnCl_2$ activation. Bioresour. Technol. **99**, 492–501 (2008)

Leyva-Ramos, R., Rangel-Mendez, J.R., Mendoza-Baron, J., Fuentes-Rubio, L., Guerrero-Coronado, R.M.: Adsorption of Cadmium(II) from aqueous solution onto activated carbon. Water Sci. Technol. **35**, 205–211 (1997)

Leyva-Ramos, R., Bernal-Jacome, L., Acosta-Rodriguez, I.: Adsorption of cadmium(II) from aqueous solution on natural and oxidized corncob. Sep. Purif. Technol. **45**, 41–49 (2005)

Machida, M., Fotoohi, B., Amamo, Y., Ohba, T., Kanoh, H., Mercier, L.: Cadmium(II) adsorption using functional mesoporous silica and activated carbon. J. Hazard. Mater. **221**, 220–227 (2012)

Mohan, D., Pittman Jr., C.U., Steele, P.H.: Single, binary and multi-component adsorption of copper and cadmium from aqueous solutions on Kraft lignin—a biosorbent. J. Colloid Interface Sci. **297**, 489–504 (2006)

Moussous, S., Selatnia, A., Merati, A., Junter, G.A.: Batch Cadmium(II) biosorption by an industrial residue of macrofungal biomass (*Clitopilus scyphoides*). Chem. Eng. J. **197**, 261–271 (2012)

O'Connell, D.W., Birkinshaw, C., O'Dwyer, T.F.: Heavy metal adsorbents prepared from the modification of cellulose: a review. Bioresour. Technol. **99**, 6709–6724 (2008)

Purkayastha, D., Mishra, U., Biswas, S.: A comprehensive review on Cd(II) removal from aqueous solution. Journal of Water Process Engineering **2**, 105–128 (2014)

Remenarova, L., Pipiska, M., Hornik, M., Rozloznik, M., Augustin, J., Lesny, J.: Biosorption of cadmium and zinc by activated sludge from single and binary solutions: mechanism, equilibrium and experimental design study. J. Taiwan Inst. Chem. Eng. **43**, 433–443 (2012)

Saha, P., Chowdhury, S.: Insight into adsorption thermodynamics. In: Tadashi, M. (ed.) Thermodynamics, pp. 349–364. InTech, Vienna, Austria (2011)

Salam, M.A., Al-Zhrani, G., Kosa, S.A.: Simultaneous removal of copper(II), lead(II), zinc(II) and cadmium(II) from aqueous solutions by multi-walled carbon nanotubes. C. R. Chim. **15**, 398–408 (2012)

Srivastava, V.C., Mall, I.D., Mishra, I.M.: Equilibrium modelling of single and binary adsorption of cadmium and nickel onto bagasse fly ash. Chem. Eng. J. **117**, 79–91 (2006)

Subha, R., Namasivayam, C.: $ZnCl_2$-modified activated carbon from biomass coir pith for the removal of 2-Chlorophenol by adsorption process. Bioremediat. J. **14**, 1–9 (2010)

Sud, D., Mahajan, G., Kaur, M.P.: Agricultural waste material as potential adsorbent for sequestering heavy metal ions from aqueous solutions–a review. Bioresour. Technol. **99**, 6017–6027 (2008)

Tsai, W.T., Chang, C.Y., Lin, M.C., Chien, S.F., Sun, H.F., Hsieh, M.F.: Characterization of activated carbons prepared from sugarcane bagasse by $ZnCl_2$ activation. Journal of Environmental Science and Health: Part B **36**, 365–378 (2001)

Waalkes, M.P.: Cadmium carcinogenesis in review. J. Inorg. Biochem. **79**, 241–244 (2000)

Wan Ngah, W.S., Hanafiah, M.A.K.M.: Removal of heavy metal ions from wastewater by chemically modified plant wastes as adsorbents: a review. Bioresour. Technol. **99**, 3935–3948 (2008)

Wang, C.P., Wang, B.L., Liu, J.T., Yu, L., Sun, H.W., Wu, J.Z.: Adsorption of Cd(II) from acidic aqueous solutions by tourmaline as a novel material. Chin. Sci. Bull. **57**, 3218–3225 (2012)

Wang, F.Y., Wang, H., Ma, J.W.: Adsorption of cadmium(II) ions from aqueous solution by a new low-cost adsorbent–bamboo charcoal. J. Hazard. Mater. **177**, 300–306 (2010)

Removal of CO_2 in a Multistage Fluidized Bed Reactor by Activated Carbon Prepared from Green Coconut Shell

Dipa Das, Debi Prasad Samal and B.C. Meikap

1 Introduction

Increase of greenhouse gases in atmosphere causes global warming. The major constituent of greenhouse gas is carbon dioxide (CO_2). Tremendous amount of CO_2 comes from emissions of combustion of fossil fuels. Carbon capture and storage (CCS) is the most important techniques to control CO_2 emission (Plaza et al. 2007). Out of different separation techniques like absorption, adsorption, cryogenic distillation and membrane separation, adsorption can be a competitive method for CO_2 capture because it requires lower energy for solvent regeneration and cost of post-combustion capture reduces (Chaffee et al. 2007). For adsorption of CO_2, different sorbents like coal, molecular sieves, zeolites, activated carbon and carbon replicas (Mastalerz et al. 2004) etc. are used. Among these adsorbents, activated carbon is the best suitable adsorbent due to its lower cost, hydrophobicity, and less energy requirements to regenerate (Himeno et al. 2005). The performance of adsorption of activated carbon depends on its surface chemistry and pore structure. (Sircar et al. 1996; Lee et al. 2010). In the present work, green coconut shell based

D. Das (✉) · D.P. Samal · B.C. Meikap
Department of Chemical Engineering, IIT Kharagpur,
Kharagpur 721302, West Bengal, India
e-mail: dipadas@che.iitkgp.ernet.in

D.P. Samal
e-mail: debiprasad.nitrkl@gmail.com

B.C. Meikap
e-mail: bcmeikap@che.iitkgp.ernet.in

B.C. Meikap
Department of Chemical Engineering, School of Engineering,
Howard College, University of Kwazulu-Natal, Durban, South Africa

I. Regupathi et al. (eds.), *Recent Advances in Chemical Engineering*,
DOI 10.1007/978-981-10-1633-2_15

activated carbon has been prepared. Due to small macropores structure, high fixed carbon content and low ash content, it is more superior than other adsorbent for the adsorption of CO_2 from flue gas. The advantages of multistage fluidized bed reactor performance is more as compared to fixed bed reactor due to rapid mixing of solid-fluid and large surface area available for reaction. The multistage fluidized bed reactor with a down comer is one of the recently developed equipment used in industry for removal of pollutants from flue gas (Mohanty et al. 2008, 2009; Mohanty and Meikap 2009). Reduction of gas bypassing is the major advantage of multistage fluidized bed reactor along with narrow residence time distribution of solids as a result of which the reactor approaches the condition of plug flow (Roy et al. 2009). This study represents the operation of a four-stage fluidized bed reactor for removal of carbon dioxide (CO_2) from flue gas by using activated carbon.

2 Material and Method

2.1 Materials

Green coconut Shells were collected from nearby local market of IIT Kharagpur and then cut into small pieces, washed with tap water to remove dirt followed by drying in the sunlight for 15–20 days till it becomes completely dry. Dried materials were kept inside the furnace at 150 °C for 24 h for removal of moisture and other volatile impurities. Dried samples were crushed with a locally made crusher and sieved to size range of 300–700 μm. The powdered precursor was chemical activated with $ZnCl_2$ with impregnation ratio (activating agent/precursor) 1:1. The slurry was properly mixed and kept for 24 h for proper soaking with $ZnCl_2$ and dried inside the oven at 100 °C for 24 h. The chemical impregnated samples were kept inside the furnace by placing the material inside galvanized iron pipe. The materials inside the furnace was heated at the rate of 10 °C min^{-1} and reaches the temperature 650 °C i.e. the carbonization temperature of under 150 cm^3 min^{-1}, nitrogen flow rate for 1 h at STP. Then it was cooled under the constant flow of nitrogen gas till the temperature reaches the room temperature. The resulting material was washed with 0.5 N HCl for 2–3 times and washed with hot distilled water and then with cold water to remove any kinds of residual organic and mineral matter, till the solution becomes neutral. Finally the samples were dried for 24 h at 100 °C inside an oven to become completely dry.

2.2 Characterization of Activated Carbon

See Tables 1 and 2.

Table 1 Physical characteristics of activated carbon

Property	Value
Particle size (μm)	300–700
Particle density (kg/m^3)	2410
Porosity (ϵ_{mf})	0.61
Sphericity (Φ_s)	0.72
Minimum fluidization velocity (U_{mf})	0.142

Table 2 Pore structure parameter for activated carbon sample

Sample	BET SSA (m^2/g)	V_{Tot} (cm^3/g)	V_{micro} (cm^3/g)	A_{micro} (m^3/g)	Avg. Pore radius(A^0)
Activated carbon	1076.678	0.5487	0.425	998.21	9.8659

3 Experimental Methods

Figures 1 and 2 shows the schematic diagram and the pictorial view of four staged fluidized bed reactor. The fluidized bed column was consisted of four stages (0.21 m height per stage and 0.095 m internal diameter). Stages were assembled together with a flanged joint. Four number of stainless steel made plate (S_1, $S_{2,}$, S_3, S_4) of 0.002 m thick were used as internal baffles for separating two stages. Hole of diameter 0.002 m was there on each plate. To avoid solids were falling down through the plate, grid plates were covered with 100 mesh size weir mesh with

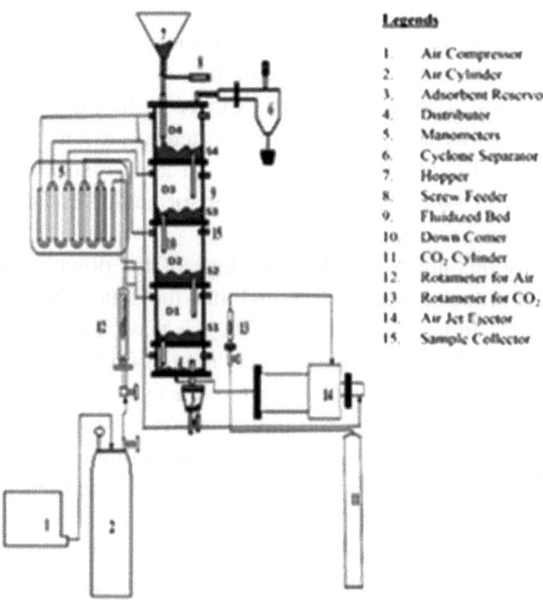

Fig. 1 Experimental set-up for four stage fluidized bed reactor (schematic diagram)

Fig. 2 Experimental set-up for four stage fluidized bedreactor (pictorial view)

openings smaller than particle size. Down comers (D_1, D_2, D_3, D_4) of 0.024 m internal diameter and height of 0.265 m and were made up of perspex material. Each section was provided with down-comers and it was further fitted with a cone of diameter 0.007 and 0.024 m height at the exit end so that up-flow of the gas through the down-comer was reduced as a result of which stable operation was maintained. The down-comers were further fitted to the plate for adjustment of required weir height i.e. considered as bed height. The material flows from stage to stage through the down-comer. There were provisions for measuring pressure drop. Gas distributor was there at the bottom of the column so that air was properly distributed to the column. A rotameter was fitted to the column for measurement of the air flow rate. For storage of the solid, a hopper was attached to the bottom of the column. A feeding funnel was there at the top of the column to hold the activated carbon particle and it was connected to the screw feeder. The speed of the motor (0.25 HP) can be controlled by variable rheostat and the screw feeder was connected to the motor. The compressor having capacity 5HP was used to supply the air for fluidization.

Solids are put to the first stage of the down-comer from the top of the funnel connected to screw feeder and then the materials were passed through perspex tube (0.011 m internal diameter). The gas leaving from the top stage was passed through 0.14 m diameter cyclone to collect the fines coming out from the fluidized bed, a bag was attached at the bottom of the cyclone. Compressed air at 2 kg/cm^2 pressure was passed through the pipe line. At the same time CO_2 from the cylinder was passed at certain flow rate. From the mixing chamber, the CO_2 and air mixture was fed into at the bottom stage of the fluidized bed reactor. Counter-current flow of the solids and the gas flow were occurred at each stage. The operating range of gas velocity for this experiments were 0.233–0.396 m/s and the solid flow rate range of 2.15–4.12 kg/hr. The weir height of the down comer was kept at 0.02 m and

0.06 m. From the literature we have studied that the flue gas containing CO_2 coming out from stack of industries are generally very high concentration. So we have taken an approximate range (low concentration to high concentration) in ppm scale varies in the range of 3000–12,000 ppm. Hence we have kept in this range. The inlets CO_2 loadings were varied from 3000 to 12,000 ppm in four stages for each gas flow rate. Activated carbon was used as adsorbent for this experiment.

4 Sampling and Analysis

The gas samples (CO_2 and air) were drawn with the help of aspiratory bottles at the inlet and outlet of the column under steady and stable operation of the reactor. The air and CO_2 mixture that was collected, were analyzed for carbon dioxide by the "Orsat Analysis" apparatus. The change in reading gives the volume % of CO_2 absorb by the KOH solution. Then it was converted to ppm. The removal of CO_2 (%) has been calculated by the following Eq. (1).

$$\%\,\text{Removal of CO}_2 = \frac{\text{CO}_{2\,\text{inlet}} - \text{CO}_{2\,\text{outlet}}}{\text{CO}_{2\,\text{inlet}}} \times 100 \quad (1)$$

The % Removal of CO_2 in stage i, can be calculated as,

$$\eta_{CO_2} = \frac{C_{i+1} - C_i}{C_{i+1}} \times 100 \quad (2)$$

Where C_i and C_{i+1} are concentrations of carbon dioxide in gas at outlet and inlet of the column.

5 Results and Discussion

5.1 *Effect of Gas Velocity on Percentage Removal of CO_2*

Figures 3 and 4 represents the percentage removal of CO_2 at different inlet CO_2 loading at different gas velocity at a constant solid flow rate. It may be seen from Figs. 3a and 4a that for inlet concentration 3000 ppm, the maximum removal efficiency of carbon dioxide was 55.39, 53.62 and 49.62 % at 0.02 m weir height with gas flow rate 0.233, 0.349 and 0.396 m/s respectively, at constant solid flow rate 2.15 kg/hr. The maximum removal efficiency of carbon dioxide for inlet concentration 12,000 ppm was 34.76, 32.19 and 30.09 % at 0.02 m weir height at same operating conditions. The maximum removal efficiency of carbon dioxide for inlet concentration 3000 ppm is 61.92, 57.83 and 55.93 % at 0.06 m weir height with gas flow rate 0.233, 0.349 and 0.396 m/s respectively at constant solid flow

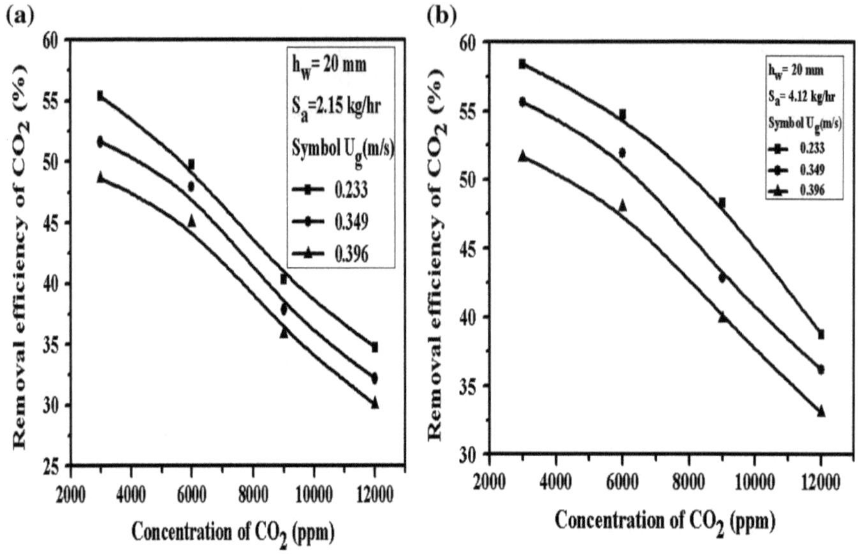

Fig. 3 Effect of gas velocity on CO_2 removal efficiency at h_w = 20 mm for S_a: **a** 2.15 and **b** 4.12 kg/hr

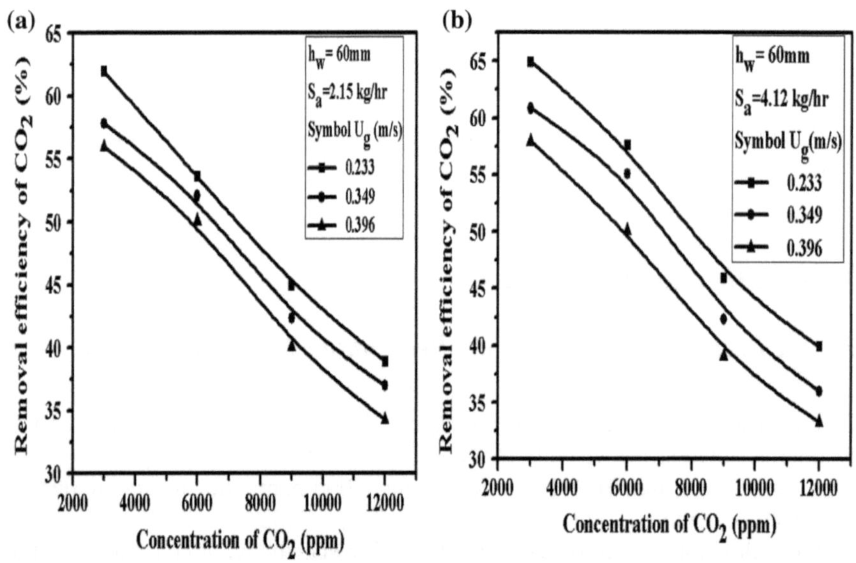

Fig. 4 Effect of gas velocity on CO_2 removal efficiency at h_w = 60 mm for S_a: **a** 2.15 and **b** 4.12 kg/hr

rate 2.15 kg/hr. The maximum removal efficiency of carbon dioxide for inlet concentration 12,000 ppm is 38.92, 36.98 and 34.28 % at 0.06 m weir height at same other operating conditions. Similar trend was shown in the for Figs. 3b and 4b for constant solid flow rate. Also this type of similarity was reported in the literature (Mohanty et al. 2009; Roy et al. 2009). The reason of such decreasing trend may be due to an increase in concentration of carbon dioxide on the surface of activated carbon particle and formation of monolayer which results in decrease of the sorbent activity. While increasing in the velocity of gas, the percentage removal of carbon dioxide also decreased since solids hold up in the bed decreased thereby decreasing the probability of diffusion of gas to activated carbon particles.

5.2 *Effect of Solid Flow Rate on Percentage Removal of CO_2*

Figures 5 and 6 represents the percentage removal of CO_2 at different inlet CO_2 loading at different solid flow rate at a constant gas velocity. It may be seen from the Figs. 5a and 6a that with increasing flow rate of solids, carbon dioxide removal efficiency increases. A similar tendency of increasing removal efficiency CO_2 also reported in the literature (Mohanty et al. 2009; Roy et al. 2009). At 0.02 m weir height, and gas velocity (0.233 m/s), the range of maximum removal efficiencies were 55.62 and 49.08 % for inlet concentration of 3000 ppm. The range of minimum removal efficiencies were 42.91 and 33.29 % for inlet concentration

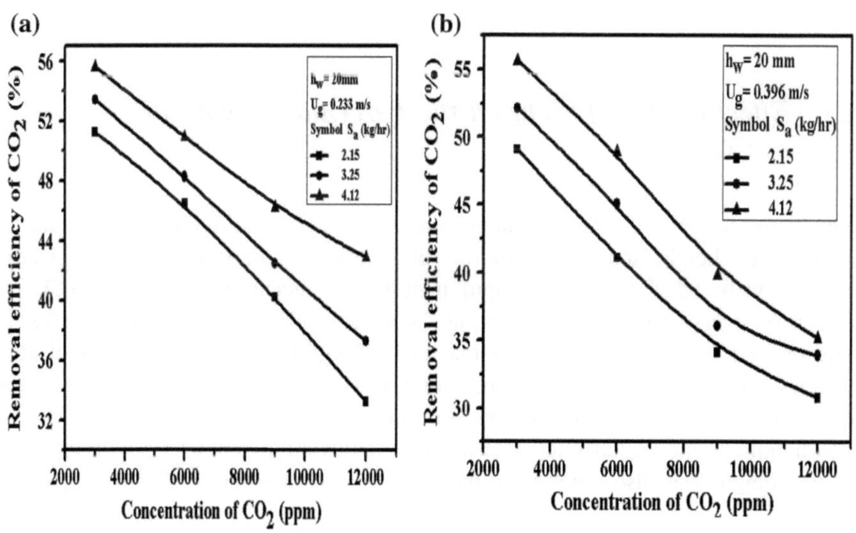

Fig. 5 Effect of solid flow rate on CO_2 removal efficiency at h_w = 20 mm for U_g: **a** 0.233 and **b** 0.396 m/s

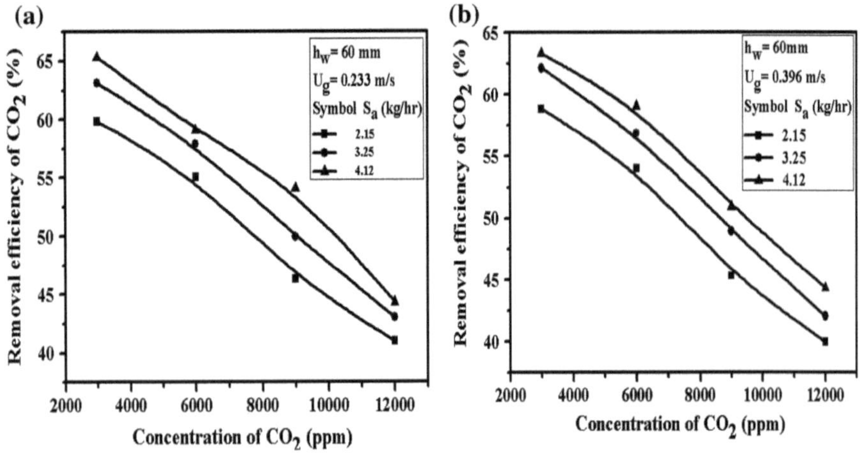

Fig. 6 Effect of solid flow rate on CO_2 removal efficiency at h_w = 60 mm for U_g: **a** 0.233 and **b** 0.396 m/s

12,000 ppm. At gas velocity 0.233 m/s and at 0.06 m weir height, the ranges of maximum removal efficiencies were 65.31 and 60.83 % for inlet concentration of 3000 ppm. The minimum removal efficiencies ranges were 44.37 and 40.98 % at 12,000 ppm. Similar trends were also observed for Figs. 5b and 6b. It was clear from the figures that increasing the flow rate of solids increases carbon dioxide removal efficiency. The results indicate that as the flow rate of solid increases, the solid hold in each stage increases. These increases in hold-up of solids augment to adsorb more quantity of CO_2 on activated carbon particle.

5.3 Effect of Weir Height on the Percentage Removal of CO_2

Figures 7a, b represents the percentage removal of CO_2 at different weir heights. At 0.06 m weir height and solid flow rate of 2.15 kg/hr, the range of removal efficiencies were 62.02 % at 3000 ppm inlet CO_2 concentration and 38.03 % for 12,000 ppm inlet CO_2 concentration at a constant velocity of gas (0.233 m/s) At 0.06 m weir height and solid flow rate of 4.12 kg/hr, the range of removal efficiencies were 65.01 % at 3000 ppm inlet CO_2 concentration and 41.93 % for 12,000 ppm inlet CO_2 concentration. Similar trends were also observed at 0.02 m weir height. A similar tendency of increasing removal efficiency CO_2 with increase in weir height was also reported in the literature (Mohanty et al. 2009; Roy et al. 2009).

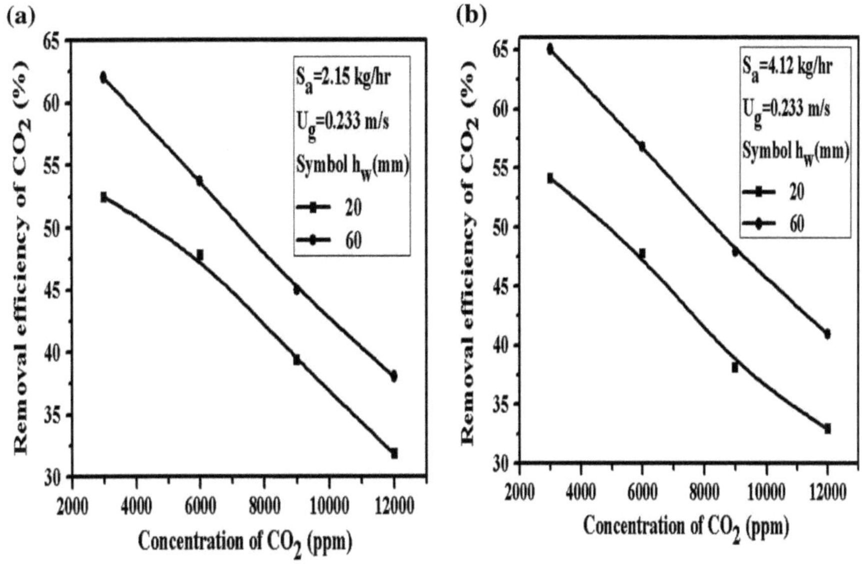

Fig. 7 Effect of weir height on CO_2 removal efficiency for U_g = 0.233 m/s at S_a: **a** 2.15 and **b** 4.12 kg/hr

6 Conclusions

From the above experiment it has been found that activated carbon was the most efficient adsorbent for removal of carbon dioxide in the four stage fluidized bed reactor and the percentage removal of carbon dioxide was affected by different operating conditions like velocity of gas, flow rate of solid and weir height. With increase in weir height carbon dioxide removal was also increases. The percentage removal of carbon dioxide was 65 % at maximum weir height of 60 mm and at high solid flow rate and low gas velocity. But at 20 mm weir height, the removal efficiency was 54 % at the same operating condition. So a multistage fluidized bed reactor can be used as a suitable equipment for capture of carbon dioxide from flue gas.

References

Chaffee, A.L., Knowles, G.P., Liang, Z., Zhang, J., Xiao, P., Webley, P.A.: CO_2 capture by adsorption: materials and process development. Int. J. Greenhouse Gas Control **1**, 11–18 (2007)

Himeno, S., Komatsu, T., Fujita, S.: High-pressure adsorption equilibria of methane and carbon dioxide on several activated carbons. J. Chem. Eng. Data **50**, 369–376 (2005)

Lee, S.W., Daud, W.M.A.W., Lee, M.G.: Adsorption characteristics of methyl mercaptan, dimethyl disulfide, and trimethylamine on coconut-based activated carbons modified with acid and base. J. Ind. Eng. Chem. **16**, 973–977 (2010)

Mastalerz, M., Gluskoter, H., Rupp, J.: Carbon dioxide and methane sorption in high volatile bituminous coals from Indiana, USA. Int. J. Coal Geol. **60**, 43–55 (2004)

Mohanty, C., Adapala, S., Meikap, B.C.: Hydrodynamics of a multistage countercurrent fluidized bed reactor with downcomer for lime-dolomite mixed particle system. Ind. Eng. Chem. Res. **47**, 6917–6924 (2008)

Mohanty, C.R., Adapala, S., Meikap, B.C.: Removal of hazardous gaseous pollutants from industrial flue gases by a novel multi-stage fluidized bed desulfurizer. J. Hazard. Mater. **165**, 427–434 (2009)

Mohanty, C., Meikap, B.C.: Pressure drop characteristics of a multi-stage counter-current fluidized bed reactor for control of gaseous pollutants. Chem. Eng. Process. **48**, 209–216 (2009)

Plaza, M., Pevida, C., Arenillas, A., Rubiera, F., Pis, J.: CO_2 capture by adsorption with nitrogen enriched carbons. Fuel **86**, 2204–2212 (2007)

Roy, S., Mohanty, C., Meikap, B.C.: Multistage fluidized bed reactor performance characterization for adsorption of carbon dioxide. Ind. Eng. Chem. Res. **48**, 10718–10727 (2009)

Sircar, S., Golden, T., Rao, M.: Activated carbon for gas separation and storage. Carbon **34**, 1–12 (1996)

Comparative Study for Removal of Cr(VI) Using Sweetlime Peel Powder and Lemon Peel Powder in A Fixed Bed Column

N.M. Rane, A.V. Kulkarni, S.P. Shewale and R.S. Sapkal

1 Introduction

Over the last century, rapid population growth and industrialization have caused degradation of a variety of water ecosystems. Consequently, the level of pollutants is found well above the permissible discharge limit of 0.05 ppm. Chromium is extensively used in various industries like electroplating, leather tanning, dye, cement, photography, paint pigments, wood treatment and so on due to its various physico-chemical properties. Among various heavy metals found in wastewater, hexavalent chromium is found to be most toxic, carcinogenic and mutagenic due to its high mobility. The chromium containing wastewater can be treated by using various techniques such as reverse osmosis, electrodialysis, ultrafiltration, membrane separation, ion exchange and adsorption. Considering the effects of high concentration of Cr(VI) in water bodies and available techniques (Paterson et al. 1975; Jorgensen 1979; Mukhopadhyay et al. 2007) for treating of it, adsorption method is found to be very popular because of its low operating and capital cost (Garg et al. 2004; Loukidou et al. 2004).

Optimisation of the adsorption parameters were done using the batch data of stirred tank reactor before the column studies. The parameters such as initial Cr(VI) concentration, pH and flow rate were studied (Baral et al. 2006, 2007a, b, 2008). In the present study removal of Cr(VI) from aqueous solutions using sweetlime peel powder (A) and lemon peel powder (B) as adsorbents in a fixed bed column carried out. Experimental work was done by varying initial Cr(VI) concentration, bed height and flow rate to evaluate the performance of selected adsorbents for Cr(VI)

N.M. Rane (✉) · A.V. Kulkarni · S.P. Shewale
Department of Chemical Engineering, MIT Academy of Engineering, Alandi(D), Pune, India
e-mail: nmrane@chem.maepune.ac.in

R.S. Sapkal
UDCT, Sant Gadge Baba Amravati University, Amravati, India

I. Regupathi et al. (eds.), *Recent Advances in Chemical Engineering*,
DOI 10.1007/978-981-10-1633-2_16

uptake capacity. Gathered data were used to determine the optimum operating parameter for adsorbent A and B. The best fit of the experimental data was also calculated using three different kinetic models.

2 Experimental Material Preparation and Characterization

2.1 *Preparation of Adsorbent*

Adsorbent A and B which contains citric acid was collected from fruit centres. The collected peels were kept for drying for a 3 week period at room temperature. After drying, the next step was grinding and screening. Using household simple mixer grinder, the dried adsorbent A and B were grinded and converted to a fine powder. After grinding, the adsorbent A and B powder were screened through the sieve shaker. 180 μ pore size powder material was selected as adsorbent for further studies (Kundu et al. 2005).

2.2 *Adsorbent Characterization*

2.2.1 SEM (Scanning Electron Microscope) Images for Adsorbent A and B Before and After Adsorption

See Figures 1, 2, 3 and 4.

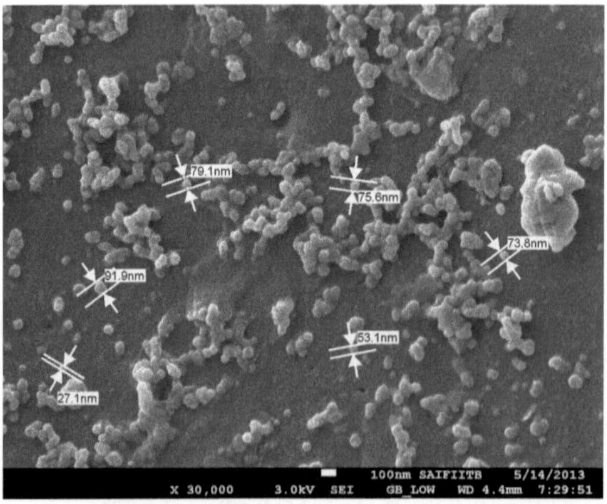

Fig. 1 SEM image of A before adsorption

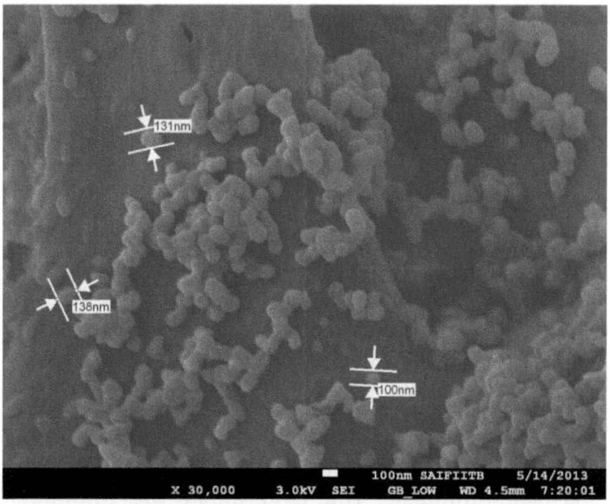

Fig. 2 SEM image of A after adsorption

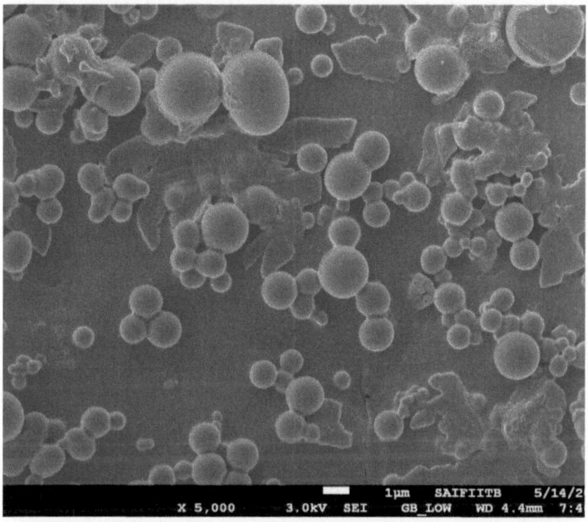

Fig. 3 SEM image of B before adsorption

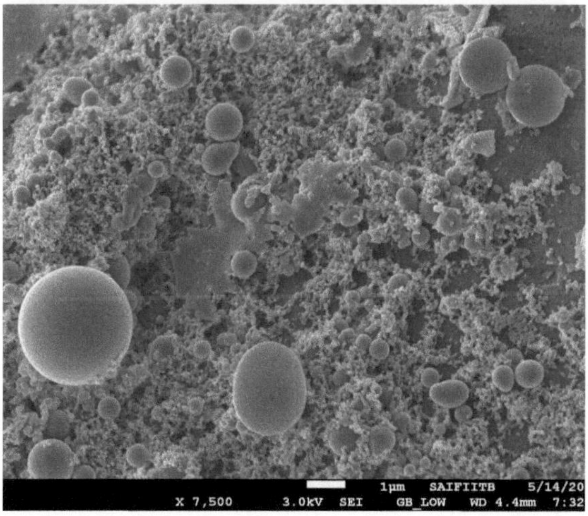

Fig. 4 SEM image of B after adsorption

2.2.2 FTIR Analysis of Sweetlime Peel Powder (A) and Lemon Peel Powder (B) Before and After Adsorption

See Figures 5, 6, 7 and 8.

3 Adsorption Study

The Cr(VI) solution of known concentration was passed with the help of peristaltic pump at the bottom of the column. The treated Cr(VI) solution was collected at regular intervals from top of the column. The flow rate, bed height and initial Cr (VI) concentration were checked regularly. The analysis of the effluent was measured spectrophotometrically by using (Thermofischer 840-210800) UV-VIS spectrophotometer. The experiment was conducted till there were no further change in Cr(VI) concentration both at inlet and outlet of the column. Repetition of the experiment was done for the change in various parameters.

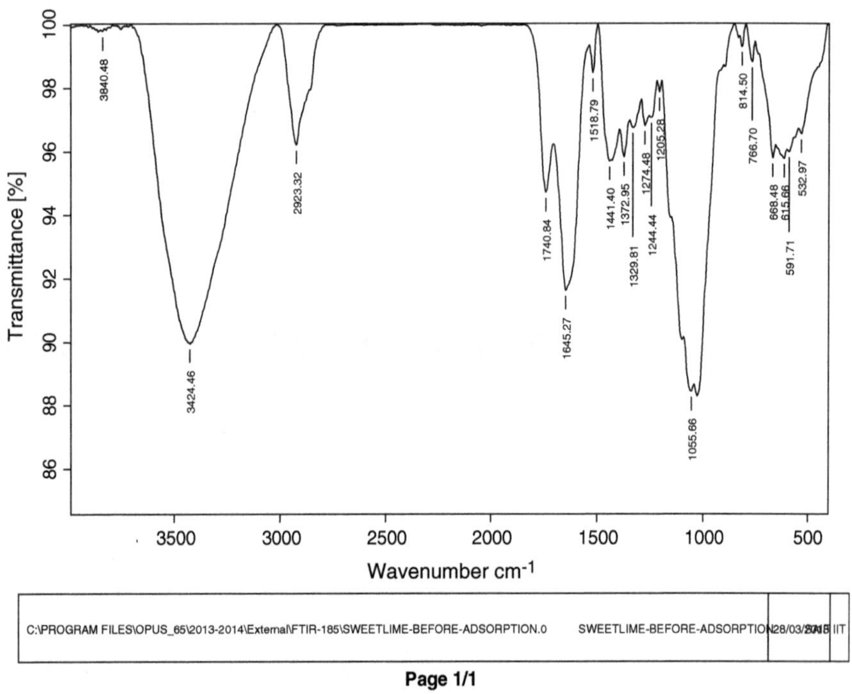

Fig. 5 FTIR for A before adsorption

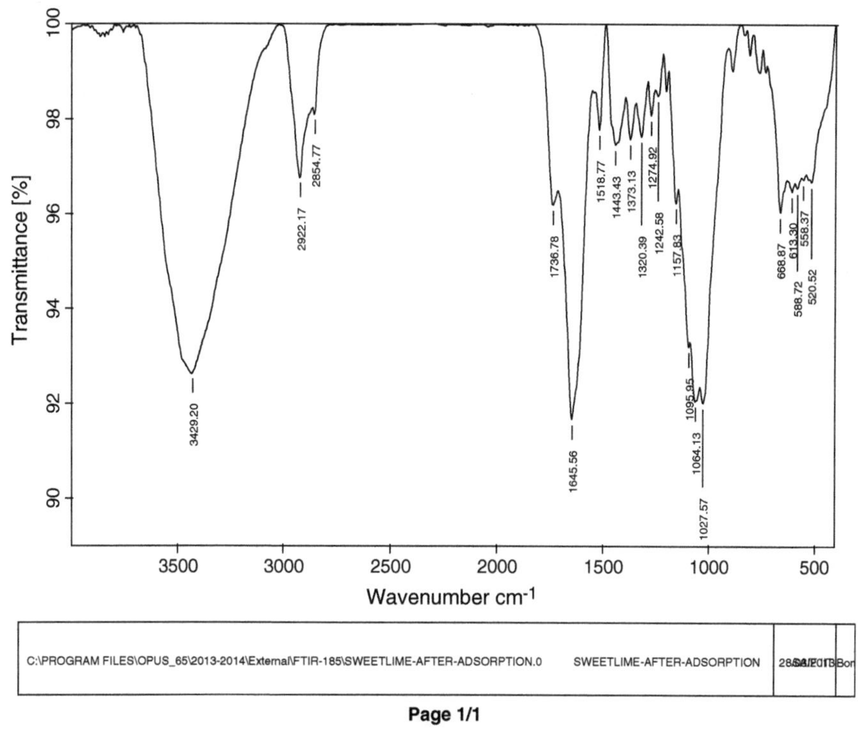

Fig. 6 FTIR for A after adsorption

4 Results and Discussion

The various parameters such as initial Cr(VI) concentration, bed height and flow rate were studied by maintaining pH at 1.25 (from previous batch studies).

4.1 Effect of Initial Cr(VI) Concentration

Figures 9 and 10 represents the performance of breakthrough curve at various initial Cr(VI) concentration using adsorbent A and B respectively. It is observed from the experimental study that with increasing inlet Cr(VI) concentration, the uptake and total Cr(VI) adsorbed were increased. From the experimental study, it was also observed that at high initial Cr(VI) concentration the mass transfer resistance is neglected due to high driving force for the desired transfer process. Among adsorbent A and B, adsorbent A shows higher Cr(VI) removal than B.

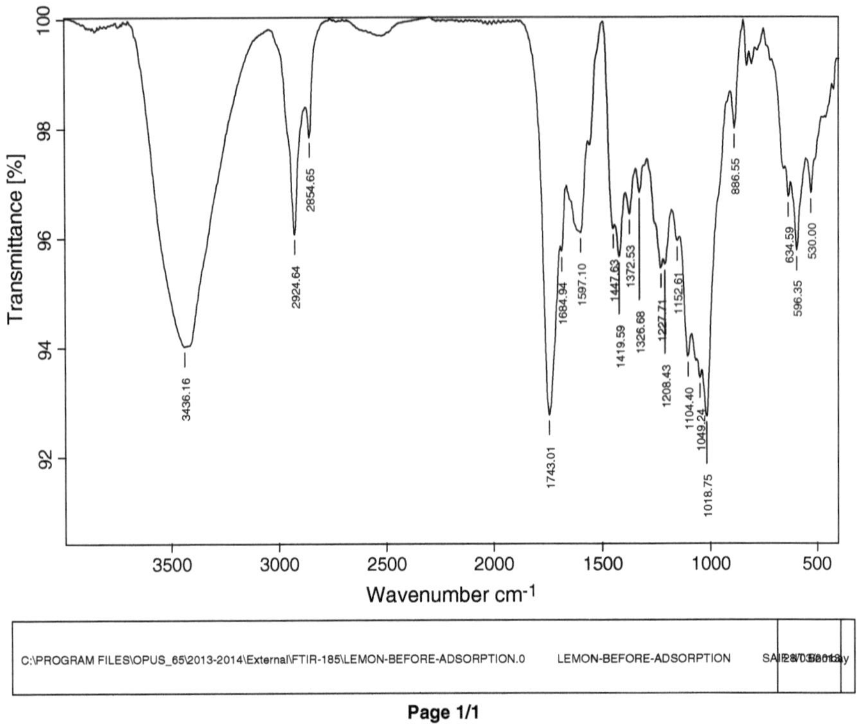

Fig. 7 FTIR before adsorption for B

4.2 *Effect of Bed Height*

Figures 11 and 12 represents the breakthrough curve performance at various bed height for A and B respectively. With the increase in bed height, metal uptake capacity increases because of increased area of the adsorbent.

4.3 *Effect of Flow Rate*

Figures 13 and 14 represents various flow rate effect on performance of break-through curve for adsorbent A and B respectively. It can be observed that adsorption efficiency increases with decrease in flow rate, because of higher residence time.

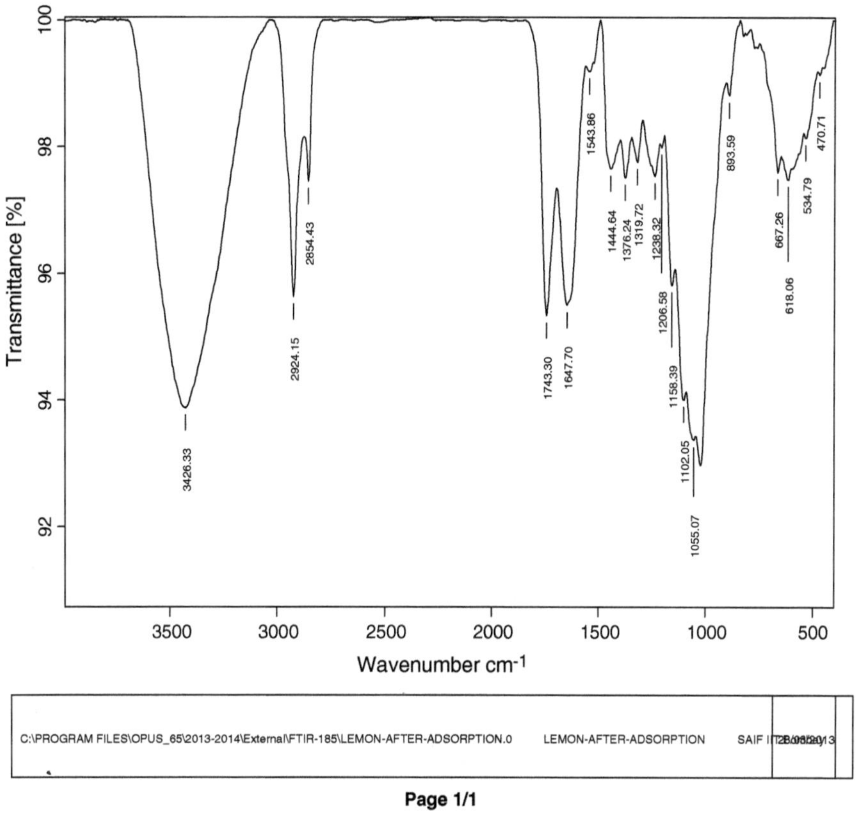

Fig. 8 FTIR after adsorption for B

Fig. 9 Effect of initial concentration for A

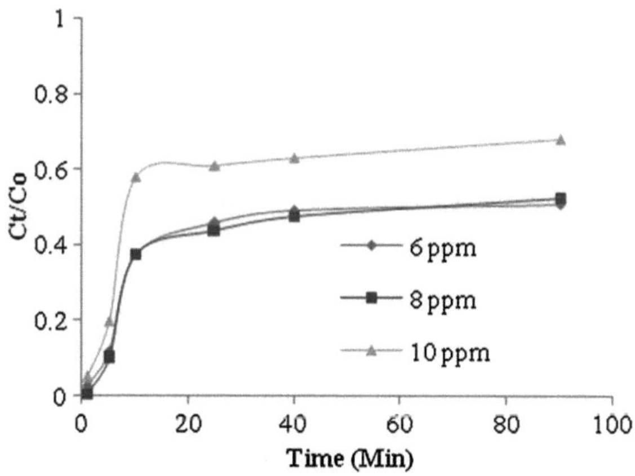

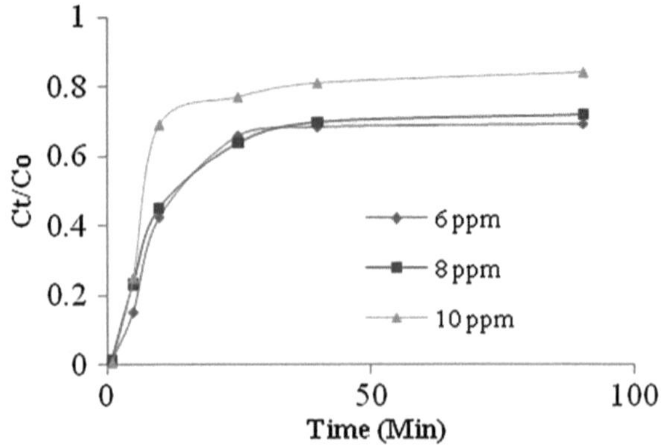

Fig. 10 Effect of initial concentration for B

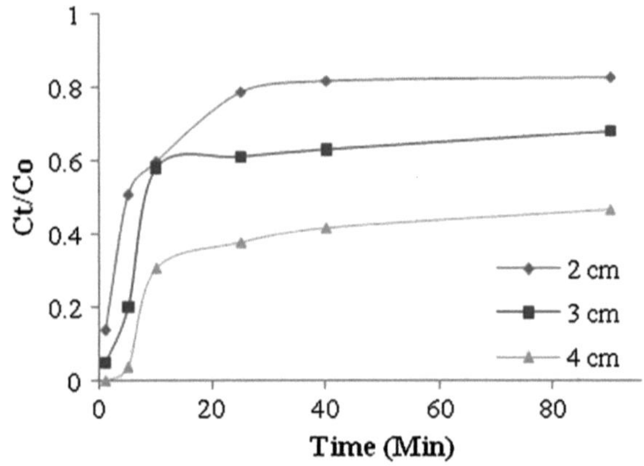

Fig. 11 Effect of bed height for A

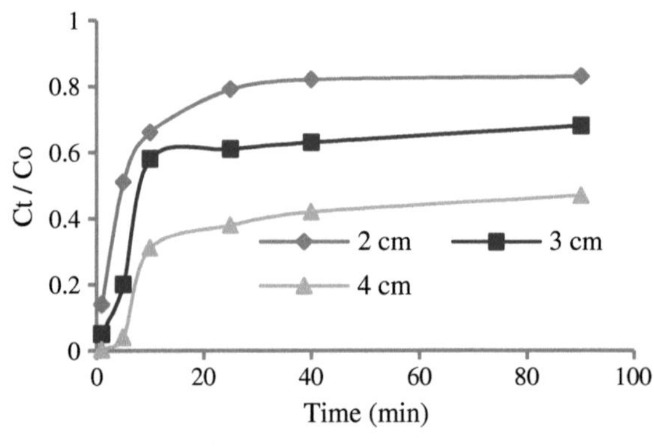

Fig. 12 Effect of bed height for B

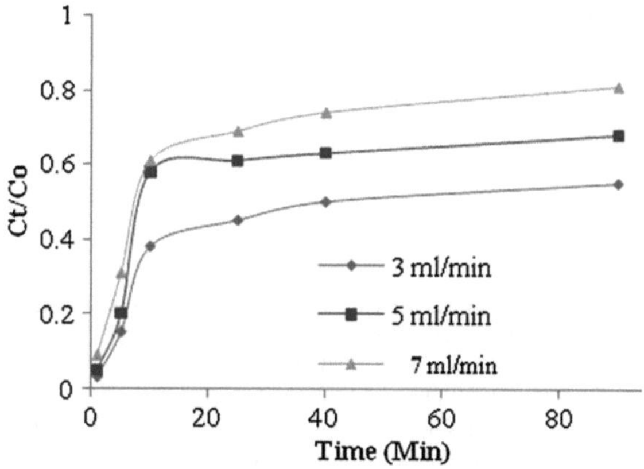

Fig. 13 Effect of flow rate for A

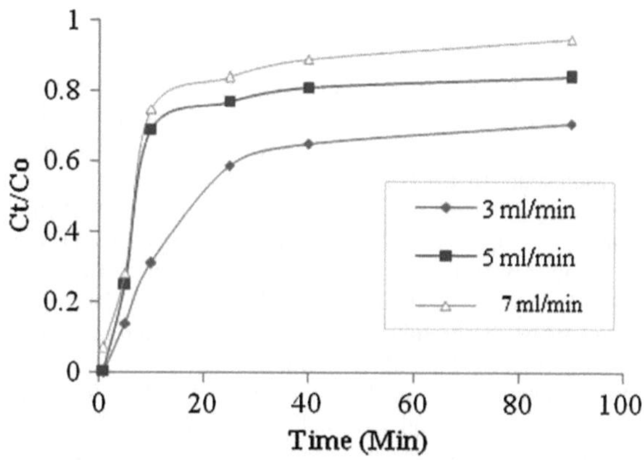

Fig. 14 Effect of flow rate for B

5 Application of Different Breakthrough Models

The experimental data gathered from the packed bed column were tested for various breakthrough curve. The dynamic response studies of fixed bed column were evaluated using important characteristics such as shape of breakthrough curve and time for breakthrough appearance. Concentration-time profiles were also studied. The various kinetic models such as Bohart-Adams, Yoon-Nelson and BDST (Baral et al. 2009; Kundu et al. 2005) were verified with experimental data obtained.

6 Conclusion

Adsorbent sweetlime peel powder (A) was found to be comparatively more efficient than lemon peel powder (B) for removal of Cr(VI) from aqueous synthetic solutions. For both A and B, it was found that adsorption capacity was increased with increase in initial adsorbent concentration and also increased with decrease in both bed height and flow rate. Bohart-Adam model fits well with experimental data in case of both A and B.

References

Baral, S.S.: Removal of Cr(VI) by thermally activated weed Salvinia Cucullata in a fixed-bed column, In: Baral, S.S., Das, N., Ramulu, T.S., Sahoo, S.K., Das, S.N., Chaudhuary G.R. J. Hazard. Mater. **161**, 1427–1435 (2009)

Baral, S.S.: Hexavalent chromium removal from aqueous solution by adsorption on treated sawdust, In: Baral, S.S., Das, S.N., Rath, P. Biochem. Eng. J. **31**, 216–222 (2006)

Baral, S.S.: Cr(VI) removal by calcined bauxite, In: Baral, S.S., Das, S.N., Rath, P., Chaudhury, G.R. Biochem. Eng. J. **34**, 69–75 (2007a)

Baral, S.S.: Removal of Cr(VI) from aqueous solution using waste weed, Salvinia Cucullata, In: Baral, S.S., Das, S.N., Rath, P., Chaudhury, G.R., Swamy, Y.V. Chem. Ecol. **23**, 105–117 (2007b)

Baral, S.S.: Adsorption of Cr(VI) using thermally activated weed Salvinia Cucullata, In: Baral, S.S., Das, S.N., Chaudhury, G.R., Swamy, Y.V., Rath, P. Chem. Eng. J. **139**, 245–255 (2008)

Garg, V.K., Gupta, R., Kumar, R.: Adsorption of Cr from aqueous solution on treated saw dust. Bioresour. Technol. **92**, 78–81 (2004)

Jorgensen, S.E.: Ind. Waste water Manage. **7**, 81–92 (1979)

Kundu, S., Gupta, A.K.: Analysis and modeling of fixed bed column operations on As(V) removal by adsorption onto iron oxide-coated cement. J. Colloid Interface Sci. **290**, 52–60 (2005)

Loukidou, M.X., Zouboulis, A.I., Karapantsios, T.D., Matis, K.A.: Equilibrium and kinetic modeling of chromium(VI) biosorption by Aeromonascaviae. Colloids Surf. A Physicochem. Eng. Aspects **242**, 93–104 (2004)

Mukhopadhyay, B., Sundquist, J., Schmitz, R.J.: Removal of Cr(VI) from Cr-contaminated groundwater through electrochemical addition of Fe(II). J. Environ. Manage. **82**, 66–76 (2007)

Paterson, J.W.: Wastewater treatment technology, In: Paterson, J.W. Ann Arbour Science, Michigan, pp. 43–58. (1975)

Adsorption Kinetics for the Removal of Chromium(VI) from Synthetic Waste Water Using Adsorbent Derived from Saw Dust, Bark and Rice Husk

Devaki Nandan Swami and S.K. Gupta

1 Introduction

Chromium(VI) is one of most important heavy metal used in various industries. It is one of the most toxic metals in waste water because of its carcinogenic, mutagenic and teratogenic characteristics. Chromium(VI) is released to the environment through different industrial process such as, metallurgy, leather tanning, pain, textiles industries, pulp production, ore and petroleum refining, metal finishing and manufacturer of products for corrosion protection. Chromium(VI) is released to the environment by leakage, poor storage and improper disposal in river. In aqueous solution chromium exists in two oxidation state trivalent and hexavalent. Cr(VI) has high mobility in aqueous solution and high accumulation in living organism where as Cr(III) is easily precipitated because of limited solubility in aqueous systems (Bayat 2002). These two Chromium oxidation state have drastically different chemical, biological and environmental characteristics. Cr(III) is less toxic and required by living organisms in small quantities as an essential trace metal because it help in carbohydrate and lipid metabolism, where as Cr(VI) is highly toxic and known to be carcinogenic, mutagenic that leads to different types of cancer diseases in living organism and increased incidence of birth and development defects among the children and other people living near to the tanneries. Since long time, extensive uses of chromium(VI) in different industries have resulted in contamination of soil,

D.N. Swami (✉) · S.K. Gupta
Department of Chemical Engineering, H.B.T.I. Kanpur, Kanpur 208002, UP, India
e-mail: devakinandan_hbti@yahoo.co.in

S.K. Gupta
e-mail: skjee@yahoo.com

I. Regupathi et al. (eds.), *Recent Advances in Chemical Engineering*,
DOI 10.1007/978-981-10-1633-2_17

surface water and ground water at production sites due to improper disposal which cause a serious aquatic biodiversity (Daneshvar et al. 2002). Due to environmental concern, discharge limits of both Cr(III) and Cr(VI) have been instituted by most industrial countries. The Ministry of Environment and Forest has set the permissible limit of Cr(VI) in industrial effluents to 0.05 mg/L (Baral et al. 2008). According to the World Health Organization (WHO) drinking water guidelines, the maximum limit for total chromium is 0.05 mg/L. whereas, according to BIS (Bureau of Indian Standards) the maximum permissible and acceptable limits for Cr(VI) in drinking water is 0.05 mg/L. Government of India has set the tolerance limit for Cr(VI) for discharge into inland surface waters is 0.1 mg/L. The recommended limits of Cr (VI) in waste water and in potable water is 0.05 mg/L (Bhattacharya et al. 2009).

Lot of conventional method have been developed for wastewater treatment, such as adsorption, chemical precipitation, ion exchange, electrochemical precipitation, solvent extraction, membrane separation, evaporation, electro-deposition, photo catalysis reduction, reverse osmosis and nano filtration. Our study is aimed at the selection of low-cost adsorbents which is more economically and can adsorb chromium(VI) from waste water. Process has been carried out in batch studies with selected adsorbent. The effect of pH, adsorbent dose and contact time is investigated with kinetic, equilibrium isotherm and thermodynamic study.

2 Materials and Methods

2.1 Materials Required

Eucalyptus (SDC), Neem (SDC), Eucalyptus Bark, Neem Bark, Rice husk carbon. These five adsorbents were used for the removal of chromium. Potassium dichromate for the preparation of stock solution, H_2SO_4, for the acid activation, double distilled water.

2.2 Methods

Batch adsorption method has been used to evaluate the maximum adsorption capacity of these adsorbent to evaluate their potential for the Cr(VI) removal from waste water prepared by stock solution with depending parameter.

2.3 Apparatus Used

Digital pH meter for measuring pH of solutions during experiment. 2 MLH magnetic stirrer for mixing of solution and heating the solution during mixing if needed. SL-218UV/VIS Spectrophotometer for analyzing wavelength at maximum absorbance, and concentration of different samples. Digital weight balance for weighing the reagents. Glass wares such as beakers of different capacities, pipette, volumetric flasks, measuring cylinders etc.

2.4 Preparation Methods of Adsorbents

Our work is planned to developed new activated carbon from these adsorbents. According to the research of various general we find that there are no of activation methods to activate the carbon prepare from different adsorbents. Such as steam activation, chemical activation with acid (H_2So_4, HCl, HNO_3, H_3PO_4, H_2SO_4 + H_2O_2, H_2SO_4 + $(NH_4)_2S_2O_8$), chemical activation with NaOH, chemical activation with $ZnCl_2$ within these acid chemical activation method give the maximum yield of carbon percentage range from 35 to 50 % depends upon the acid used for the activation. According to the acid activation study H_2SO_4 give the maximum percentage yield of activated carbon (Babu and Gupta 2008; Bansal et al. 2009). So we used chemical activation method with H_2SO_4 for the preparation of activated carbon from different adsorbent. These different methods for activation give the different characteristics of same adsorbent.

2.5 Preparation of Adsorbents

Saw dust of Neem and Eucalyptus used in this study was procured from local timber house. Bark was collected from a local source and Rice husk is collected from a local rice mill. Impurities from adsorbents removed manually. Then saw dust was washed with distilled water and dried under the sun (Bishnoi et al. 2004). Then bark was cut into small pieces and ground to pass through mesh screen. After screening it again we washed it with distilled water several times to remove dust and then dried under the sun and similarly with rice husk carbon. Finally the adsorbents was dried separately in an oven at 105 °C for 6 h. The sample was activated with H_2SO_4 (50 %) in a ratio of 2:1 (acid volume: weight of saw dust) and allowed to get carbonized at 150 °C in hot air oven for 24 h (Dutta et al. 2011).

Table 1 Characteristics of different adsorbent activated from acid (H_2SO_4)

Characteristics	E.B.	E. (SDC)	N.B.	N. (SDC)	RHC
Surface area (M^2/g)	0.59	0.64	280.39	280.88	600.20
Bulk density (g/cm^3)	0.25	0.27	0.95	0.96	0.63
Moisture	10.1	9.94	6.31	6.12	0.4
Ash	19.0	19.5	3.60	3.69	19.2
Volatile	65.7	65.9	–	–	–
Fixed carbon	15.3	15.8	15.2	15.3	15.7
Carbon (%)	43.68	44.98	43.23	43.28	38.7
Hydrogen (%)	8.14	8.10	4.39	4.17	5.0
Nitrogen (%)	0.43	0.42	0.93	0.58	0.7
Oxygen (%)	47.75	46.35	38.2	38.8	36.0

2.6 Characteristics of Derived Adsorbent

The size distribution of particle and surface morphology was investigated with the Scanning Electron Machine (SEM-LEO 430) operated at 25 kV, magnification kx56.99 (Table 1).

2.7 Preparation of Stock Solution of Chromium (VI)

Stock solution (1000 mg L^{-1}) of chromium (VI) was prepared by dissolving 2 gm of potassium dichromate in double distilled water. The aqueous solution was diluted with distilled water to obtain the desired concentrations of Cr(VI) for batch study. pH of the solutions was adjusted using 0.01 M NaOH/HCl using pH meter.

2.8 Cr(VI) Removal Batch Studies

In this work, batch experiments were conducted to remove chromium (VI) from Stock solution (1000 mg L^{-1}) by using activated carbon prepared from the adsorbents. The study is examine the effects of different reaction parameter. The Cr (VI) solution of 50, 100, 200, 250 ppm is prepared for experimental study by diluting the stock solution with double distilled water. For each batch experiment, 100 ml Cr(VI) solution of 250 mg L^{-1} concentration was used. All adsorption experiment were carried out at room temperature. The experiments were carried out at different pH range (2–4), adsorbent dose (5–10 g L^{-1}) and contact time of 180 min with stirring speed of 180 rpm. After adding 5 gm of adsorbent, pH was adjusted to desired value and then the mixture was agitated on orbital shaker. At different time interval the mixture was filtered to separate the adsorbent from supernatant. The residual concentration of chromium(VI) in supernatant was checked. All batch experiment is done at least 3 times for each adsorbent potential and the results were average.

2.9 *Reduction of Cr(VI) by Adsorbents*

The chromium(VI) removal percentage (R%) is calculated using the formula given by Eq. (1):

$$R(\%) = \left[\frac{C_i - C_e}{C_i}\right] \times 100 \quad (1)$$

where C_i and C_e were the initial and final concentration of chromium(VI) (mg L^{-1}) in the solution after treatment.

The chromium(VI) removal capacity (qe) is calculated using the formula given by Eq. (2):

$$q_e(\text{mg g}^{-1}) = \left[\frac{C_i - C_e}{M}\right] \times V \quad (2)$$

where C_i and C_e were the initial and final concentration of chromium(VI) (mg L^{-1}) in the solution after treatment. V is the volume of solution and M is the mass of adsorbent used for the study.

3 Results and Discussion

3.1 *Effect of pH*

The effect of removal potential of chromium(VI) from waste water using SDC (Eucalyptus), EB, SDC(Neem), NB and RHC, was found to be pH dependent as shown in Fig. 1. Chromium(VI) adsorption was favoured by acidic pH range 1.5–2.0. and maximum adsorption potential was observed at pH 2.0. At pH 1.0 the percentage removal of Cr(VI) was very less and as increasing the pH from 1 to 1.5 the percentage removal of Cr(VI) has been increased very low and as again increasing in pH from 1.5 to 2.0 the percentage removal of Cr(VI) was very high then again as we increased the pH from 2.0 to 2.5 the percentage removal of Cr(VI) was decrease. So the study results that maximum Cr(VI) removal was at pH 2.0 and so this was taken as optimum pH value for further adsorption studies.

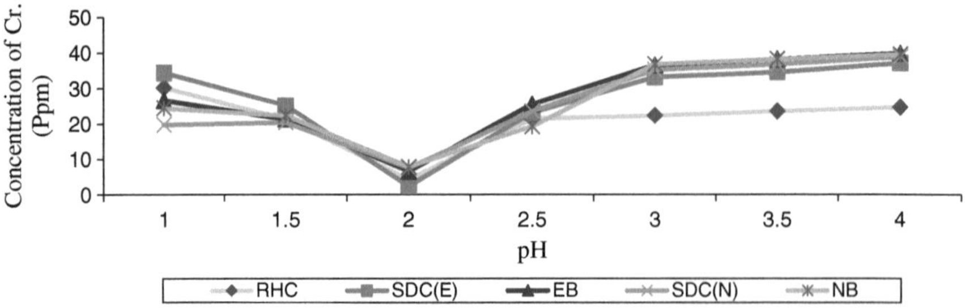

Fig. 1 Effect of pH on removal of Cr(VI) by the adsorbents

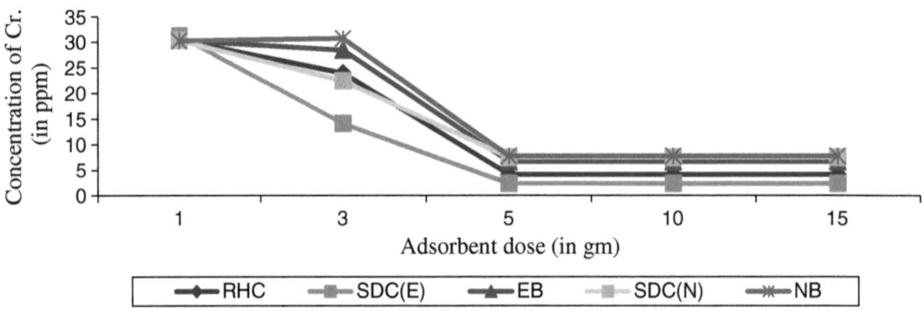

Fig. 2 Effect of Adsorbent dose on removal of Cr(VI) by the adsorbents

3.2 *Effect of Adsorbent Dose*

The effect of removal potential of chromium(VI) from waste water using SDC (Eucalyptus), EB, SDC (Neem), NB and RHC, was also found to adsorbent dose dependent as shown in Fig. 2. The removal potential of chromium(VI) by these adsorbents was studied by varying adsorbent doses (1, 3, 5, 10 gm) at pH 2.0. The percentage removal of Cr(VI) was very less at adsorbent dose of 1 gm and as we increased the adsorbent dose from 1 to 3 gm the percentage removal of Cr(VI) was increased very low and again as increasing the adsorbent dose from 3 to 5 gm the percentage removal of Cr(VI) has increased at its maximum value and again as increasing the adsorbent dose from 5 to 10 gm the result of percentage removal of Cr(VI) was same as adsorbent dose of 5 gm. So this study results that the maximum percentage removal of Cr(VI) was at adsorbent dose of 5 gm and at pH of 2.0.

3.3 *Effect of Contact Time*

The effect of removal potential of chromium(VI) from waste water using SDC (Eucalyptus), EB, SDC(Neem), NB and RHC, was also studied at fixed adsorbent

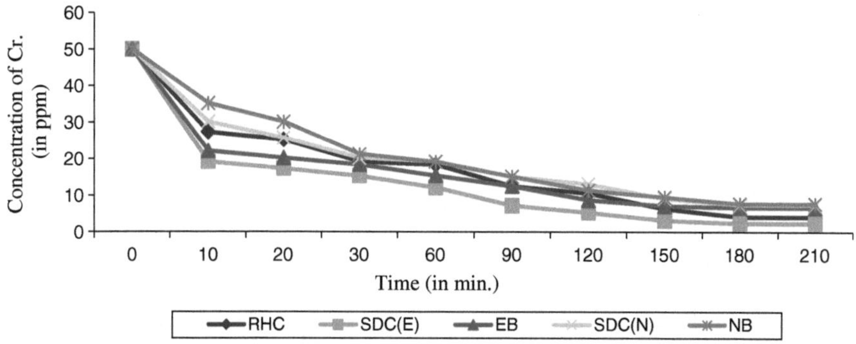

Fig. 3 Effect of Contact time on removal of Cr(VI) at pH 2 and adsorbent at dose 5 gm

dose of 5 gm and at fix pH (2). The study found to be the contact time dependent also as shown in Fig. 3. The removal potential of chromium(VI) by these adsorbents was studied at different time interval. First three results were observed at the time interval of 10 min. and then at the time interval of 30 min. It had been established that optimum time required to attain the equilibrium between Cr(VI) adsorbed on these adsorbent was with in 10 min. and saturation reached for adsorption after 180 min.

3.4 *Adsorption Kinetics*

Several Kinetics models are available to examine the adsorption mechanism of Cr (V1) from aqueous solution but we use only pseudo-first order and pseudo-second order kinetic equation to explain the adsorption mechanism rate based on adsorption capacitive of Cr(V1) from aqueous solution (Argun et al. 2007). The linear form of Lagergren's first order kinetic equation is

$$\ln(qe - q_t) = \ln(qe) - K_1 t \tag{3}$$

where qe is the amount of Cr(V1) adsorbed onto the adsorbent at equilibrium condition (mg/g). q_t is the amount of Cr(V1) adsorbed onto the adsorbent at any time t (mg/g). K_1 is the rate constant for the first order kinetic equation. The Linear form of pseudo-second order equation is-

$$\frac{t}{q_t} = \frac{1}{K_t q_e^2} + \frac{1}{q_t} \cdot t \tag{4}$$

where K_2 is the rate constant for second order kinetics equation.

These two Kinetics models are used to examine the adsorption mechanism of Cr (V1) from aqueous solution but only pseudo-first order kinetic plot as Fig. 4 shows

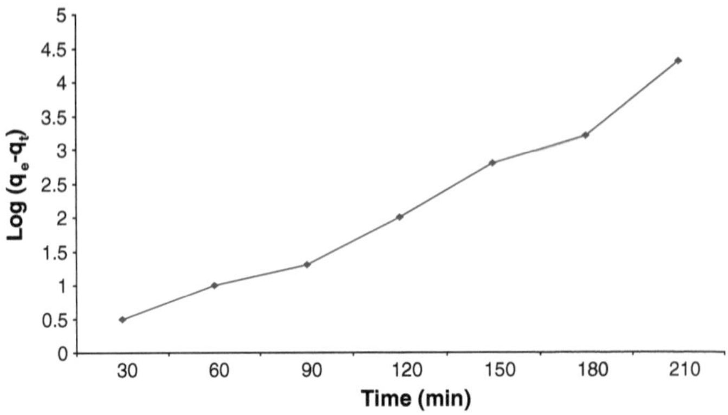

Fig. 4 Pseudo first order kinetic plot for Cr(VI) on Eucalyptus (SDC)

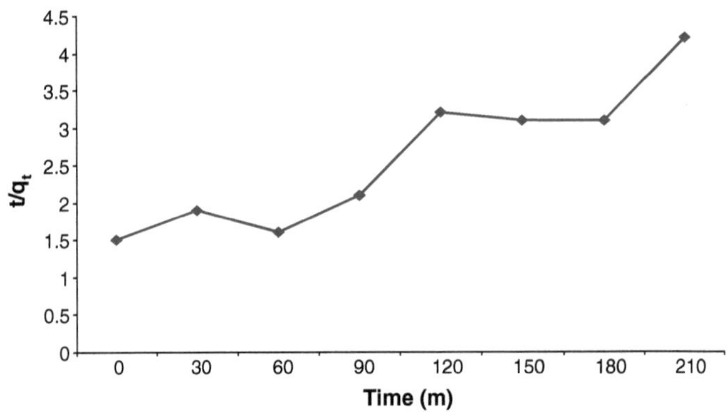

Fig. 5 Pseudo first order kinetic plot for Cr(VI) on Eucalyptus(SDC)

linearity whereas pseudo-second order kinetic equation plot Fig. 5 shows not linearity. So the data follow the Pseudo first order kinetic equation to explain the adsorption mechanism rate based on adsorption capacitive of Cr(V1) from aqueous solution.

3.5 *Adsorption Isotherm*

There are several equilibrium model available to analyze and investigate the suitable adsorption isotherm such as Langmuir, Freundlich, Redlichpeterson, Koblecorrigan, Tempkin, Dubinin-Redushkevich and Generalized isotherm (Abdullah and Prasad 2009). But only Langmuir and Freundlich are the most common equilibrium model. So we use there two models for the isotherm study and compare which one is the best fitted model for the adsorption isotherm. A well known linear farm of Langmuir eg^n is expressed as

$$C_e/q_e = C_e/q_m \quad + 1/q_m b \tag{5}$$

where q_e is the amount of Cr(V_1) adsorbed (mg/g), Ce is the equilibrium concentration of the adsorbate qm and b are the Langmuir constant related to maximum adsorption capacity (mg/g) and energy of adsorption (L/mg). A well known linear farm of Freundlich eq^n is expressed as-

$$\log q_e = \log k_f + 1/nC_e \tag{6}$$

where q_e is the amount of Cr(V1) adsorbed at equilibrium, K_f is the Freundlich constant, 1/n is the heterogeneity factor related to capacity and intensity of adsorption C_e is the equilibrium concentration (mg/L). The value of K_f and 1/n can be calculated from the slope and intercept of graph for this (Figs. 6 and 7).

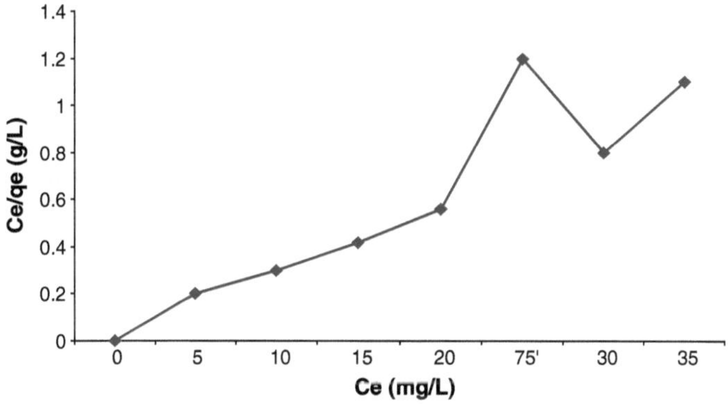

Fig. 6 Langmuir plot for Cr(VI) adsorption SDC(E)

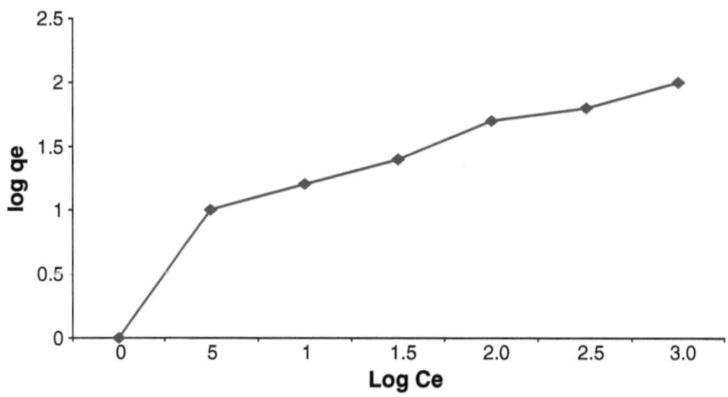

Fig. 7 Freundlich plot for Cr(VI) adsorption on SDC(E)

Both isotherm data is used to check which is best filled model but the graph no 7 plot log q_e versus log Ce shows as linear for the adsorption. Where as the graph no. 6 plot Ce/qe versus Ce is not linear for the adsorption. So the isotherm study results that Freundlich adsorption isotherm is best titled model for this study.

4 Conclusion

- The kinetics study of Cr(V1) adsorption follow only first order kinetics mechanism.
- The isotherm study of Cr(V1) adsorption is best fitted and satisfactions explain only by Freundlich isotherm model.
- The study, proved that the SDC (Eucalyptus and Neem), Bark (Eucalyptus and Neem) and RHC can be used for the Cr(V1) effluent treatment.
- Cr(V1) adsorption process strongly dependent on pH of aqueous solution, adsorbent dose and contact time. Maximum Cr(V1) adsorption was observed at pH 2.0, adsorbent dose 5 gm and contact time 180 min.
- Efficiencies of adsorbent for Cr(V1) removal is follow as Eucalyptus (SDC) is 94.33 %, RHC is 91.75 %, EB is 86.7 %, Neem (SDC) is 85 % and of NB is 84.5 %.
- The best study of the experimental also state the SDC of (Eucalyptus and Neem) is best as compare to Bark of (Eucalyptus and Neem). So the data of this study can be used by different tanneries and other industries for the Cr(V1) effluent treatment.

References

Abdullah, M.A., Prasad, A.G.D.: Kinetic and equilibrium studies for the biosorption of Cr(VI) from aqueous solutions by potato peel waste. Int. J. Chem. Eng. Res. **1**, 51–62 (2009)

Argun, M.E., Dursun, S., Ozdemi, C., Karatas, M.: Heavy metal adsorption by modified oak sawdust: thermodynamics and kinetics. J. Hazard. Mater. **141**, 77–85 (2007)

Babu, K.V., Gupta, S.: Adsorption of Cr(VI) using activated neem leaves: kinetic studies. Adsorption **14**, 85–92 (2008)

Bansal, M., Singh, D., Garg, V.K.: A comparative study for the removal of hexavalent chromium from aqueous solution by agriculture wastes'carbons. J. Hazard. Mater. **171**, 83–92 (2009)

Baral, S.S., et al.: Removal of hexavalent chromium by varying various parameters using saw dust. Bioresour. Technol. **91**, 305–507 (2008)

Bayat, B.: Comparative study of adsorption properties of fly ashes. The case of chromium (VI) and cadmium (II). J. Hazard. Mater. **B95**, 275–290 (2002)

Bhattacharya, A.K., et al.: Removal of Cr(VI) from aqueous solution by batch adsorption technique using different low-cost adsorbents. J. Hazard. Mater. **171**, 83–92 (2009)

Bhatti, I., Qureshi, K., Kazi, R.A., Ansari, A.K.: Preparation and characterization of chemically activated almond shells by optimization of adsorption parameters for removal of chromium VI from aqueous solutions. World Acad. Sci. Eng. Technol. **34**, 199–204 (2007)

Bishnoi, K., Bajaj, M., Sharma, N., Gupta, A.: Adsorption of Cr(VI) on activated rice husk carbon and activated alumina. Bioresource Technol. **91**, 305–307 (2004)

Candela, M.P., Martinez, M., Macia, R.T.: Chromium removal with activated carbons **29**, 2174–2180 (1995)

Dakiky, M., Khami, A., Manassra, A., Mereb, M.: Selective adsorption of chromium(VI) in industrial wastewater using low cost abundantly available adsorbents. Adv. Environ. Res. **6**(4), 533–540 (2002)

Daneshvar, N., Salari, D., Aber, S.: Chromium adsorption and Cr(VI) reduction to trivalent chromium in aqueous solutions by soya cake. J. Hazard. Mater. B **94**, 49–61 (2002)

Demirbas, E., Kobyab, M., Senturk, E., Ozkan, T.: Adsorption kinetics for the removal of chromium (VI) from aqueous solutions on the activated carbons prepared from agricultural wastes. Water S. A. **30**, 533–540 (2004)

Dikshit, V.P.: Removal of chromium (VI) by adsorption using sawdust. Natil. Acad. Sci. Lett. **12** (12), 419–421 (1989)

Dubey, S.P., Gopal, K.: Adsorption of chromium(VI) on low cost adsorbents derived from agricultural waste materials: a comparative study. J. Hazard. Mater. **145**, 465–470 (2007)

Dutta, S., Bhattacharyya, A., Ganguly, A., Gupta, S., Basu, S.: Application of response surface methodology for preparation of low-cost adsorbent from citrus fruit peel and for removal of methylene blue. Desalination **275** (2011)

Mohan, D., Singh, K.P., Singh, V.K.: Removal of hexavalent chromium from aqueous solution using low-cost activated carbons derived from agricultural waste materials and activated carbon fabric cloth. Ind. Eng. Chem. Res. **44**, 1027–1042 (2005)

Pehlivan, E., et al.: Removal of Cr(VI) ion from aqueous solutions through biosorption using, the shells of Walnut (WNS). J. Hazard. Mater. **145**, 465–470 (2007)

Utilization of an Agro Waste, Groundnut Shell Ash, for Removal of 2,4-Dichlorophenoxyacetic Acid

Nikhilesh S. Trivedi and Sachin A. Mandavgane

1 Introduction

The global groundnut seed production averages approximately 45,654 metric tons (Source United Nations Food and Agriculture Organization (FAO) 2013). According to the 2013 Food and Agricultural Organization report, groundnut is the most produced oil seed, out of all it is first major oil seed produced in India. Groundnut has been cultivated as a major food crop for several thousands of years now. The seeds of the groundnut crop termed "peas" are used as the raw material for preparing several foodstuffs worldwide, starting from oil various verities are made by pea. Consequently, an equally large volume of biomass is produced. This huge volume of biomass above the groundnuts termed as shell remaining after the production of seeds is considered a waste, and is locally used in different industries as a source of energy. It is ignited in situ for reducing the waste in farmlands. This produces a large amount of biomass ash, which can be utilized for a variety of purposes by applying scientific methods. One such use of this groundnut shell ash (GSA) is as an adsorbent for 2,4-D.

2,4-D has been used as a herbicide in agricultural sector. It acts as an antiweed for farm crops such as palm oil sugarcane, cotton, cocoa and rubber. It is a low-cost herbicide and is selective in activity. It is considered moderately toxic and its maximum allowable concentration in drinking water is 100 ppb. However, 2,4-D is poorly biodegradable and has been frequently detected in water bodies in various regions around the world (Bekbolet and Yucel 1999).

N.S. Trivedi · S.A. Mandavgane (✉)
Department of Chemical Engineering, Visvesvaraya National Institute of Technology, Nagpur, India
e-mail: sam@che.vnit.ac.in

N.S. Trivedi
e-mail: nst2011@gmail.com

I. Regupathi et al. (eds.), *Recent Advances in Chemical Engineering*,
DOI 10.1007/978-981-10-1633-2_18

GSA has been reported for its use as an adsorbent (Malik 2006; Malik et al. 2007), groundnut shell as a lignin source (Khan and Ashraf 2006), partial replacement to ordinary Portland cement (Alabadan 2005), previously. In addition, groundnut shells are used for production of energy (Malik 2006).

Several methods have been suggested for removal of 2,4-D, including adsorption (Hameed 2009; Deokar et al. 2015; Maryam Khoshnood 2012), photocatalytic degradation, electrochemical decomposition, combined photo-Fenton and biological oxidation, advanced oxidation processes, aerobic degradation, nanofiltration membranes, ozonation, coagulation, fluid extraction, and solid-phase extraction (Ahmad et al. 2010). Among these methods, adsorption is the best method to reduce 2,4 D concentration and economically feasible.

In this work, our attempts to use GSA as an adsorbate, source of micronutrient supply to the soil, and a rich silica source are described. Chemical constituents in GSA (K, Ca, and P) act as a soil micronutrient. The groundnut crop is grown all over the world, and thus, during its removal of peas, a huge volume of waste is generated in the form of shell remains (or their ashes).

2 Materials and Methods

Groundnut shells were dried after removing peas completely for 2 weeks in summer and ignited in open air under natural conditions prepare the groundnut shell ash. The thus-prepared ash was allowed to cool down to 30 °C and is collected in a plastic bag to prevent open-air contact. The plant remains were manually cleaned to remove dust and soil traces (to avoid contamination). The GSA was then sieved through a BSS 25-mesh sieve to separate bigger sizes and unburned particles. GSA was rinsed almost 6 times with Milli-Q water, until the water used for washing becomes clear, and dried in an oven at 110 °C for 24 h. 98 % pure technical-grade 2,4-D (Fig. 1.) (Sigma-Aldrich (M), Malaysia) was used as the adsorbate. Milli-Q water was used for the preparation of all solutions.

Properties of 2,4-D (Trivedi et al. 2016)

Molecular formula	$C_8H_6Cl_2O_3$
Molecular weight	221 g/mol
Solubility in water	0.90 g/l

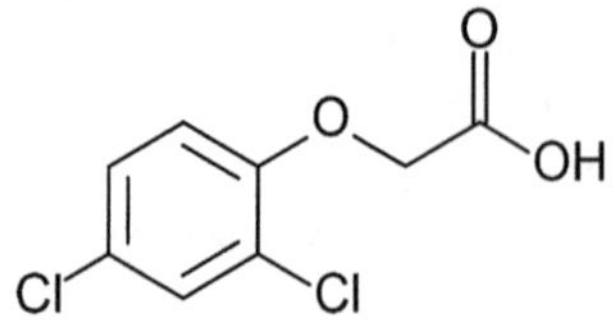

Fig. 1 Structure of 2,4-dichlorophenoxyacetic acid

3 Batch Adsorption Study

Stock solution was prepared by adding 400 mg of 2,4-D in 1000 ml of Milli-Q water. Further dilutions are prepared using this stock solution. Batch sorption studies were carried out by adding different amount of GSA into flasks containing initial concentrations (100 mg/l) of the 2,4-D solution. For initial concentration study an optimized dose of GSA was added into flasks containing different initial concentrations (50–400 mg/l) of the 2,4-D solution. All the flasks were agitated in an isothermal water bath shaker at a constant speed (30 °C) for 20 h. Initial and equilibrium concentration of 2,4-D were determined by measuring the absorbance of the solution using a double-beam UV/VIS spectrophotometer (Shimadzu, Japan) at 283 nm (Njoku and Hameed 2011). Based on this value, the concentration of 2,4-D was computed using the standard calibration curve. The amount of adsorption equilibrium was calculated using the following equation:

$$qe = \frac{(\mathrm{C0} - \mathrm{Ce})\mathrm{v}}{\mathrm{W}}, \quad (1)$$

where C_0 and C_e (mg/l) are the initial and equilibrium liquid-phase concentrations of 2,4-D, respectively; v(L) is the volume of the solution and w(g) is the mass of dry adsorbent used. For the kinetic study, the concentration (C) of 2,4-D at time t is denoted as q_t (mg/g) and calculated as follows:

$$qt = \frac{(\mathrm{C0} - \mathrm{C})\mathrm{v}}{\mathrm{W}} \quad (2)$$

The percentage of 2,4-D adsorbed (% 2,4-D adsorbed) is calculated using the following equation:

$$\%\,2,4 - \mathrm{D\,adsorbed} = \frac{100(\mathrm{Co} - \mathrm{Ce})}{Co} \quad (3)$$

The adsorption study was carried out and the following parameters were studied: optimization of adsorbent dose, variation of initial concentration of 2,4-D, and contact time. Based on the experimental data, kinetics and equilibrium models were studied and respective results were obtained.

4 Characterization Study

The XRF analysis of GSA was performed to obtain the elemental composition using an XRF spectrometer (PAN analytical Model no. PW 2403). Scanning electron microscopy (SEM) was used to study the surface morphology of GSA

(JSM-6380, JEOL, Japan). The Brunauer–Emmett–Teller (BET) surface area was measured using the BET surface analyzer by the nitrogen adsorption technique SMART SORB 92/93 surface area analyzer.

5 Results and Discussion

The XRF characterization study results indicated SiO_2 as a major constituent (28.01 %), with Al_2O_3 (10.26 %), CaO (11.2 %) K_2O (5.73 %), P_2O_5 (3.32 %) and Fe_2O_3 (14.62 %) as the minor constituents (Table 1). Loss on ignition (LOI) was found to be 18.05, which is due to incomplete combustion of groundnut shell and presence of unburned carbon and other volatiles. The ash composition showed the presence of K_2O, CaO, and P_2O_5, all of which act as micronutrients to the soil (Demeyer et al. 2001). Ash acts as a rich source of K, Ca, P, and Mg similar to wood ash (Voundi Nkana et al. 1998). GSA solution acts as a neutralizer to the acidic soil as its pH is found to be 10.25. The pH of ash solution was measured after stirring 1 g ash in 50 ml of Milli-Q water for 24 h. Addition of ash contributes toward the water-holding capacity of the soil as well as improving the texture, aeration, and salinity of the soil (Demeyer et al. 2001). Being a rich source of silica, it also improves content of silica in the soil.

The BET surface area of GSA is 22 m^2/g. Results of proximate analysis (Table 2) show a large value for fixed carbon due to the incomplete combustion of biomass. Morphology of GSA was studied by using SEM image. The high-resolution SEM image (Fig. 2) shows surface irregularities. On this image, pores are distinctly visible, which confirms the surface characteristics that are responsible for the binding of 2,4-D onto the GSA surface. This result confirms the possibility of 2,4-D to trap and adsorb onto the surface of GSA or into its pores.

Table 1 XRF properties of GSA

	Na_2O	MgO	Al_2O_3	SiO_2	P_2O_5	SO_3	K_2O	CaO	TiO_2	MnO	Fe_2O_3	CuO	LOI
GSA	0.97	5.44	10.26	28.01	3.32	1.98	5.73	11.2	0.2	0.25	14.63	0.05	18.05

Table 2 Properties of GSA

Parameter	Value
(BET) surface area	22 m^2/g
C	2.5
H	0.32
N	0.17
S	0.66
Fixed carbon	0.57
Volatile matter	18.7
Ash content	78.5
Moisture content	2.3
LOI	18.05

Fig. 2 SEM image of GSA at 2500×

5.1 Effect of Adsorbent Dosage

The effect of adsorbent dosage was studied by varying the GSA dosage from 1 to 6 g. The adsorbate (i.e., 2,4-D) concentration was fixed at 100 ppm. Experimental results indicated that with the increase in adsorbent dose, the adsorption amount increases. This is due to the increase in the availability of more vacant sites for adsorption with the increase in dosage. However, when the adsorbent dose is above 4 g/25 ml of 100 ppm 2,4-D solution, the 2,4-D is not adsorbed, probably because the equilibrium may have been established between the adsorbate and adsorbent molecules. The graph showing the % removal of 2,4-D with adsorbent dosage is presented in Fig. 3.

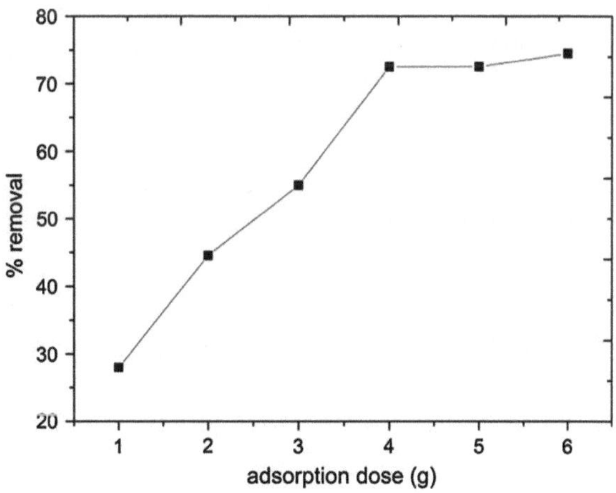

Fig. 3 Effect of adsorbent dose

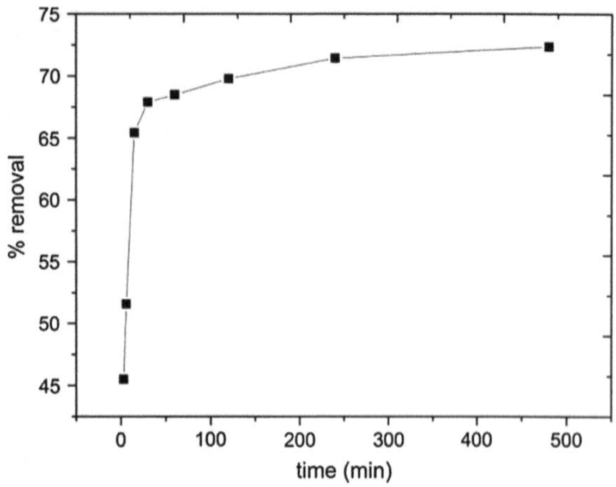

Fig. 4 Effect of adsorption time

Table 3 Pseudo-first-order and pseudo-second-order rate constants

Model	Parameter	Value
Pseudo first order	$q_{e,cal}$ (mg/g)	0.23
	k_1 (min^{-1})	0.006
	R^2	0.78
Pseudo second order	$q_{e,cal}$ (mg/g)	2.73
	k_2(g/mg/min)	0.104
	R^2	0.99

5.2 *Effect of Contact Time*

Batch study of the optimized dose, that is, 4 g/25 ml of 100 ppm 2,4-D solution, was carried out and the effect of contact time was studied (Fig. 4). The batch adsorption study was carried out in separate flasks and the effect was reported after 3, 6, 10, 30, 60, 120, 240, 480 min, and infinite time. The equilibrium time was found to be 60 min. We also carried out a kinetic study and data obtained were fitted to the pseudo-first-order and pseudo-second-order kinetics models (Table 3). The pseudo-second-order model was found to show the best fit with the experimental data (R^2 = 0.99).

5.3 *Effect of Initial Concentration of Adsorbate*

The initial concentration variation was studied by varying the concentration of 2,4-D from 400 to 50 mg/l. Batch studies were carried out by treating the optimum dose of GSA (i.e., 4 g) with 25 ml of different concentrations of the 2,4-D solution. Results of adsorption study (Fig. 5) indicated that with the increase in the initial

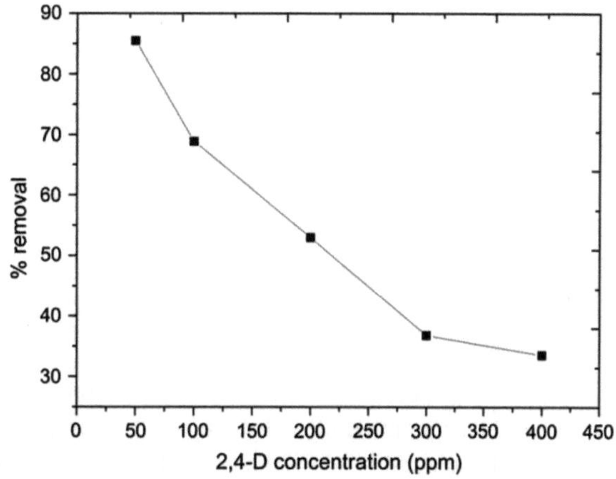

Fig. 5 Effect of initial concentration on adsorption

Table 4 Langmuir, freundlich and temkin isotherm parameters for GSA adsorption

Isotherm	Parameters	Values
Langmuir	q_0	0.87
	B	0.037
	R^2	0.98
Freundlich	K_f	0.43
	N	3.21
	R^2	0.96
Temkin	A	0.86
	B	0.35
	R^2	0.95

concentration of the 2,4-D solution, the % removal decreases, that is, the number of adsorption sites is constant and the activity is limited. Isotherms were fitted to the experimental data (Table 4).

5.4 GSA Utilization in Farmland

The Central Insecticides Board and Registration Committee (CIBRC) (Major Uses of Pesticides 2014), India, has provided guidelines regarding the use of approved registered herbicides. 2,4-D in salt and ester forms is recommended for wheat, maize, sugarcane, potato, rice, sorghum, citrus, etc., to control different weed species. The sprayed pesticide runs off from crop and leaches into groundwater though soil. To avoid this, based on the batch study, it is recommended to spread 574 kg of GSA/hectare in case of wheat and maize. According to the composition, 574 kg of GSA will supplement 64 kg of CaO, 32 kg of K_2O, and 18 kg of P_2O_5, which would subsequently improve crop yield (Demeyer et al. 2001). Thus, GSA

Table 5 GSA dose per hectare required

Type of crop	2,4-D/hectare required (kg) (MOA, GOI 2014)	GSA proposed to be added to soil per hectare (kg)
Maize, wheat	0.5	574
Transplanted rice, sorghum, citrus	1	1148
Sugarcane	1.2	1377
Grapes	2	2296
Aquatic weeds	2.5	2870

acts as a soil-protecting layer and as a barrier preventing the contact between 2,4-D and soil. The recommended dosage of GSA is based on the assumption that the entire 2,4-D sprayed is washed away. The quantity of GSA spread will form a layer on the soil surface, which will act as an adsorbent of 2,4-D. Table 5 lists the recommended dosages of GSA dispersal/hectare according to this study for different crops and fruits being cultivated in India. It is also recommended that GSA should be spread before each spraying of 2,4-D.

6 Conclusion

Our study results show that GSA serves dual purposes: as a source of both silica and micronutrient and as an adsorbent for the removal of 2,4-D from aqueous solution over a wide range of concentrations. The high-resolution SEM image showed that the GSA surface is highly porous and rough, which favors the adsorption of 2,4-D. The XRF study indicated CaO, K_2O, and P_2O_5 as the major constituents in GSA, and thus, the adsorbent could also enhance the composition of microelements in the soil. Addition of ash contributes toward the water-holding capacity of the soil as well as improving the texture, aeration, and salinity of the soil. The BET surface area was 22 m^2/g. The pseudo-second-order model showed the best fit with the experimental data with an R^2 value over 0.99, suggesting that the adsorption process depended on the availability of sites on the GSA surface. The Langmuir adsorption isotherm model also showed the best fit to the experimental data, indicating monolayer adsorption over ash surface. The dosages of GSA/hectare have been recommended, so that the dual properties of GSA (i.e., as an adsorbent and a soil additive) can be utilized effectively.

Acknowledgments Authors are thankful to the Science and Engineering Research Board, India for providing research grant (SB/S3/CE/077/2013) to undertake the work.

References

Ahmad, T., Arniza Ghazali, M.R., Sulaiman, O., Hashim, R., Ahmad, A.: Removal of pesticides from water and wastewater by different adsorbents: a review. J. Environ. Sci. Health Part C **28**, 231–271 (2010)

Alabadan, B.A., Olutoye, M.A., Abolarin, M.S., Zakariya, M.: Partial replacement of ordinary portland cement (OPC) with bambara groundnut shell ash (BGSA) in concrete. Leonardo Electron. J. Practices Technol. (6), 43–48 (2005)

Bekbolet, M.Y.O., Yucel, I.: Sorption studies of 2,4-D on selected soils. Water Air Soil Pollut. **111**, 75–88 (1999)

Demeyer, A., Nkana, J.C.Voundi, Verloo, M.G.: Characteristics of wood ash and influence on soil properties and nutrient uptake: an overview. Bioresour. Technol. **77**, 287–295 (2001)

Deokar, S.K., Mandavgane, Sachin A.: Rice husk ash for fast removal of 2, 4-dichlorophenoxyacetic acid from aqueous solution. Adsorpt. Sci. Technol. **33**(5), 429–440 (2015)

Hameed, B.H., Salman, J.M., Ahmad, A.L.: Adsorption isotherm and kinetic modeling of 2,4-D pesticide on activated carbon derived from date stones. J. Hazard. Mater. **163**, 121–126 (2009)

Khan, M.A., Ashraf, S.M.: Development and characterization of groundnut shell lignin modified phenol formaldehyde wood adhesive. Indian J. Chem. Technol. **13**, 347–352 (2006)

Major uses of pesticides: Central insecticides board and registration committee. Ministry of Agriculture, Government of India (2014)

Malik, R., Ramteke, D.S., Wate, S.R.: Physico-chemical and surface characterization of adsorbent prepared from groundnut shell by $ZnCl_2$ activation and its ability to adsorb color. Indian J. Chem. Technol. **13**, 319–328 (2006)

Malik, S.R., Ramteke, R., Wate, D.S.: Adsorption of malachite green on groundnut shell waste based powdered activated carbon. Waste Manag. **27**(9), 1129–1138 (2007)

Maryam Khoshnood, S.A.: Adsorption of 2,4-dichlorophenoxyacetic acid pesticide by graphitic carbon nanostructures prepared from biomasses. J. Ind. Eng. Chem. **18**, 1796–1800 (2012)

Njoku, V.O., Hameed, B.H.: Preparation and characterization of activated carbon from corncob by chemical activation with H_3PO_4 for 2,4-dichlorophenoxyacetic acid adsorption. Chem. Eng. J. **173**, 391–399 (2011)

Source United Nations Food and Agriculture Organization (FAO) (2013)

Trivedi, NS., Mandavgane, S.A., Kulkarni, B.D.: (2016) Mustard plant ash: a source of micronutrient and an adsorbent for removal of 2,4-dichlorophenoxyacetic acid. Environ. Sci. Pollut. Res. 1–13

Voundi Nkana, J.C., Verloo., A.D.M.G.: Chemical effects of wood ash on plant growth in tropical acid soils. Bioresour. Technol. **63**, 251–260 (1998)

Part IV
Photooxidation

Synthesis of ZnO Based Nanocatalyst with Palash Flower Powder for Degradation of Phenol

Rajani Bharati and S. Suresh

1 Introduction

Phenols and its derivatives have been used in many production processes in industries. Phenol and its derivatives affect badly environment, water reservoir and health of people because of their harmful effect and these are poisonous to liver, kidneys, lungs and vascular systems (ATSDR 1998). Conventional methods for treatment of phenolic compounds have been found costly and inadequate so it increases researchers' attention to develop more effective treatment methods and technology (Labbe et al. 2014). Researchers found application of nanocatalyst more effective to remove phenolic derivatives from wastewater by catalytic oxidation method than other conventional treatment methods (Labbe et al. 2014). It has been found that use of leaf residues to synthesize nanoparticles (NPs) is cheap and eco-friendly approach (Banerjee et al. 2014).

No work has been done to synthesize ZnO nanocatalyst by Palash flower powder (PFP). ZnO nanocatalyst have been synthesized by taking PFP as reducing and stabilizing agent by one step green synthesis processes, it is biosynthesis process to make ZnO nanocatalyst (Varshney et al. 2010). It has been concluded from literature that plants or plants extract have been immerged as more effective and fast reducing and stabilizing agents for synthesis of nanoparticles than chemicals. So it has been found that biosynthesis method has immerged as a good alternatives in place of chemical methods (Gadekar and Ghoshal 2012). In this paper PFP has been used as reducing and stabilizing agent because PFP contents many biomolecules which convert Metals Oxides in nano range. Synthesized nanocatalyst has been characterized by energy dispersive X-ray spectroscopy (EDAX), Field Emission Gun-Scanning Electron Microscopy (FEG-SEM), and X-ray diffraction (XRD) and

R. Bharati · S. Suresh (✉)
Green Catalysis and Process Technology Research Laboratory, Department of Chemical Engineering, Maulana Azad National Institute of Technology, Bhopal, India
e-mail: sureshpecchem@gmail.com; sureshs@manit.ac.in

I. Regupathi et al. (eds.), *Recent Advances in Chemical Engineering*,
DOI 10.1007/978-981-10-1633-2_19

Fourier Transform Infrared Spectrophotometer (FTIR) to determine their size, shape and particle distribution and it has been used for phenol removal from aqueous solution of phenol in annular photo reactor.

2 Materials and Methods

In this work all chemicals have been used of analytical grade (AR). ZnO was purchased from Molychem Company, Mumbai. Phenol was purchased from Ranbaxy fine chemicals, New Delhi, India which was used for the preparation of synthetic aqueous solution. PFP was used as stabilizing and reducing agent to synthesize ZnO nanocatalyst by grinding the dried palash flower and then screening the flower powder by a 34 mesh sieve. For the synthesis of ZnO nanoparticles, 0.5 g dried Palash flower powder was added to 50 mL of 1 mM aqueous ZnO solution and heated at 60 °C for 1 h. ZnO is non soluble in water but by adding some drops of 1 N HCl, it is soluble in water. The nanocatalyst of ZnO precipitated at the bottom of the conical flask after 1 h of reaction time. A change in color of the solution from brownish yellow to dark brownish occurred, confirming green synthesis of ZnO. The experiment has been performed at pH 9.3 by adding 1 N NaOH solution to get better results. After the precipitation, the suspension above the precipitate was centrifuged at 10,000 rpm to separate ZnO nanocatalyst from biomass residue. After separation, ZnO nanocatalyst was dried at 60 °C in oven. The samples of ZnO nanocatalyst were characterized by FEG-SEM, EDAX, XRD and FTIR Instruments.

2.1 Experimental Setup

In this batch reactor, two coaxial cylinders are there. The inner with reaction zone and lamp is placed on the symmetry axis. The total volume of the reactor is 1 L and working volume is 0.7 L. Practically, all the photons emitted by UV lamp reach the reaction medium if the outer walls of the reactor are reflective. This geometry has been selected for the photoreaction used in our study and offers a number of advantages viz. efficient photon captures, practical to fluidize the photocatalyst in a cylindrical geometry and an outer jacket can be used for meeting the cooling requirements of UV lamp. At the bottom of reactor the magnetic stirrer is fitted for uniform mixing. The temperature maintained around 28 ± 2 °C.

For this experiment, 500 mL aqueous solution of phenol with 100 mg/L initial concentration of phenol was taken in a 1000 mL of concentric annular reactor containing 1.0 g/L mass of the ZnO with PFP nanocatalyst. 500 mL aqueous solution of phenol has been treated in a 1000 mL of total volume of reactor. The mixture solution was placed into an annular reactor and started the UV light (6 W) to study the degradation or removal of phenol for duration of 6 h. The temperature was maintained at 30 °C. The samples were withdrawn from annular reactor periodically in an interval of 0.5, 1, 1.5, 2, 3, 4, 5, 6 h. After appropriate degradation, the samples were filtered then obtained remaining concentrations of phenol have been found out by using a high performance liquid chromatography (HPLC) C-18 Column (Waters Pvt. Ltd., India).

The removal of pollutant Phenol (%) have been found by relationships given by Eq. (1):

$$\text{Percentage degradation of Phenol} = 100\left(C_{i(0)} - C_{f(t)}\right)/C_{i(0)} \quad (1)$$

First orders kinetic model was fitted for understanding the degradation mechanism of phenol.

First-order rate model

$$\ln\left(C_{f(t)}/C_{i(o)}\right) = -kt \quad (2)$$

where, $C_{i(0)}$ is the initial phenol concentration (mg/L) at time 0 and, $C_{f(t)}$ is the final concentration (mg/L) of phenol at time t.

3 Results and Discussion

Chemical analysis of sample was done by EDAX test in SAIF, IIT Mumbai and it has been shown in Fig. 1 and composition is: C = 74.38 %, O = 20.62 %, Na = 0.96, Mg = 0.47 %, S = 0.35 %, K = 0.96 %, Ca = 1.65 % and Zn = 0.61 %.

In EDAX images, Zn has been found at 1 keV (Gayen et al. 2011), So from EDAX test, it has been confirmed that ZnO nanoparticles have been synthesized, % of carbon has been found more than ZnO in sample, because during separation of nanoparticles from PFP, some flower powder was remained in the sample.

The particle size, shape and surface morphology of the ZnO nanocatalyst have been analyzed by FEG-SEM image Fig. 2. The FEG-SEM test has been done in SAIF, IIT Mumbai. Mean particle size and particle size range have been calculated from FEG-SEM image shown in Fig. 2.

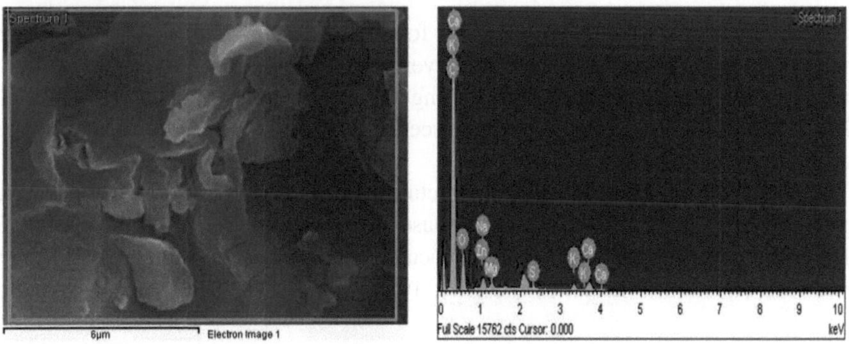

Fig. 1 EDAX images of ZnO with palash flower powder nanocatalyst

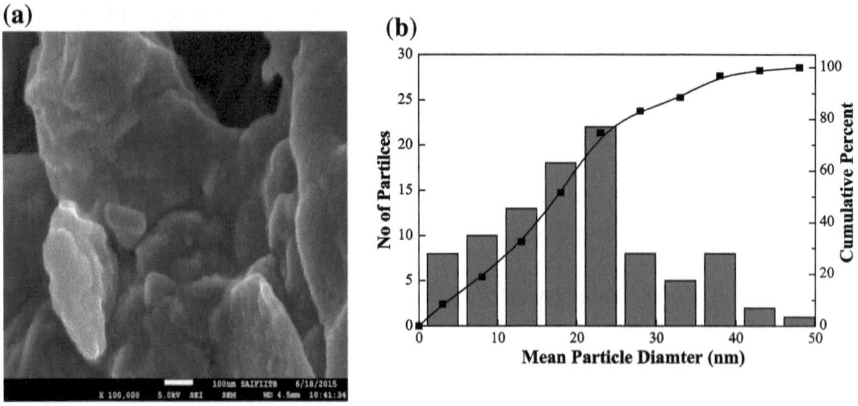

Fig. 2 FEG-SEM images of ZnO/palash flower powder nanocatalyst **a** FEG-SEM Image, **b** mean particle size from FEG-SEM image

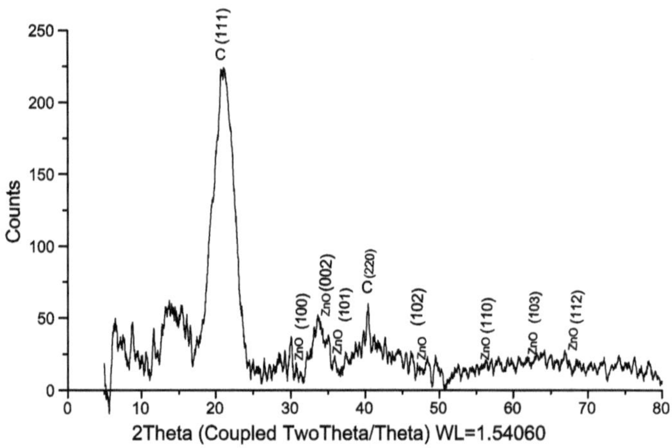

Fig. 3 XRD pattern of ZnO nanoparticles synthesized with palash flower powder

The range of nanocatalyst has been found from 3 to 40 nm and the average particle (mean diameter) is 17.6 nm (Gayen et al. 2011). From FEG-SEM image it has been found that nanoparticles are angular and irregular in shape (Khan et al. 2011). From FEG-SEM image it can be seen that at some places some nanoparticles of ZnO are aggregated.

The particle dimeter and crystal structure of ZnO nanocatalyst in sample have been found by using XRD test and its phase identification has been shown in Fig. 3 The mean particles size of the ZnO nanocatalyst has been determined using Debye Sherrer's equation (Alexander and Klug 1950).

$$d = 0.94\,\lambda/B\cos\theta \quad (3)$$

The XRD analysis has been done in Central laboratory facility of North Maharashtra University, Jalgaon.

Here d = mean particles diameter, λ = wavelength of X-Ray in this case (1.54060 Å), B = line broadening at half the maximum, and θ = ½ (Bragg angle; 2θ), the mean particle diameter of ZnO nanocatalyst is 17.4 nm has been calculated from this equation.

From XRD results we found prominent peaks of Carbon (C) and weak and clear peaks of ZnO nanocatalyst. XRD peaks confirmed the presence of ZnO corresponding to PDF No. 800075. Lattice Constance (parameters) a and c were calculated to be 3.25 and 5.209 Å respectively (Khan et al. 2011). From XRD results it has been found that ZnO has been found in hexagonal Crystal structure, Peaks at 31.744° (100), 34.419° (002), 36.233° (101), 47.525° (102), 56.558° (110), 62.879° (103), 67.998° (112) has been found for ZnO (Kumar et al. 2013) and peaks at 19.866° (100), 20.895° (100), 21.542° (80), 23.937° (40), 40.237° (40) have been found for Carbon.

FTIR spectroscopy test has been done to find functional groups present in synthesized ZnO nanocatalyst. The FTIR test has been done in Central laboratory facility of North Maharashtra University, Jalgaon. The FTIR spectra of ZnO with PFP have been given in Table 1 FTIR spectrum of ZnO with PFP nanocatalyst has been shown in Fig. 4.

Table 1 Infrared spectroscopy analysis of ZnO with Palash flower powder nanocatalyst

Frequency (cm^{-1})	Strength of bond	Bond
449.43	Moderate	ZnO stretching (Khan et al. 2011)
648.10, 681.86	Medium	ZnO stretching (Kumar et al. 2013)
886.32	Medium	ZnO stretching (Geyen et al. 2011)
1084.99	Strong	C=C Stretching (Geyen et al. 2011)
1248.95, 1256.67, 1312.60	Strong	–C–O Stretching (Rao 1963)
1367.58	Strong	C=O stretching (Dutta and Ganguly 2012)
1423.51	Strong	C–O stretching (Mitra and Mondal 2013)
1447.62	Strong	C-O Stretching (Gayen et al. 2011)
1665.59	Strong	C=O (Dutta and Ganguly 2012)
2357.09	Strong	O=C=O (Khan et al. 2011)
2574.09,2722.61,2767.94	Medium	O–H Stretching (Khan et al. 2011)
2886.57	Medium	C–H Stretching (Dutta and Ganguly 2012)
3034.13	Medium	ZnO stretching (NIST 2009)
3082.35	Medium	C–H Stretching (Ravichandrika et al. 2012)
3195.19	Medium	O–H Stretching (Khan et al. 2011)
3214.48	Strong, broad	O–H Stretching (Rao 1963)
3227.98	Strong, broad	O–H Stretch (Ravichandrika et al. 2012)
3428.58, 3537.57	Medium	ZnO stretching (NIST 2009)
3616.65	Strong, Sharp	O–H Stretching (Khan et al. 2011)

NIST National Institute of Standards and Technology

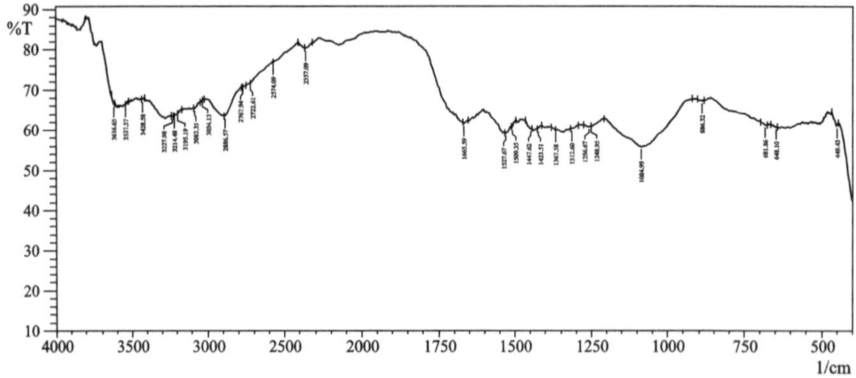

Fig. 4 FTIR spectra of ZnO with Palash flower Powder

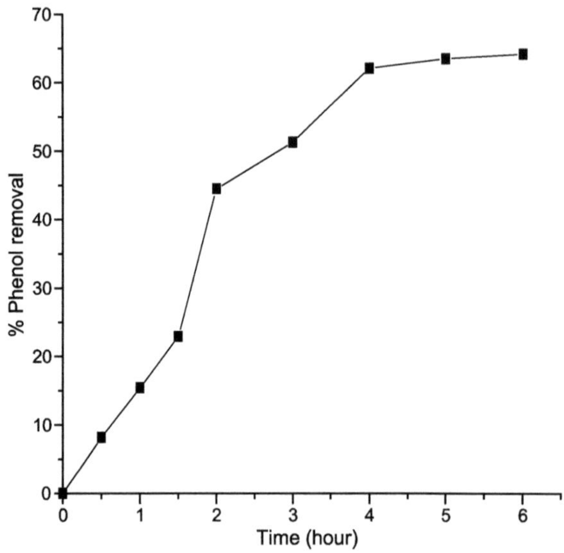

Fig. 5 Photocatalytic degradation of phenol by ZnO nanocatalyst with palash flower powder

Phenol degradation is done in annular reactor. 100 mg/L initial concentration of phenol has been taken for treatment. After treatment for 6 h final phenol concentration was 35.8 mg/L. The Optimum phenol degradation was 62 % at 4 h, reaction time, 30 °C reaction temperature and mass loading of ZnO nanocatalyst is 1.0 g/L for aqueous solution of phenol. Figure 5 Shows the effect of time on degradation of phenol from aqueous solution. The result indicates that with increasing of time percent removal of phenol has also increased. At 4 h time percent removal of phenol tends to constant (at 62 %). Kinetic study of photocatalytic degradation of phenol has been shown in Fig. 6. Here k (Experimental rate constant) for first order is found to be 0.0034 min^{-1}.

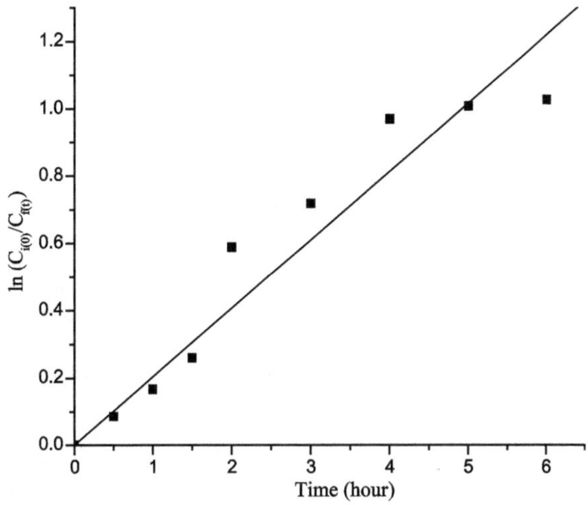

Fig. 6 Kinetic study of photocatalytic degradation of phenol

4 Conclusion

We have synthesized ZnO nanocatalyst by green synthesis route in this method we have used PFP as reducing and stabilizing agent to synthesized ZnO nanocatalyst in place of chemicals so it is eco-friendly and nontoxic method. During synthesis of ZnO nanocatalyst we have used PFP which are easily available in India at very low cost and we have used simple mixing and grinding techniques to synthesized ZnO nanocatalyst so it is simple and cost effective method in comparison to other methods. PFP contains some metals naturally; it can be seen by EDAX analysis so it increases the photocatalytic activity of ZnO nanocatalyst. By FEG-SEM image and Debye Sherrer's equation for XRD we have calculated particle size of ZnO nanocatalyst and it has been found in nano range. The mean particle diameter is 17.6 nm and particles are in 3–40 nm range. The EDAX and XRD results confirmed that particles are ZnO nanocatalyst. FTIR shows stretching for ZnO and various functional groups which are naturally present in PFP and this functional groups work as capping and stabilizing agent for ZnO nanocatalyst. The optimum photocatalytic degradation of phenol is 62 % at 4 h reaction time, 30 °C reaction temperature and mass loading of ZnO nanocatalyst is 1.0 g/L for aqueous solution of phenol. After 6 h reaction time phenol concentration has been reduced from 100 to 35.8 mg/L.

Acknowledgment The authors would like to thank the Department of Chemical Engineering, Maulana Azad National Institute of Technology (MANIT) Bhopal, India, for providing necessary facilities and established the Green Catalysis and Process Technology Research laboratory through Research grant. One of author (Rajani Bharati) would like to special thank the Ministry of Human Resource Development (MHRD), Government of India for giving financial support to research work.

References

Agency for Toxic Substances and Disease Registry (ATSDR) Atlanta, GA: U.S. Department of Health and Human Services. Public Health Service (1998)

Alexander, L., Klug, H.P.: Determination of crystallite size with the X-ray spectrometer. J. Appl. Phys. **21**, 137 (1950)

Banerjee, P., Satapathy, M., Mukhopahayay, A., Das, P.: Leaf extract mediated green synthesis of Silver nanoparticles from widely available Indian plants: synthesis, characterization, antimicrobial property and toxicity analysis. Bioresources Bioprocess SpringerOpen J. 1–3 (2014)

Dutta, S., Ganguly, B.N.: Characterization of ZnO nanoparticles grown in presence of Folic acid template. Dutta and Ganguly J. Nanotechnol. 10–29 (2012)

Gadekar, G.P., Ghoshal, K.P.: Green synthesis of silver nanoparticles from Butea Monosperma and their antimicrobial activities. Bionano Frontier ISSN **5**(2), 7–11 (2012)

Gayen, R.N., Sarkar, K., Hussain, S., Bhar, R. Pal, A.K.: ZnO films prepared by modified Sol-gel technique. Indian J. Pure Appl. Phys. **49**, 470–477 (2011)

Khan, Z.R., Khan, M.S., Zulfequar, M., Khan, M.S.: Optical and structural properties of ZnO thin films fabricated by Sol-gel method. Materials Sciences and applications **2**, 340–345 (2011)

Kumar, H., Rani, R.: Structural and optical characterization of ZnO nanoparticles synthesized by microemulsion route. Int. Lett. Chem., Phys. Astron. **14**, 26–36, 2299–3843 (2013)

Labbe, M.C., Shewa, W.A., Lalman, J.A., Shanmugam, S.R.: Photocatalytic degradation of phenol and phenol derivatives using a nano-TiO_2 catalyst. Integration quantitative and qualitative factors using response surface methodology. Water **6**, 1785–1806 (2014)

Mitra, P., Mondal, S.: Structural and morphological characterization of ZnO thin films synthesized by SILAR. Prog. Theoret. Appl. Phys. **1**, 17–31 (2013)

NIST-National Institute of Standards and Technology, Data from NIST Standard Reference Database 69: NIST Chemistry WebBook (2009)

Rao, C.N.R.: Chemical Applications of Infrared Spectroscopy. Academic press, New York and London (1963)

Ravichandrika, K., Kiranmayi, P., Ravikumar, R.V.S.S.N.: Synthesis, characterization and antibacterial activity of ZnO nanoparticles. Int. J. Pharm. Pharm. Sci. **4**(4), 0975–1491 (2012)

Varshney, R., Bhadauria, S., Gaur, M.S.: Biogenic synthesis of silver nanocubes and nanorods using sundried stevia rebaudiana leaves. Adv. Mater. Lett. **1**(3), 232–237 (2010)

Calculation of Hydroxyl Radical Concentration Using an Indirect Method-Effect of pH and Carbonate Ion

Snigdha Khuntia, Manish Kumar Sinha, Subrata Kumar Majumder and Pallab Ghosh

1 Introduction

The use of hydroxyl radicals in oxidation is known as advanced oxidation process, which is an emerging method in the water and wastewater treatment (Legrini et al. 1993). The oxidation potentials of O_3 (molecular ozone) and hydroxyl radical (•OH) are 2.07 and 2.80 V, respectively (Beltrán 2004). The reaction with molecular ozone is selective to certain organic and inorganic compounds such as the compounds having C=C bonds, and the functional groups containing sulfur, phosphorous, nitrogen and oxygen (Oppenländer 2003). The hydroxyl radicals are able to mineralize many organic and inorganic compounds through various reaction mechanisms such as, radical–radical reactions, hydrogen abstraction, electron transfer and electrophilic addition. This often leads to the complete mineralization of the pollutant compounds (Oppenländer 2003).

S. Khuntia (✉)
Institute of Engineering and Technology-Ahmedabad University, Commerce Six Roads, Navrangpura, Ahmedabadd, India
e-mail: snigdha.khuntia@ahduni.edu.in

M.K. Sinha
Marwadi Education Foundation, Rajkot, India
e-mail: mksinha1985@gmail.com

S.K. Majumder · P. Ghosh
Indian Institute of Technology Guwahati, Guwahati 781039, Assam, India
e-mail: skmaju@iitg.ernet.in

P. Ghosh
e-mail: pallabg@iitg.ernet.in

I. Regupathi et al. (eds.), *Recent Advances in Chemical Engineering*,
DOI 10.1007/978-981-10-1633-2_20

The hydroxyl radicals concentration [•OH], generated from ozone is an important parameter to measure the rate constant of the reaction. [•OH] can be measured by various methods such as, electrochemical, optical, spectrophotometric and colorimetric techniques (Langlais et al. 1999). However, quantitative measurement of [•OH] is rather difficult due to its low concentration in water. A few computer models have been developed for the prediction of concentration of hydroxyl radicals (Chelkowska et al. 1992). Elovitz and von Gunten (1999) have developed an indirect method to determine the concentration of hydroxyl radicals generated from ozone by using a probe compound (i.e. p-Chlorobenzoic acid).

In this work, [•OH] generated from ozone microbubbles has been determined. The effects of pH and carbonate ions on the generation of hydroxyl radicals has been studied. Oxidation of phenol using O_3 and [•OH] has been investigated at pH 3–10. The contribution of hydroxyl radicals on the degradation of phenol has been calculated.

2 Materials and Methods

All the chemicals used in this work are of analytical grade. The details of the experimental setup are given in our earlier work (Khuntia et al. 2013). Oxygen was isolated from air [make: Oz-Air (India), model: HG 03]. An ozonator [make: Oz-Air (India), model: ISM 10 Oxy EC] converted the oxygen to ozone by the corona-discharge method. The ozone generator worked in the range of 0–3 mg s^{-1}. The flow rate of the oxygen and ozone mixture coming out of the ozonator was measured by a rotameter, which had the range of 8–80 cm^3 s^{-1}. The percentage of ozone in the gas mixture was varied in the range of 0.7–2 %. The gas mixture was passed into the microbubble generator [make: Riverforest Corporation (USA), model: AS MK-III] where the dissolution of gas in water was achieved by applying a high pressure. The microbubbles (mean diameter ~25 μm) were generated by releasing the pressure. The gas intake capacity of the microbubble generator was 1.7 cm^3 s^{-1}. A polycarbonate reactor of 20 L capacity was used for conducting ozonation. The microbubbles were continuously passed into the reactor from the microbubble generator by recirculation the aqueous phase.

The [O_3] was measured by using the Indigo Colorimetric Method (Eaton et al. 2005). The PCBA concentration was measured by using a HPLC [make: Shimadzu (Japan), model: LC-20AD] equipped with a ternary pump delivery system and UV detector. 0.3 cm^3 sample was injected into the C18 column [make: Varian (The Netherlands), model: Hypersil ODS, 250 mm length, 4.6 mm ID, 5 μm particle size]. The mobile phase consisted of 55 % methanol and 45 % water containing 10 mol m^{-3} phosphoric acid. PCBA was detected at 234 nm. With 0.3 cm^3 injection, the detection limit was 0.05 μmol dm^{-3}. The concentration of phenol was measured by HPLC using the C18 column at 280 nm wavelength. The mobile phase was constituted of water and acetonitrile (with 0.2 % phosphoric acid, pH ~ 3) in 50:50 volume ratio.

The concentration of hydroxyl radical was measured by using PCBA as the radical probe. 0.05 μmol dm^{-3} PCBA was added to the aqueous phase in the reactor after it was saturated with ozone. At certain time intervals, 20 cm^3 samples were withdrawn, out of which 2 cm^3 was used for measuring the concentration of PCBA and the rest was used for measuring [O_3]. The effect of carbonate on the generation of hydroxyl radicals was studied by adding 1 and 2 mmol dm^{-3} sodium carbonate. After the saturation of ozone in water, PCBA was added, which was followed by calcium carbonate. All the experiments were performed at 298 K.

The hydroxyl radicals contribution on phenol degradation was determined at pH 3–10. 0.53 mmol dm^{-3} phenol and 0.05 μmol dm^{-3} PCBA were added after the saturation of the aqueous phase by ozone. The procedure for determining the concentrations of PCBA and phenol was similar to that described earlier.

3 Results and Discussion

3.1 Determination of [•OH]

Direct determination of the concentration of hydroxyl radicals is rather difficult because they have a very short life time (of the order of nanoseconds), and they react at the site of their formation. Therefore, an indirect method was proposed by Elovitz and von Gunten (1999) to measure the concentration of hydroxyl radicals. In this method, a probe compound (e.g. PCBA) is used, which readily reacts with •OH, but remains almost indifferent towards O_3 (Elovitz and von Gunten 1999). To illustrate, the rate constants of the reaction of PCBA with •OH and O_3 are 5×10^9 and 0.15 $dm^3\ mol^{-1}\ s^{-1}$, respectively. When the probe compound and the scavenger are both present in water, the consumption of •OH will be due to both of them, as illustrated in Fig. 1.

When PCBA is present at very low concentrations, the contribution of the probe on •OH scavenging is negligible, and therefore, k_h[PCBA] ≪ k_s[S]. As PCBA shows almost no reaction with O_3 as compared to •OH, the rate equation for PCBA can be written as (Elovitz and von Gunten 1999)

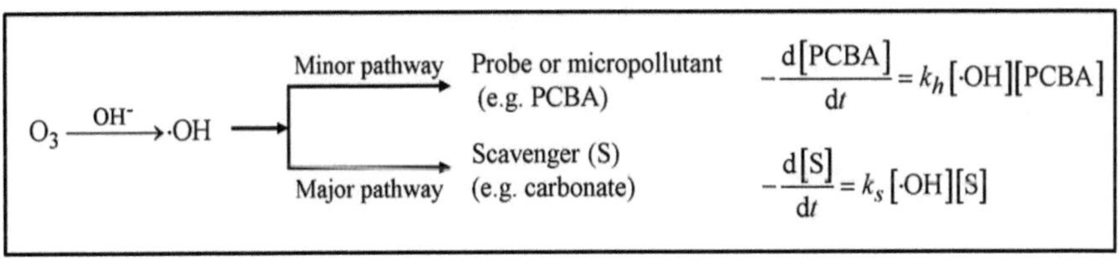

Fig. 1 Mechanism of reaction of probe and scavenger with •OH

$$-\frac{\mathrm{d}[\mathrm{PCBA}]}{\mathrm{d}t} = k_h[\cdot\mathrm{OH}][\mathrm{PCBA}] \quad (1)$$

Rearranging and integrating Eq. (1), we get

$$\ln\left(\frac{[\mathrm{PCBA}]}{[\mathrm{PCBA}]_0}\right) = -k_h \int [\cdot\mathrm{OH}]\mathrm{d}t \quad (2)$$

where, $\int[\cdot\mathrm{OH}]\mathrm{d}t$, is called the time-integrated [•OH], or *·OH exposure*. From Eq. (2) it is observed that, the decrease in [PCBA] at any time, t, is an indirect measurement of the •OH exposure for the reaction time. Therefore, a term, R_{ct}, is defined as

$$R_{ct} = \frac{\int [\cdot\mathrm{OH}]\,\mathrm{d}t}{\int [\mathrm{O}_3]\,\mathrm{d}t} \quad (3)$$

The O_3 exposure or $\int[\mathrm{O}_3]\mathrm{d}t$, can be calculated from the integration of the ozone concentration profile, as shown in Fig. 2a. From Eqs. (2) and (3) we get

$$\ln\left(\frac{[\mathrm{PCBA}]}{[\mathrm{PCBA}]_0}\right) = -k_h R_{ct} \int [\mathrm{O}_3]\,\mathrm{d}t \quad (4)$$

From Eq. (4) R_{ct} can be determined by measuring the concentration profile of PCBA and O_3. The concentration profile of PCBA with time is shown in Fig. 2b. The •OH generated from ozone is consumed by PCBA, and therefore, the concentration of PCBA decreases. However, the ozone dissolution in water increases with time, which results the increase in the ozone exposure. From the above linear equation [i.e. Eq. (4)], the O_3 exposure, R_{ct} can be calculated from the slope of the plot.

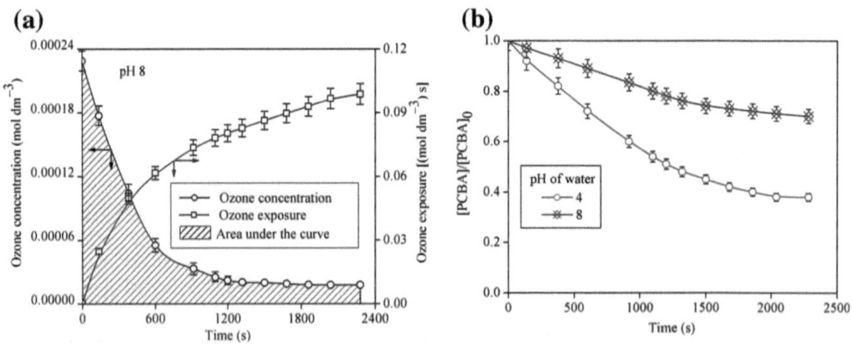

Fig. 2 Variation of **a** O_3 concentration and O_3 exposure, and **b** PCBA concentration, with time

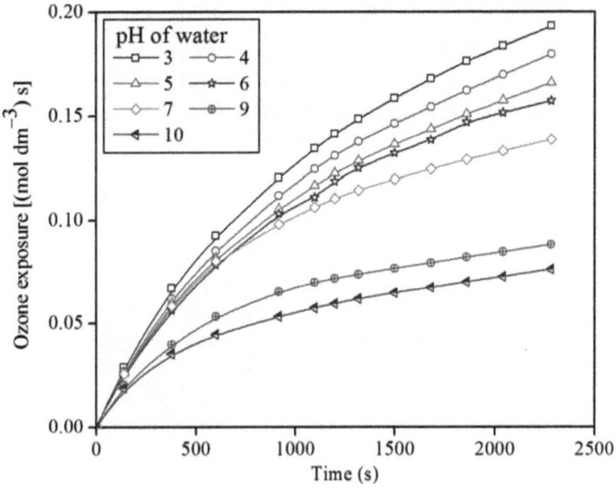

Fig. 3 O_3 exposure profile at various pH

Figure 3 depicts the O_3 exposure profiles at various pH. The value of O_3 exposure was higher at the acidic pH than that at the alkaline pH. The decomposition of ozone increases with increasing pH. Therefore, at acidic pH, the molecular ozone is more stable. The relationship between $\ln([\mathrm{PCBA}]/[\mathrm{PCBA}]_0)$ and the O_3 exposure shows that, R_{ct} is a constant and independent of the reaction time. So, $R_{ct} = [\cdot\mathrm{OH}]/[O_3]$ at any time t during the reaction (Elovitz and von Gunten, 1999), from which the value of [•OH] can be calculated.

The variation of •OH exposure, O_3 exposure and R_{ct} with pH is shown in Fig. 4. The variation of ozone exposure with time at different pH is shown in Fig. 5. At acidic pH, the value of •OH exposure was more than that at the alkaline pH. This was the result of the higher •OH generation from the ozone microbubbles.

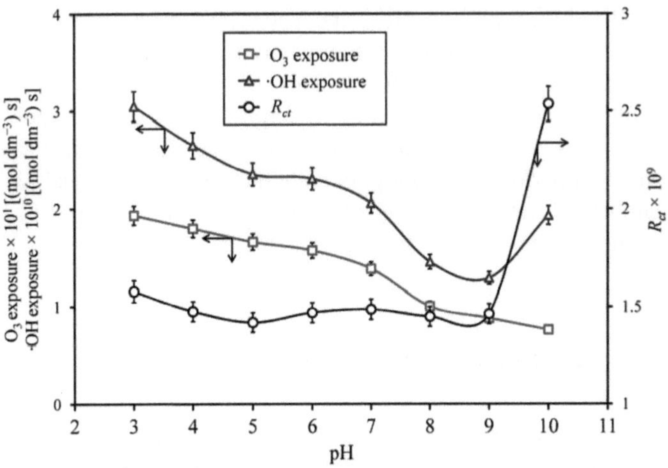

Fig. 4 Comparison of •OH exposure, O_3 exposure and R_{ct} at different pH at 2400 s

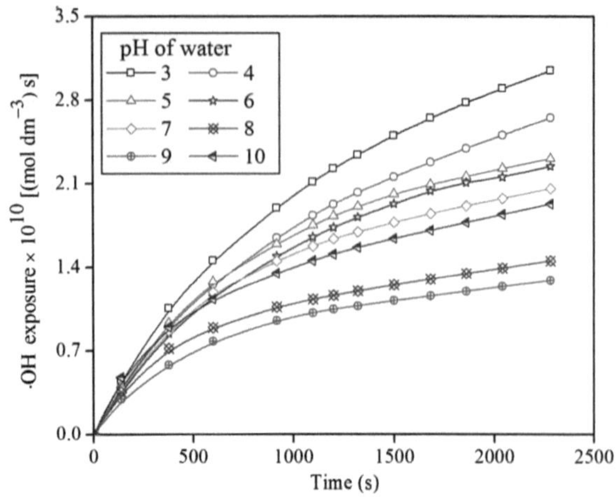

Fig. 5 Variation of •OH exposure at different pH

However, the ozone decomposition was less at pH < 7, which increased the O_3 exposure. Therefore, the value of R_{ct} was less at acidic pH. At alkaline pH, the decomposition of ozone was more, and therefore, the O_3 exposure was decreased. The generation of •OH was due to the chain reactions triggered by the OH^{-}. However, a significant amount of hydroxyl radicals was generated in the acidic conditions. Although the •OH exposure was less at pH < 9, but R_{ct} was similar to that of the acidic conditions. At pH 10, the •OH exposure and the ozone decomposition rapidly increased due to the presence of a large amount of OH^{-} ions. Thus, the O_3 exposure noticeably decreased and R_{ct} increased.

Elovitz and von Gunten (1999) have reported that R_{ct} increased with pH, whereas the values of •OH exposure were very close to each other for pH 6–9 (viz. 5×10^{-14} mol dm^{-3} s). This happened due to the higher ozone decomposition and lower •OH generation in this pH range. However, they have not reported the values of [•OH], [O_3] and R_{ct} at pH < 6. Haag and Yao (1993) have reported that the value of R_{ct} was almost constant at pH 7–8.2 for a variety of water samples ranging from clean surface water to wastewater. In our study, higher depletion of PCBA at acidic pH clearly shows that the generation of •OH from the ozone microbubbles in the acidic medium was more than that in the alkaline medium. Studies at pH < 3 or pH > 10 were not performed in our pilot plant due to the operational limitations of the microbubble generator. The range of the pH could not be increased, as it would damage the internal materials of the microbubble generator.

3.2 *Degradation of Phenol by •OH*

Hydroxyl radicals and molecular ozone simultaneously react with the pollutants. Therefore, the depletion rate of the pollutant (D) with ozone may be written as

Fig. 6 Concentration profiles of phenol

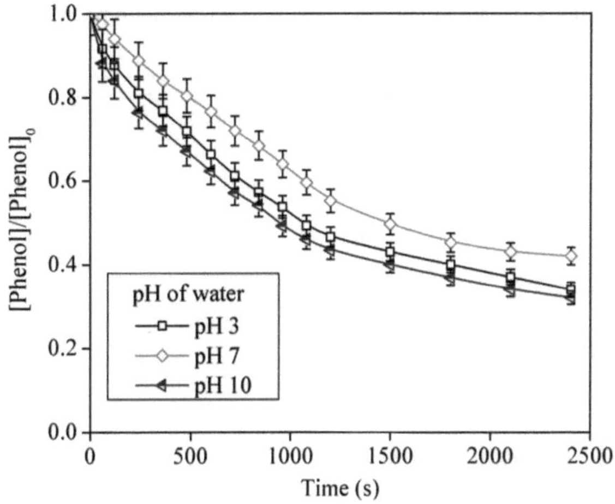

$$-\frac{d[D]}{dt} = k_h[\cdot OH][D] + k_{O_3}[O_3][D] \quad (5)$$

$$-\frac{d[D]}{dt} = (k_h R_{ct} + k_{O_3})[O_3][D] \quad (6)$$

where k_h and k_{O_3} are the second-order rate constants for reaction of micropollutant, D, with •OH and O_3, respectively. The values of k_h and k_{O_3} can be calculated from Eq. (6). The degradation of phenol using ozone microbubbles at different pH is shown is Fig. 6. The rate constant of the reaction of phenol with ozone and •OH is pH dependent. The O_3 exposure was slightly less in presence of phenol due to the consumption of ozone. The degradation of phenol was more effective at acidic and alkaline pH than at pH 7. By using the PCBA probe, the vales of R_{ct} were calculated. The rate constants of the reaction of phenol with •OH and O_3 (Table 1) were determined by using Eq. (6). The values of k_{O_3} and k_h are similar to that reported by Hoigné and Bader (1983), and Land and Ebert (1967), respectively.

Furthermore, the values of k_{O_3} are of the same order at all pH. The R_{ct} parameter is a measure of the contributions of OH and O_3 for the degradation of a micropollutant in water. The fraction of D degraded by •OH and O_3 can be expressed as

$$f_{\cdot OH} = \frac{\text{Amount of D reacted with } \cdot \text{ OH}}{\text{Amount of D reacted with } \cdot \text{ OH} + \text{Amount of D reacted with } O_3} \quad (7)$$

$$f_{\cdot OH} = \frac{k_h[\cdot OH][D]}{k_h[\cdot OH][D] + k_{O_3}[O_3][D]} = \frac{k_h R_{ct}}{k_h R_{ct} + k_{O_3}} \quad (8)$$

Using Eq. (8), the fraction of phenol reacted with •OH was calculated. These values are given in Table 1.

Table 1 Contribution of O_3 and •OH on phenol oxidation

pH	$R_{ct} \times 10^{10}$	k_{O_3} (dm^3 mol^{-1} s^{-1})	$k_h \times 10^{-10}$ (dm^3 mol^{-1} s^{-1})	R^2	$f_{\cdot OH} \times 10^3$
3	11.4	1203	1.49	0.97	14.00
4	10.2	1202	1.48	0.99	12.50
5	6.9	1262	1.50	0.96	8.08
6	5.5	1260	1.50	0.97	6.46
7	5.3	1359	1.51	0.98	5.78
8	5.2	1420	1.51	0.96	5.43
9	5.2	1420	1.49	0.97	5.43
10	7.9	1420	1.51	0.97	8.22

4 Conclusions

The indirect method of determination of [•OH] is promising. The depletion rate of PCBA was high in the acidic condition, and it decreased with increasing pH. But at pH 10, the PCBA depletion rate increased. From the rate of depletion of PCBA and ozone, the values of O_3 exposure, •OH exposure and R_{ct} were calculated. At acidic pH, the generation of •OH was due to the thermal dissociation of ozone. However, ozone was more stable in the acidic condition than in the alkaline condition. Therefore, both O_3 and •OH exposures were higher in the acidic media than the alkaline conditions. At alkaline pH, ozone decomposition increased due to the presence of OH^-. However, the generation of •OH was more at the acidic pH. At pH 10, •OH exposure strongly increased due to the high concentration of OH^-. As a result, the O_3 exposure increased. The degradation of phenol by •OH was higher at acidic pH than that at the alkaline pH. The rate constant of reaction of phenol with •OH was much higher than that of O_3 and it was constant at all pH.

References

Beltrán, F.J.: Ozone Reaction Kinetics for Water and Wastewater System. CRC Press, Boca Raton, Florida (2004)

Chelkowska, K., Grasso, D., Fabian, I., Gordon, G.: Numerical simulations of aqueous ozone decomposition. Ozone Sci. Eng. **14**, 33–49 (1992)

Eaton, A.D., Clesceri, L.S., Rice, E.W., Greenberg, A.E.: Standard Methods for the Examination of Water & Wastewater. American Public Health Association, New York (2005)

Elovitz, M.S., von Gunten, U.: Hydroxyl radical/ozone ratios during ozonation processes. I. the R_{ct} concept. Ozone Sci. Eng. **21**, 239–260 (1999)

Haag, W.R., Yao, C.C.D.: Ozonation of U.S. drinking water sources: HO concentration and oxidation-competition values, in: Eleventh Ozone World Congress. San Francisco, California, pp. 119–126 (1993)

Hoigné, J., Bader, H.: Rate constants of reactions of ozone with organic and inorganic compounds in water-I. Water Res. **17**, 173–183 (1983)

Khuntia, S., Majumder, S.K., Ghosh, P.: Removal of ammonia from water by ozone microbubbles. Ind. Eng. Chem. Res. **52**, 318–326 (2013)

Land, E.J., Ebert, M.: Pulse radiolysis studies of aqueous phenol. water elimination from dihydroxycyclohexadienyl radicals to form phenoxyl. Trans. Faraday Soc. **63**, 1181–1190 (1967)

Langlais, B., Reckhow, D.A., Brink, D.R.: Ozone in Water Treatment: Applications and Engineering. Lewis Publishers, Boca Raton, Florida (1999)

Legrini, O., Oliveros, E., Braun, A.M.: Photochemical processes for water treatment. Chem. Rev. **93**, 671–698 (1993)

Oppenländer, T.: Photochemical Purification of Water and Air. Wiley-VCH, Weinheim, Germany (2003)

Granular Activated Carbon Supported Titanium Dioxide Photocatalytic Process for Carbofuran Removal

M.A. Vishnuganth, Mathava Kumar and N. Selvaraju

1 Introduction

In the past, the usage of pesticides has been encouraged significantly to increase the yields in agriculture and also to control the diseases transformed to human by the insects (Rekha et al. 2006). It is vital to increase the food productivity to cater the need of rapidly increasing population (Agoramoorthy et al. 2008). There are wide variety of options available for increasing the productivity and the simplest being the efficient pest control. However, pest control is one of the difficult tasks in tropical countries owing to the favorable speedy production of pests in the very hot and humid conditions (Lakshmi 1993). The conventional chemical oxidation techniques are ineffective in mineralizing refractory organic materials but it is economically suitable for the removal of pollutants at high concentrations. On the other hand, the biological methods have the following main weaknesses: slow reaction rate, problem with disposal of sludge and the need for severe control of suitable pH and temperature. In this situation, heterogeneous photocatalysis has many benefits for the removal of pollutants:

(1) A broad range of organic composites could be destroyed (Hoffmann et al. 1995)

M.A. Vishnuganth · N. Selvaraju
Department of Chemical Engineering, National Institute of Technology Calicut, Calicut, Kerala, India
e-mail: vishnuchem12@gmail.com

N. Selvaraju
e-mail: selvaraju@nitc.ac.in

M. Kumar (✉)
Department of Civil Engineering, Indian Institute of Technology Madras, Chennai, India
e-mail: mathav@iitm.ac.in

I. Regupathi et al. (eds.), *Recent Advances in Chemical Engineering*,
DOI 10.1007/978-981-10-1633-2_21

(2) The target compounds which are organic in nature will be converted to simple end-products such as carbon-di-oxide, acids and water (Alberici and Jardim 1997)
(3) The overall operation cost and energy requirement of photocatalysis could be minimized by combining with solar energy input (Malato et al. 2001)

In this work, TiO_2 was coated on the GAC surface by a predetermined standard protocol and subsequently, the TiO_2-GAC was used as a catalyst for the removal of carbofuran under the photocatalytic process. Moreover, the research work was to display that carbofuran could be degraded CO_2 and H_2O in the photodegradation process and to compare the effect of various process conditions such as TiO_2 concentration and reaction pH on carbofuran removal.

2 Materials and Methods

2.1 Chemicals and Reagents

Carbofuran (98 %), Titanium tetra isopropoxide (99 %), reagent grade hydrochloric acid (37 %), ≥99 % pure ethanol and GAC were purchased from commercial suppliers. The TiO_2 catalyst was coated on GAC surface as per the standardized protocols (Gao et al. 2011). The prepared catalyst was analyzed using the scanning electron microscopy (SEM) and the corresponding elemental composition was determined by electron dispersive spectroscopy (EDS) (Asha and Kumar 2015).

2.2 Experimental Set-up and Procedures

A cylindrical double walled batch reactor to maintain 1.9 L volume, provision to incorporate 4 UV lamps and overhead stirrer with sample collection ports was designed and used for the experiments. A 500 mg L^{-1} of carbofuran solution was prepared (i.e. stock solution) and subsequently, diluted to cater the experimental requirements. Exactly 1.9 L of carbofuran solution (as per predetermined concentration) was added into reactor and the pH was adjusted to predetermined value using 0.1 N NaOH or 0.1 N H_2SO_4. A predetermined amount of TiO_2 was supplemented to the reactor, the UV irradiation was provided by UV lamp and immediately the electronic overhead stirrer was switched ON (mixed continuously at 200–300 rpm). At regular period, samples were withdrawn from the reactor, filtered and analyzed for carbofuran in the high performance liquid chromatography (HPLC) (Shimatzu, Japan). During the investigation, the pH and temperature of the reactor was measured using calibrated pH and temperature meters (Hanna-3121, Germany).

3 Results and Discussion

3.1 *Effect of Initial PH and Temperature*

The initial pH and temperature of reaction mixture are the most important parameters in the appraisal of photocatalytic reactions. The adsorption of organic compounds on the TiO_2 surface is mainly influenced by the reaction pH, which is considered as vital in the photocatalytic reaction (Bahnemann et al. 2007). The profiles of pH and temperature under the batch photocatalytic experiments conducted at 250 mg L^{-1} carbofuran concentration with increasing TiO_2 concentration of 5 and 75 mg L^{-1} at various pH conditions are shown in Fig. 1a–e. The temperature of reaction constantly increased owing to the UV irradiations, which reflected later in the increase of reaction rate. At the start of the experiment average temperature of the reaction components was 24°C and it was increased to 63°C at the end of the experiments, i.e. in 4 h. Figure 2a, b show carbofuran removal at pH 3 and 250 mg L^{-1} carbofuran with increasing TiO_2 concentration, i.e. 5 and 75 mg L^{-1}. It can be observed in the figure that a maximum carbofuran removal of 77 and 74 %, respectively, at the TiO_2 concentration of 5 and 75 mg L^{-1}. However, a significant increase in carbofuran removal (84 and 83 % respectively) can be noticed in Fig. 2c, d (experiments conducted at fixed pH value of 7 and carbofuran concentration of 250 mg L^{-1} with increasing the TiO_2 concentration of 5 and 75 mg L^{-1}). The increase in pH value from 7 to 11, maximized the carbofuran to 100 % under both the TiO_2 concentrations investigated (Fig. 2e, f). The formation of hydroxyl radicals is mainly affected by the reaction pH. The cause being the reaction between photo-induced holes on the TiO_2 surface and the hydroxide ions present in the solution. It was reported earlier that positive photo-induced holes and hydroxyl radicals were responsible for oxidation under low pH and neutral/alkaline pH, respectively (Shifu et al. 2005). The improved carbofuran removal at pH 7 and 11 in this investigation could be due to the presence of higher hydroxyl radicals, which is logically enhanced under the alkaline pH (Wu et al. 2009b).

3.2 *Effect of Photocatalyst Dosage*

To recognize the optimum TiO_2 quantity required for photocatalytic degradation of carbofuran, the experiments were conducted at fixed carbofuran concentration (250 mg L^{-1}) with varying the pH 3, 7 and 11 and varying TiO_2 dosages (5 and 75 mg L^{-1}) as mentioned earlier. Table 1 shows the carbofuran removal under all the experimental conditions investigated in the current work. The carbofuran removal efficiency at the beginning of the photocatalytic process was higher at 75 mg/L compared to 5 mg/L TiO_2 dosage. However, the overall removal efficiency was slightly lesser in case of higher TiO_2 dosage owing to (1) scattering of light by the GAC-TiO_2 composite and (2) shielding of the target compound

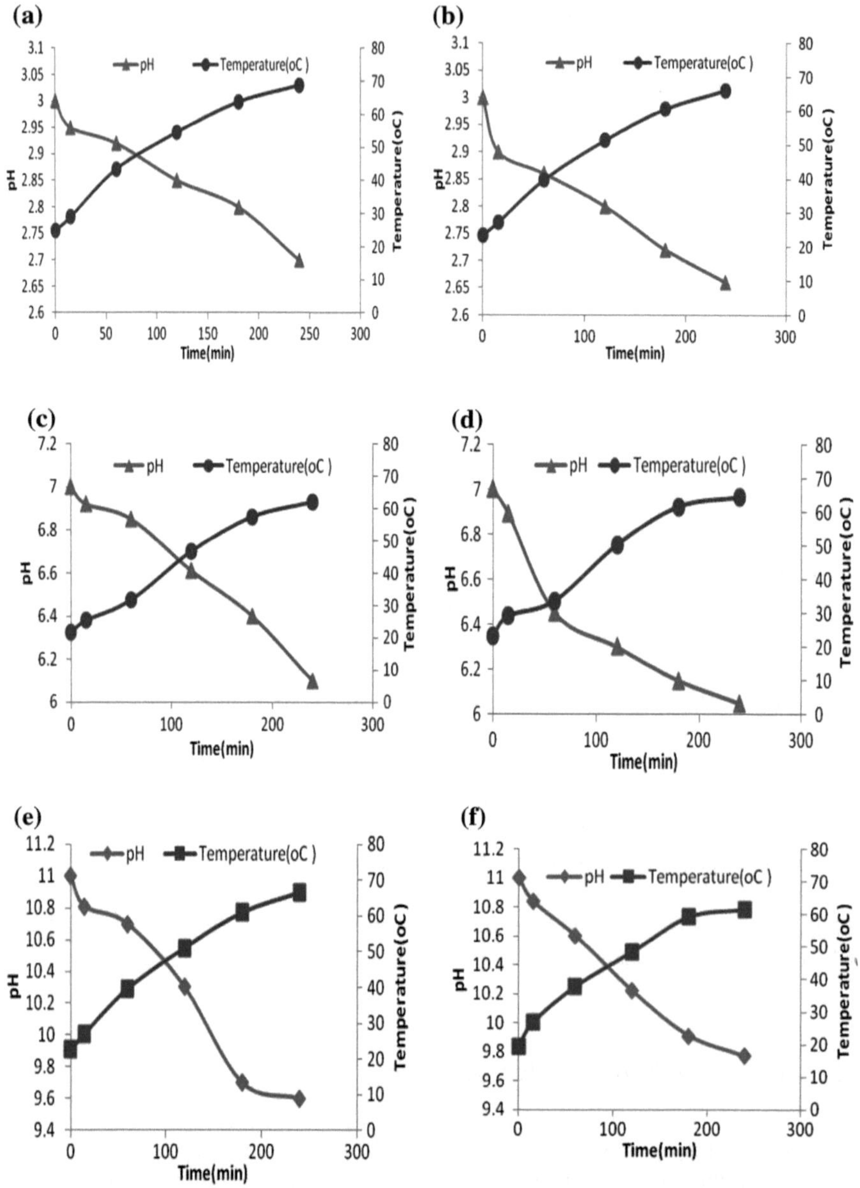

Fig. 1 **a** and **b** pH and temperature profiles on photocatalytic degradation of carbofuran at fixed carbofuran concentration (250 mg L^{-1}), pH 3 and with 5 and 75 mg L^{-1} of TiO_2, respectively. **c** and **d**. pH and temperature profiles on photocatalytic degradation of carbofuran at fixed carbofuran concentration (250 mg L^{-1}), pH 7 with 5 and 75 mg L^{-1} of TiO_2, respectively. **e** and **f** pH and temperature profiles on photocatalytic degradation of carbofuran at fixed carbofuran concentration (250 mg L^{-1}), pH 11 with 5 and 75 mg L^{-1} of TiO_2, respectively

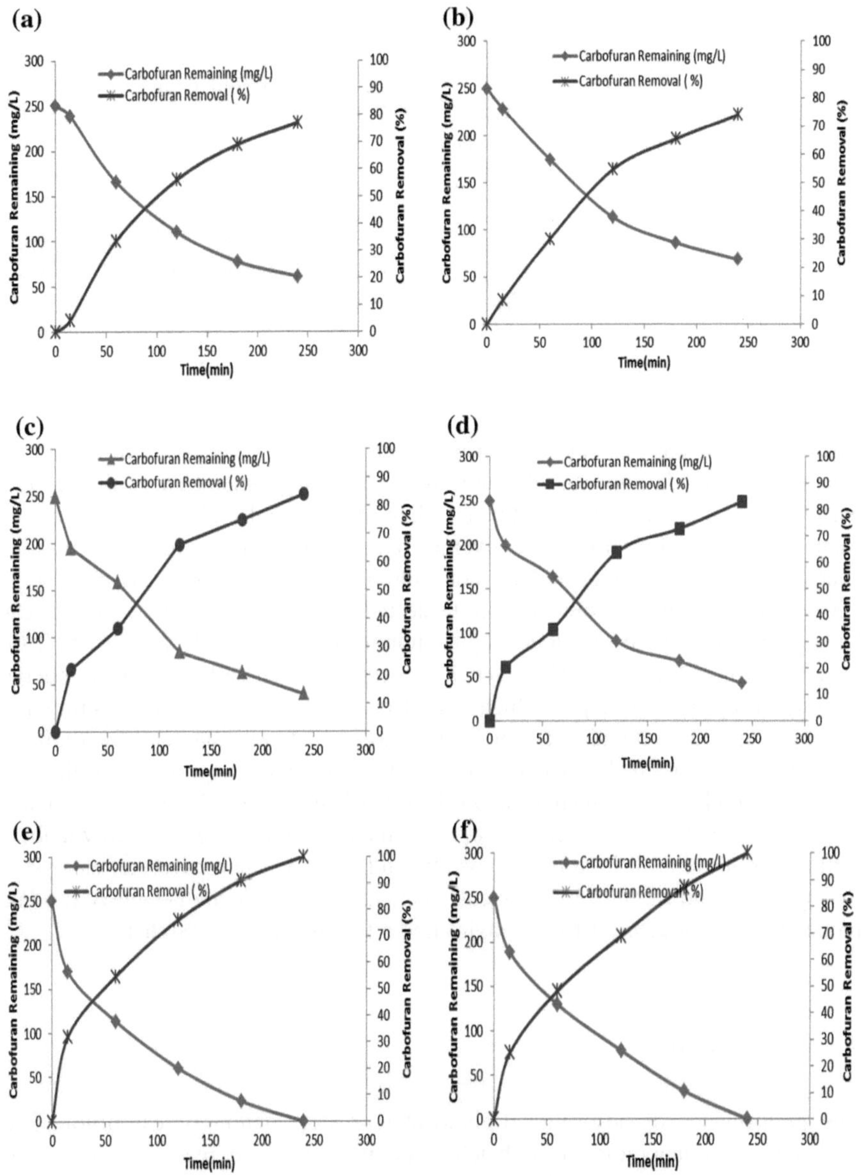

Fig. 2 **a** and **b** Profile of Carbofuran removal at fixed carbofuran concentration (250 mg L^{-1}), pH 3 with 5 and 75 mg L^{-1} of TiO_2 at 240 min, respectively. **c** and **d** Profile of Carbofuran removal at fixed carbofuran concentration (250 mg L^{-1}), pH 7 with 5 and 75 mg L^{-1} of TiO_2 at 240 min, respectively. **e** and **f** Profile of Carbofuran removal at fixed carbofuran concentration (250 mg L^{-1}), pH 11 with 5 and 75 mg L^{-1} of TiO_2 at 240 min, respectively

Table 1 Comparison of results for photocatalytic degradation of carbofuran experiments

Exp. No.	Experimental conditions			Carbofuran removal (%) at various time (min)				
	Carbofuran concentration (mg L^{-1})	TiO_2 concentration (mg L^{-1})	pH	15	60	120	180	240
1	250.0	5	3	4.5	33.6	56	69.0	77.0
2	250.0	75	3	8.5	30.2	54.7	65.4	74.0
3	250.0	5	7	22.0	36.5	66.1	75.0	84.0
4	250.0	75	7	20.3	34.8	63.7	72.8	83.0
5	250.0	5	11	32.0	54.6	76.0	91.0	100
6	250.0	75	11	25.0	48.3	69.0	87.4	100

adsorbed on the catalyst surface. On the other hand, the decrease in removal efficiency at higher dosage was also attributed to the agglomeration of the catalyst particles, which might have resulted a decline in the overall light absorption on the catalyst surface. Therefore, the overall photocatalytic removal of carbofuran decreased at higher photocatalytic concentration (Adesina 2004; Shankar et al. 2004). The experimental outcomes indicate that the photocatalytic carbofuran can be successfully carried out under higher initial pH value of 11, with lower TiO_2 concentration of 5 mg L^{-1}. However, it was reported that carbofuran can undergo hydrolysis under alkaline pH, which may decrease the carbofuran concentration.

At present, further studies are in progress to estimate the quantity of carbofuran disappeared to intermediates under the alkaline pH and also to quantify the carbofuran degradation intermediates. On the other hand, the GAC-TiO_2 catalyst was easily separated from the batch-reactor (after the treatment process) by gravitational settling for 30 min followed by a conventional filtration. The recovery of catalyst in this system could minimize the overall treatment cost and simultaneously it could reduce the risk of catalyst to the environment through effluent discharge.

4 Conclusions

The performance of adsorptional-photocatalytic degradation of carbofuran was evaluated in a laboratory-scale batch-mode photocatalytic reactor system. The influence of parameters like TiO_2 concentration and pH were analysed over the carbofuran degradation process. TiO_2 concentration and initial pH have significant effects on carbofuran degradation. The experimental outcomes indicated that 5 mg L^{-1} TiO_2 concentration and pH 11 are the favorable conditions for complete carbofuran removal. Further studies are required to characterize the carbofuran intermediates formed at the alkaline pH. As a whole, the adsorption photocatalysis is better suitable for carbofuran contaminated water treatment owing to high carbofuran removal and superior catalyst separation from the effluent stream.

References

Adesina, A.A. (2004) Industrial exploitation of photo catalysis: progress, perspectives and prospects. Catalys. Surv. Asia **8** (4), 265–273

Agoramoorthy, G.: Can India meet the increasing food demand by 2020? Futures **40**, 503–506 (2008)

Alberici, R.M., Jardim, W.F.: Photocatalytic destruction of VOCs in the gas-phase using titanium dioxide. Appl. Catal. B **14**, 55–68 (1997)

Asha, C.R., Kumar, M.: Photocatalytic degradation of poultry wastewater using activated carbon-supported titanium dioxide. Desalin. Water Treat. **54**, 3279–3290 (2015)

Bahnemann, W., Muneer, M., Haque, M.M.: Titanium dioxide-mediated photocatalysed degradation of few selected organic pollutants in aqueous suspensions. Catal. Today **124**, 133–148 (2007)

Gao, B., Yap, P.S., Lim, T.M., Lim, T.T.: Adsorption-photocatalytic degradation of acid red 88 by supported TiO2: effect of activated carbon support and aqueous anions. Chem. Eng. J. **171**, 1098–1107 (2011)

Hoffmann, M.R., Martin, S.T., Choi, W., Bahnemann, D.W.: Environmental applications of semiconductor photocatalysis. Chem. Rev. **95**, 69–96 (1995)

Lakshmi, A.: Pesticides in India: risk assessment to aquatic ecosystems. Sci. Total Environ. **134**, 243–253 (1993)

Malato, S., Blanco, J., Fernandez-lbanez, P., Caceres, J., (2001). Treatment of 2,4-dichlorophenol by solar photocatalysis: comparison of coupled photocatalytic-active carbon versus active carbon. J. Sol. Energy-T.ASME 123, 138–142

Rekha, S.N., Naik, R., Prasad, : Pesticide residue in organic and conventional food–risk analysis. Chem. Health Safety **13**, 12–19 (2006)

Shankar, M.V., Anandan, S., Venkatachalam, N., Arabindoo, B., Murugesan, V.: Novel thin-film reactor for photocatalytic degradation of pesticides in aqueous Solutions. J. Chem. Technol. Biotechnol. **79**, 1279–1285 (2004)

Shifu, C., Gengyu, C.: Photocatalytic degradation of pesticides using floating photocatalyst TiO_2 SiO_2 beads by sunlight. Sol. Energy **79**, 1–9 (2005)

Wu, R., Chena, C., Chen, M., Lu, C. (2009) Titanium dioxide-mediated heterogeneous
Photocatalytic degradation of terbufos: parameter study and reaction pathways. J. Hazard. Mater. **162**, 945–953

Electrodeposition of TiO_2/Ni-P Composite Electrodes for Efficient Water Electrolysis

Liju Elias and A. Chitharanjan Hegde

1 Introduction

The water electrolysis has attracted more attention as one of the most important industrial electrochemical reactions due to its promising outcomes towards the future source of energy (Ju et al. 2015; Aal and Hassan 2009). The hydrogen gas produced at the cathode has been deemed to be a promising alternative and renewable energy source that may take the place of fossil fuels in future (Ju et al. 2015). The advantages of water electrolysis like; hydrogen production with high purity, manufacturing capability, small size and portability, lack of dependence on fossil fuel sources and simplicity over other conventional methods for hydrogen production making it more attractive. However, the high over voltage at both anode and cathode, leading to huge energy consumption, has been hindering water splitting from industrial applications (Wei et al. 2007). In order to reduce the energy consumption, it is essential to develop more active electrode materials for HER and (OER) with low cost (Wei et al. 2007). To date, the most effective electrocatalysts capable of catalyzing HER at a significant rate with almost no overpotential are based on Pt-group metals. However, the wide applications of these noble metals have hindered due to their scarcity and high costs, hence extensive research has already been devoted in this area for the development of non-Pt materials as electrocatalysts (Hou et al. 2015). As per the literatures, at present, Ni is one of the active non-precious metal as HER catalysts, but its HER activity is limited so far and hence it is very important to enhance the catalytic activity of Ni either through proper alloying or through other methods to increase its surface area.

L. Elias · A.C. Hegde (✉)
Electrochemistry Research Lab, Department of Chemistry, National Institute of Technology Karnataka, Srinivasnagar, Surathkal 575025, India
e-mail: acrhegde@gmail.com

I. Regupathi et al. (eds.), *Recent Advances in Chemical Engineering*,
DOI 10.1007/978-981-10-1633-2_22

The present research work mainly focused on the development of electrodeposited Ni-P alloy coatings as catalysts for alkaline water electrolysis and to testify the enhancement in electrocatalytic efficiency through the development of TiO_2/Ni-P composite coating, using composite electrodeposition method.

2 Experimental

2.1 Development of Ni-P Alloy Coatings

The nanocrystalline Ni-P alloy coatings were electrodeposited on copper substrate from an optimal citrate bath containing $NiSO_4$. $6H_2O$ (28.0 g L^{-1}), $NaPO_2$ H_2. H_2O (51.0 g L^{-1}), $Na_3C_6H_5O_7$. $2H_2O$ (54.0 g L^{-1}), NH_4Cl (20 g L^{-1}), H_3BO_3 (10 g L^{-1}), $C_3H_8O_3$ (20 mL L^{-1}). All the reagents used for electrodeposition and electrochemical measurements were prepared in double distilled water and the deposition processes were conducted at pH = 8. Surface and compositional studies of the electrodeposited coatings were made, using an effective exposed surface area 1 cm^2. All depositions were carried out using optimized bath, keeping anode and cathode parallel 5 cm apart in a conventional PVC cell of 250 mL capacity. Prior to plating, the mirror polished copper surface was electrocleaned, and then pickled in 0.5 M HNO_3. For comparison purpose all electrodeposition were carried out galvanostatically from the same bath for the same duration (600 s) using power source (DC Power Analyser, Agilent, N6705A).

The microstructure and composition of the Ni-P alloy coatings were characterized using, Scanning Electron Microscope (SEM) (JSM–7610F from JEOL, USA). Energy dispersive spectroscopy (EDS) analysis and X-ray diffraction (XRD) (JDX–8P, JEOL, Japan, with CuK_λ radiation (λ = 1.5418 Å) as the X–ray source) techniques were used for elemental composition and phase structure analysis, respectively.

2.2 Synthesis of TiO_2/Ni-P Composite Electrodes

The TiO_2/Ni-P composite electrodes were prepared on copper by composite electrodeposition technique from the same optimal Ni-P plating bath loaded with TiO_2 nanoparticles (0.5 g L^{-1}). The codeposition of the nanoparticles along with the metal ions were achieved using same plating setup as used for the development of alloy electrodes. The TiO_2 nanoparticles used in this study (P < 25, anatase) was procured from Sigma-Aldrich Company, St. Louis, MO, United States of America. Since its solubility is very less in the plating bath, the solution was stirred over night for homogeneous particle dispersion using a magnetic stirrer. The composite coating was developed at an applied current density (c.d.) of 4.0 A dm^{-2} for the same deposition time of 600 s.

The surface topography and composition of the composite coating was observed using field emission scanning electron microscopy (Neon 40 Crossbeam, Carl Zeiss, Oberkochen, Germany) and energy dispersive spectroscopy (EDS-Oxford) analysis tool.

2.3 Electrochemical Measurements

Electrocatalytic characterization of both Ni-P alloy and TiO_2/Ni-P composite coatings, developed on copper rod of 1 cm^2 cross sectional area, were made by using a custom made three-electrode tubular glass cell as explained elsewhere (Elias et al. 2015; Elias and Hegde 2015). Electrocatalytic behavior of the coatings towards HER was evaluated in 1.0 M KOH medium using cyclic voltammetry and chronopotentiometry techniques. Further, the electrocatalytic efficacy of the test electrodes for HER were confirmed by measuring the amount of H_2 gases liberated during the analysis.

3 Results and Discussion

3.1 Morphological and Structural Characterization of Ni-P Alloy Coatings

The heterogeneous catalytic activity of the working electrode is greatly dependant on its surface morphology and phase structure. The SEM images of Ni–P alloy coatings deposited at different c.d.'s (2.0, 4.0 and 6.0 A dm^{-2}) were taken, and shown in Fig. 1a. The corresponding EDX results shows that P content in the

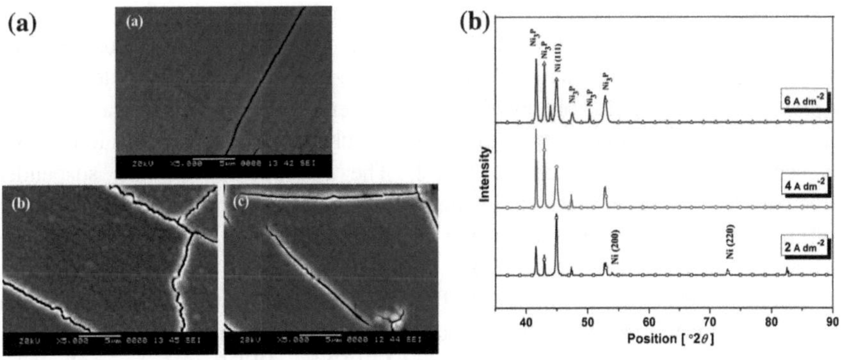

Fig. 1 **a** SEM images of Ni-P alloy coatings developed at: ***a*** 2.0 A dm^{-2}, ***b*** 4.0 A dm^{-2}, ***c*** 6.0 A dm^{-2}, and **b** The corresponding X-ray diffraction patterns of the coatings, deposited from same bath

coatings were found to be increased with deposition c.d., and is attributing to the significant morphological changes in the coatings with deposition c.d. The micro cracks observed in the coatings with increase in c.d. i.e., with increase in phosphorous content, is ascribed to the stress developed during plating (Pillai et al. 2012). The surface morphology appears to be changing from smaller grains to larger nodules, by the aggregation of small crystallites with increasing c.d. and was found to be crystalline from XRD analysis. The obtained diffraction pattern confirms the equilibrium existence of Ni and Ni_3P phase in the deposits as shown in Fig. 1b. The increased stress in the deposit due to hydrogen embrittlement during deposition, at higher c.d., resulted in the precipitation of Ni_3P phase with increase in concentration of p in the deposit (Pillai et al. 2012; Wei et al. 2007).

3.2 Hydrogen Evolution Reaction

Electrocatalytic efficacy of the Ni-P alloy electrodes (developed at c.d.'s 2.0, 4.0 and 6.0 A dm^{-2}) examined using cyclic voltammetry and chronopotentiometry techniques revealed that, coating developed at 4.0 A dm^{-2} as the most HER active under test conditions. Further, to enhance the HER efficiency of Ni-P alloy coating, TiO_2/Ni-P composite electrode was developed, at an applied c.d. of 4.0 A dm^{-2}, which was found to be the best for HER among the tested Ni-P coatings, and was taken for the electrocatalytic study in 1.0 M KOH.

3.3 Mechanism of Composite Electrodeposition

Composite electrodeposition is an attractive method for the development of wide variety of composite coatings with enhanced physical and electrochemical properties as compared with the metal or alloy coatings (Gomes et al. 2011). The codeposition of nanoparticles along with the metal ion reduction from the plating bath, during electrodeposition, follows several steps to obtain the nanoparticle incorporated metal/alloy matrix. The convective (in the bulk electrolyte) and diffusive (within the diffusion layer) mass transfer of the suspended nanoparticles along with the metal ions after attaining surface charge through aqueous ionic cloud is schematically represented in Fig. 2. The codeposition of the suspended nanoparticles can takes place simultaneously with the growing metal/alloy layer by tailoring the plating conditions using suitable complexing agents or surfactants (Gomes et al. 2011; Low et al. 2006).

Akin the development of TiO_2/Ni-P composite coating was achieved from the same optimal bath loaded with TiO_2, and the obtained composite coating at an applied c.d. of 4.0 A dm^{-2} was characterized using FESEM and EDS analyses as shown in Fig. 3b. TiO_2 particles was found to be incorporated to the alloy matrix and also some

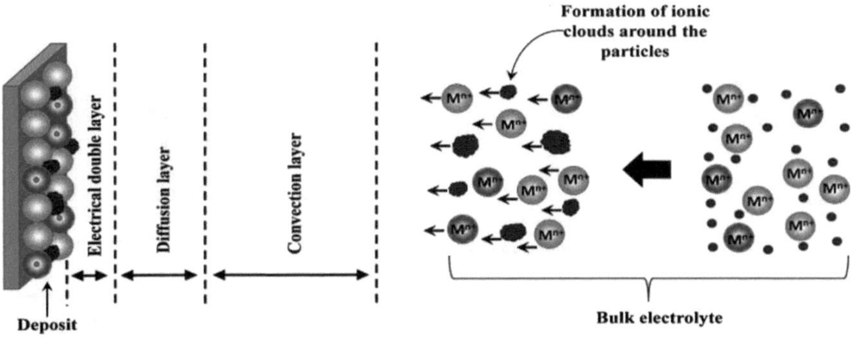

Fig. 2 Schematic representation of composite electrodeposition

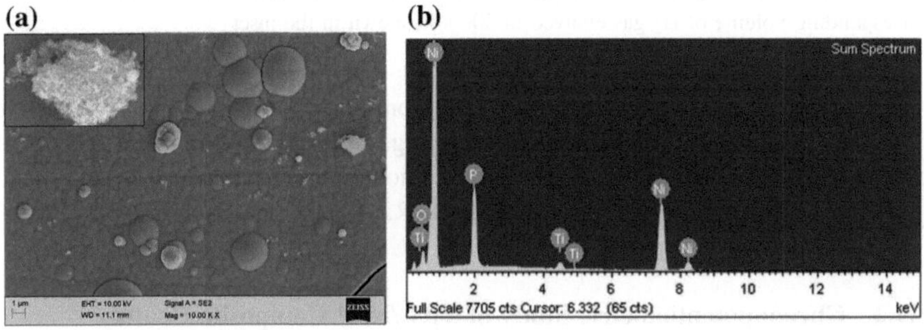

Fig. 3 **a** Surface morphology of TiO_2 incorporated Ni-P alloy coating developed at 4.0 A dm^{-2}. **b** EDX spectra of TiO_2/Ni-P composite coating

amount of agglomerated TiO_2 nanoparticles were found to be on the surface of the coatings developed, as shown in Fig. 3a.

3.4 Electrocatalytic Study of Composite Coatings

The composite coating (developed at 4.0 A dm^{-2}) was tested for its electrocatalytic activity towards HER in 1.0 M KOH using cyclic voltammetry and chronopotentiometry techniques as before.

3.4.1 Cyclic Voltammetric Study of TiO_2/Ni-P Composite Coating

Cyclic voltammetric study of the TiO_2/Ni-P composite test electrode (developed at 4.0 A dm^{-2}) was carried out in a potential window of 0 to −1.6 V for 50 cycles, at a scan rate of 50 mV s^{-1}. The obtained results as compared with the activity of

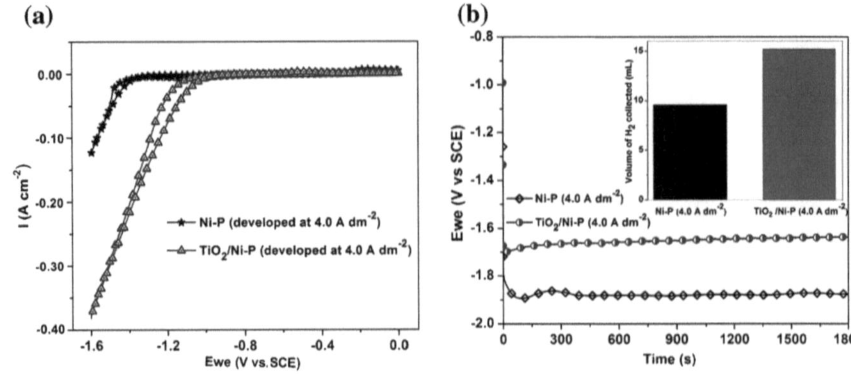

Fig. 4 Electrocatalytic activity study results of Ni-P and TiO_2/Ni-P coatings: **a** Comparison of cyclic voltammetric curves, **b** Comparison of chronopotentiometry curves along with the corresponding volume of H_2 gas evolved in 300 s is shown in the inset

normal Ni-P alloy coatings shows that the composite electrodes are more active towards HER. The cyclic votammograms in Fig. 4a shows the improved activity of composite electrodes with an increased cathodic peak c.d. (0.38 A cm^{-2}) and lowest onset potential (−1.19 V) towards HER.

3.4.2 Chronopotentiometric Study of TiO_2/Ni-P Composite Coating

The electrocatalytic behaviour of the composite coating (developed at 4.0 A dm^{-2}) for HER was evaluated by quantifying the amount of H_2 gas liberated in first 300 s during chronopotentiometric analysis. The nature of chronopotentiograms for the TiO_2 incorporated composite coating along with that of conventional Ni-P alloy coating are shown in Fig. 4b, and the corresponding volume of hydrogen liberated are also shown in the inset. A marked difference in the potential and the amount of evolved H_2 can be evidenced from the Fig. 4b. This improvement in the catalytic activity of the composite coating for HER is attributed to the increased surface area and the increase in the number of active sites for hydrogen adsorption, through incorporation of TiO_2 nanoparticles.

4 Conclusions

From the overall study and after analyzing experimental results on the catalytic efficiency of the Ni-P alloy electrodes and TiO_2/Ni-P composite electrode, prepared by electrodeposition technique, the following conclusions are established:

1. Ni-P alloy coatings electrodeposited from optimal bath were characterized, and the deposit at 4.0 A dm^{-2} was found to be the best towards HER.
2. TiO_2/Ni-P composite electrodes were synthesized successfully by composite electrodeposition technique to increase the electrocatalytic efficiency of Ni-P alloy coatings towards HER.
3. The HER efficiency of Ni-P alloy electrode was found to be improved remarkably through the development of TiO_2/Ni-P composite electrode.
4. Drastic increase in the catalytic activity of the composite coating is attributed to the increased active sites, due to increase in surface area through incorporation of TiO_2 nanoparticles.

References

Aal, A.A., Hassan, H.B.: Electrodeposited nanocomposite coatings for fuel cell application. J. Alloy. Compd. **477**(1), 652–656 (2009)

Elias, L., Scott, K., Hegde, A.C.: Electrolytic synthesis and characterization of electrocatalytic Ni-W alloy. J. Mater. Eng. Perform. **24**(11), 4182–4191 (2015)

Elias, L., Hegde, A.C.: Electrodeposition and electrocatalytic study of Ni-W alloy coating. In: Materials Science Forum, pp. 651–654. Trans. Tech. Publ. (2015)

Gomes, A., Fernández, B., Pereira, I., Pereiro, R.: Electrodeposition of Metal Matrix Nanocomposites: Improvement of the Chemical Characterization Techniques, pp. 24–126. INTECH Open Access Publisher (2011)

Hou, D., Zhou, W., Liu, X., Zhou, K., Xie, J., Li, G., Chen, S.: Pt nanoparticles/MoS_2 nanosheets/carbon fibers as efficient catalyst for the hydrogen evolution reaction. Electrochim. Acta **166**, 26–31 (2015)

Ju, H., Li, Z., Xu, Y.: Electro-catalytic activity of Ni–Co-based catalysts for oxygen evolution reaction. Mater. Res. Bull. **64**, 171–174 (2015)

Low, C.T.J., Wills, R.G.A., Walsh, F.C.: Electrodeposition of composite coatings containing nanoparticles in a metal deposit. Surf. Coat. Technol. **201**(1), 371–383 (2006)

Pillai, A.M., Rajendra, A., Sharma, A.K.: Electrodeposited nickel–phosphorous (Ni–P) alloy coating: an in-depth study of its preparation, properties, and structural transitions. J. Coat. Technol. Res. **9**(6), 785–797 (2012)

Wei, Z.D., Yan, A.Z., Feng, Y.C., Li, L., Sun, C.X., Shao, Z.G., Shen, P.K.: Study of hydrogen evolution reaction on Ni–P amorphous alloy in the light of experimental and quantum chemistry. Electrochem. Commun. **9**(11), 2709–2715 (2007)

Photocatalytic Degradation of Congo Red Dye Using Silver Doped TiO_2 Nanosheets

Abhinav K. Nair, B. Vinay Kumar and P.E. Jagadeeshbabu

1 Introduction

Titania nanomaterials have been the most researched photocatalysts for the last few decades (Lakshmi et al. 1995). Enormous amount of research work has been done on degradation of organic pollutants in water using TiO_2 (Akpan and Hameed 2009). Abundance and non toxicity of TiO_2 were the most favoring aspects for water treatment. Although nanomaterials have various morphologies the particulate form of TiO_2 is the widely used due to the ease of synthesis and size control (Kulkarni et al. 2015). One could find in literature various methods to enhance the photocatalytic properties of TiO_2 nanoparticles (TNP) using various dopents like nobel metals, transition metals, carbon nanotubes, graphene oxide, etc. (Fujishima et al. 2008). Although TiO_2 is generally considered non toxic, on a nanoscale things are different. Recent researches in nanotoxicology have revealed that TiO_2 anatase nanoparticles are cytotoxic and genotoxic (Shi et al. 2013). In such an alarming situation using TiO_2 nanoparticles for water treatment needs to be reconsidered. More than photocatalytic property enhancement, catalyst recovery needs to be emphasized up on. Catalytic efficiency of nanoparticles depends on their size, smaller the size better the specific surface area and better the catalytic efficiency. But the smaller the particle gets the tougher it becomes to recover. Centrifugation techniques that are used in laboratories can't be used for real life applications as the quantity of water to be processed is huge. Hence, nanoparticles although effective in photocatalysis have a serious problem with recovery.

TiO_2 nanosheets (TNS) are 2D nanostructures with thickness of a few nanometers and length and bread exceeding nanoscale. The thin sheet structure provides very good specific surface area and the larger size enables easy recovery

A.K. Nair · B. Vinay Kumar · P.E. Jagadeeshbabu (✉)
Department of Chemical Engineering, National Institute of Technology Karnataka, Surathkal, Mangalore, India
e-mail: dr.jagadeesh@yahoo.co.in

I. Regupathi et al. (eds.), *Recent Advances in Chemical Engineering*,
DOI 10.1007/978-981-10-1633-2_23

via filtration. These nanostructures are difficult to synthesize with effective size control. Solvothermal methods have recently shown to produce good nanosheets (Hu et al. 2014). Very few papers could be found on their applications in degradation of organic pollutants (Nakata and Fujishima 2012). Little work has been done on their doping to enhance catalytic activity. In the present work we report the synthesis of silver doped TNS (Ag-TNS). A comparative study of photocatalytic activity of TNS and Ag doped TNS with TNP is done. Congo red dye has been used as the model pollutant.

2 Experimental

2.1 *Materials*

Tetrabutyltitanate (TBT, 98 %), TiO_2 anatase nanoparticles (Dia < 21 nm) were purchased from Sigma-Aldrich Co, Bangalore, India. Hydrofluoric acid (HF, 40 %), congo red dye, silver nitrate ($AgNO_3$), ammonium hydroxide (NH_4OH) were purchased from Nice chemicals Pvt Ltd, Cochin, India.

2.2 *Synthesis of TiO_2 Nanosheets*

Solvothermal method using HF was done as reported in literature (Chen et al. 2014). 3 mL of HF was taken in a Teflon-lined stainless steel autoclave. 10 mL of TBT was added drop wise under constant stirring. The autoclave was subject to heat treatment at 200 °C for 24 h. The residue obtained was washed thrice with distilled water and then with ethanol and dried in a hot air oven at 60 °C.

2.3 *Synthesis of Ag-TiO_2 Nanosheets*

Silver doping of TNS was done using procedure reported in literature for doping TNPs (Sowmya and Meenakshi 2014). 320 mg of $AgNO_3$ was dissolved in 100 mL of distilled water. 750 mg of TNS was added to the $AgNO_3$ solution and dispersed via a magnetic stirrer. Few drops of ammonia solution were added to adjust pH in the range of 10–11. The mixture was kept stirring and irradiated with UV light for 4 h. The dark purple colored product obtained was washed thrice with distilled water and then with ethanol and dried in a hot air oven at 60 °C.

2.4 *Photocatalytic Studies*

The photocatalytic activity was assessed via congo red dye degradation under optimum conditions as reported by Muhammad (2015). The dye concentration for study was 15 ppm, catalyst loading of 25 mg and neutral pH. To 100 mL dye solution 25 mg of catalyst was added and dispersed via stirring. The solution was continuously aerated via air spurge. The solution was irradiated by two UV Lamp (18 W). 2 mL samples were collected at 30 min interval and centrifuged to separate the powder catalyst. The samples were then subjected to UV-Visible spectroscopy (Hitachi, U-2900) for estimating dye concentration. Dye adsorption studies were also carried out for a period of 2 h, dye concentration was 15 ppm and catalyst loading 25 mg.

3 Results and Discussion

3.1 *Characterization*

Transmission electron microscope (JEOL, JEM-2100) image of TNS is shown in Fig. 1. Ultra-thin nanosheets with length and breadth around 0.1–0.2 μm can be seen. Agglomeration was observed since sonication is not preferable during sample preparation of fragile nanostructures.

X-ray diffraction (XRD) patterns of TNS (Fig. 2) was obtained using X-ray Diffractometer (JEOL, Dx-GE-2P). Major XRD peaks were observed at 25.8°, 37.6°, 48.6° and 55.4° which correspond to the anatase phase of titania (Gorska

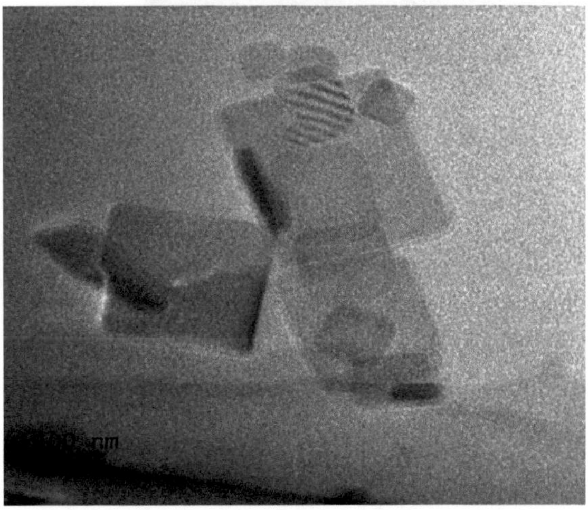

Fig. 1 Transmission electron microscope image of TNS

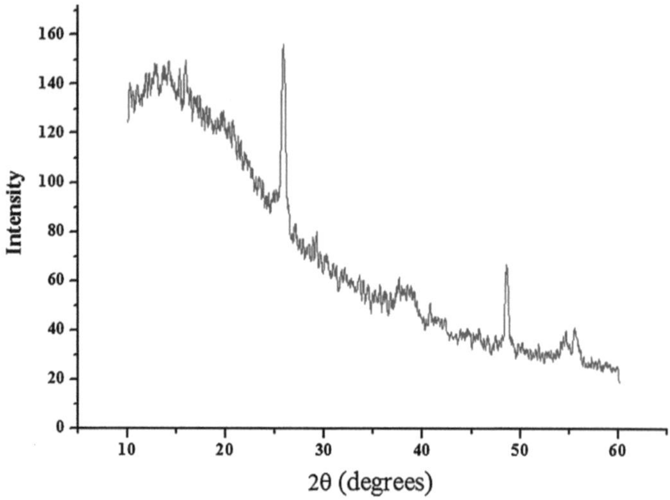

Fig. 2 XRD Pattern of TNS

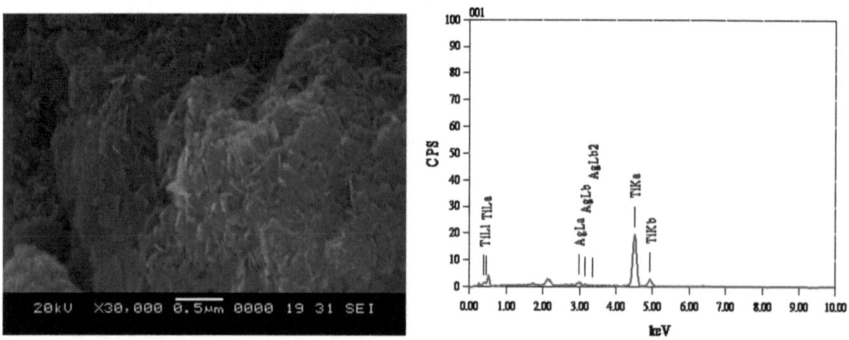

Fig. 3 Scanning electron microscope image and EDX of TNS

et al. 2008). The peaks revealed superior crystallinity and phase purity of the synthesized TNS.

Scanning electron microscope (Jeol JSM-6380LA) was used for the energy-dispersive X-ray spectroscopy (EDX) analysis of Ag-TNS. EDX analysis showed the presence of 3.44 % (atomic percentage) silver deposition (Fig. 3). Scanning electron microscope images also revealed the flake like structure of the nanosheets.

3.2 Photocatalytic Activity

Results obtained after dye degradation studies are shown in Fig. 4. It was observed that TNS and TNP showed comparable photocatalytic activity, although TNS showed a greater rate of dye degradation initially. Ag-TNS showed enhanced photocatalytic activity, 83 % of the dye got degraded in the initial 30 min itself. Complete degradation using Ag-TNS was achieved in 90 min whereas TNP and TNS took 120 min. This improved photocatalytic activity of Ag-TNS is due to the enhanced charge separation efficiency achieved via doping of silver on titania. Silver serves as electron trap and delays electron hole recombination (Liu et al. 2015).

Dye adsorption studies indicated that nanosheets have superior dye adsorption capabilities than nanoparticles (Fig. 5). Ag-TNS showed a remarkable 43 % absorption of dye in 2 h. Nanosheets have greater specific surface area owing to their ultrathin structure with a thickness of few nanometers when compared to the spherical nanoparticles. The greater specific surface area along with the greater surface energy of the defect less sheet structure results in a better tendency to adsorb dye. This greater adsorption capacity facilitates quicker adsorption of reactants on to the catalyst surface thereby improving catalytic activity.

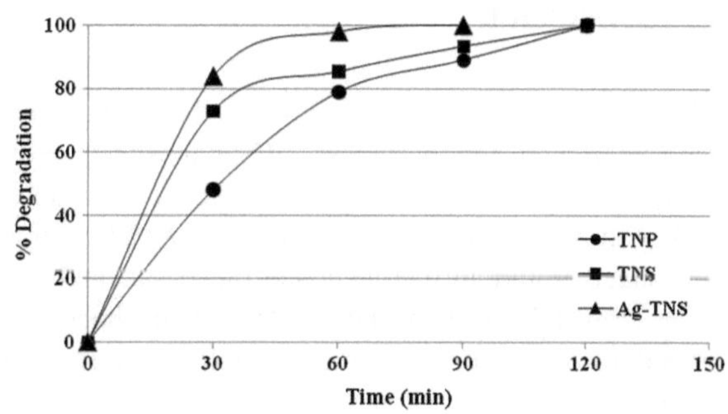

Fig. 4 Dye degradation studies of various photocatalysts

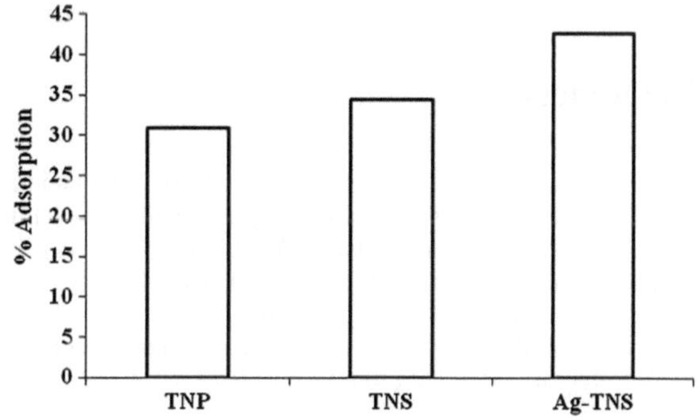

Fig. 5 Dye adsorption by various catalysts

Fig. 6 TNS recovered and clear permeate obtained after filtering the reaction mixture

Nanosheets have nanoscale thickness, but the other two dimensions are much larger in size when compared to nanoparticles and hence nanosheets can be filtered out easily. The reaction mixture containing the catalyst was poured into a filtration assembly with a 0.2 μm cellulose acetate membrane. TNS catalyst was easily filtered out from the reaction mixture leaving behind a clear permeate (Fig. 6).

4 Conclusion

Silver doped TiO_2 nanosheets showed superior photocatlytic activity than TiO_2 nanosheets and nanoparticles. TiO_2 nanosheets proved to be an efficient photocatalyst, with degradation performance comparable with anatase nanoparticles. TiO_2 nanosheets also exhibited better dye adsorption properties. Doping of TiO_2 nanosheets could further enhance the photocatalytic property as seen in case of silver doped TiO_2 nanosheets. TiO_2 nanosheets can be doped using the same procedures reported for nanoparticles. Unlike nanoparticles the larger size nanosheets could be filtered out from the reaction mixture, enabling easy catalyst recovery. It is evident from the studies on silver doped TiO_2 nanosheets that TiO_2 nanosheets are a potential alternate for nanoparticles.

References

Akpan, U.G., Hameed, B.H.: Parameters affecting the photocatalytic degradation of dyes using TiO_2-based photocatalysts: a review. J. Hazard. Mater. **170**, 520–529 (2009)

Chen, K., Jian, Z., Qin, J., Jiang, Y., Li, R., Tang, H., Yang, X.: Synthesis and improved photocatalytic activity of ultrathin TiO_2 nanosheets with nearly 100 % exposed (001) facets. Ceram. Int. **40**, 16817–16823 (2014)

Fujishima, A., Zhang, X., Tryk, D.A.: TiO_2 photocatalysis and related surface phenomena. Surf. Sci. Rep. **63**, 515–582 (2008)

Gorska, P., Zaleska, A., Kowalska, E., Klimczuk, T., Sobczak, J.W., Skwarek, E., Janusz, W., Hupka, J.: TiO_2 photoactivity in vis and UV light: the influence of calcination temperature and surface properties. Appl. Catal. B: Environ. **84**, 440–447 (2008)

Hu, C., Zhang, X., Li, W., Yan, Y., Xi, G., Yang, H., Li, J., Bai, H.: Large-scale, ultrathin and (001) facet exposed TiO_2 nanosheet superstructures and their applications in photocatalysis. J. Mater. Chem. A **2**, 2040–2043 (2014)

Kulkarni, R.M., Malladi, R.S., Hanagadakar, M.S., Doddamani, M.R., Bhat, U.K.: Ag-TiO_2 nanoparticles for photocatalytic degradation of lomefloxacin. Desalin. Water Treat. (2015). doi:10.1080/19443994.2015.1076352

Lakshmi, S., Renganathan, R., Fujita, S.: Study on TiO_2-mediated photocatalytic degradation of methylene blue. J. Photochem. Photobiol. A Chem. **88**, 163–167 (1995)

Liu, S., Wang, N., Zhang, Y., Li, Y., Han, Z., Na, P.: Efficient removal of radioactive iodide ions from water by three-dimensional Ag_2O-Ag/TiO_2 composites under visible light irradiation. J. Hazard. Mater. **284**, 171–181 (2015)

Muhammad, A.S.K.: Photocatalytic degradation of congo red dye using TiO_2, PANI and PANI/ TiO_2 nanoparticles, M.Tech. Thesis, NITK, Surathkal, Karnataka, India (2015)

Nakata, K., Fujishima, A.: TiO_2 photocatalysis: design and applications. J. Photochem. Photobiol. C: Photochem. Rev. **13**, 169–189 (2012)

Shi, H., Magaye, R., Castranova, V., Zhao, J.: Titanium dioxide nanoparticles: a review of current toxicological data. Part. Fibre Toxicol. (2013). doi:10.1186/1743-8977-10-15

Sowmya, A., Meenakshi, S.: Photocatalytic reduction of nitrate over Ag-TiO_2 in the presence of oxalic acid. J. Water Process Eng. (2014). doi:10.1016/j.jwpe.2014.11.004

Effect of TiO_2/MgO Nanocomposites on the Photocatalytic Activity Towards Removal of Organic Dyes

A. Dhanya and K. Aparna

1 Introduction

Water is one of the essential enablers of life on earth. Today water scarcity and water pollution are the two important issues being faced by the world. The industries that use large amounts of water for the processing techniques will pollute the environment in a large stretch. Textile industries are one of the most polluting industries, as the consumption of water for various processes such as sizing, scouring, bleaching, dyeing, printing and other finishing processes is about 3.2 % of the total consumption of water in the industry, and the dyes lost with waste water during synthesis and processing comes to be around 10–15 % of the dyes used in the industry.

The dye, which makes its way to the environment by the disposal of the effluents is highly toxic and is a great threat to the environment. The dyes have complex chemical structures which makes them stable and difficult to degrade (Farzana and Meenakshi 2014).

In the recent years, heterogeneous photocatalysis has been a great concern because, they not only convert photon energy into chemical energy but also has a great potential for purifying air and water (Dhanavel et al. 2014).

Photocatalytic degradation process has a great potential in controlling and degrading the contaminants in the pollutant dye (Hauas et al. 2001). Photocatalysis using semiconductor is a method widely used for environmental protection (Lezner et al. 2012). One of the most commonly used photocatalysts is titanium dioxide. TiO_2 is a semiconductor with wide bandgap (3.2 eV), high mechanical strength, refractive index and chemical stability (Chen and Caruso 2013). The main difficulty while using TiO_2 nanoparticles for photocatalytic activity is that, the photo excited electron hole pair recombination rate is high in the irradiated particles

A. Dhanya (✉) · K. Aparna
National Institute of Technology Calicut, Calicut, India
e-mail: dhanyaarikal@yahoo.com

I. Regupathi et al. (eds.), *Recent Advances in Chemical Engineering*,
DOI 10.1007/978-981-10-1633-2_24

(Ambrus et al. 2007). MgO nanoparticles are of high interest due to its low heat capacity, chemical inertness, optical transparency and high thermal stability (Gao et al. 2012). The photocatalytic performance of TiO_2 can be enhanced by combining an insulator (MgO or Al_2O_3) with TiO_2 (Bandara et al. 2006; Fujishima et al. 2011). By using MgO or SiO_2 with TiO_2 shows good degradation activity for cationic surfactants (Fujishima et al. 2011). Compared to TiO_2 nanoparticles, TiO_2/MgO nanocomposites show enhanced photocatalytic activity as MgO plays the role of a barrier for charge recombination in the TiO_2 nanoparticles (Bayal and Jeevanandam 2014).

Methyl Orange is a basic azo dye having azo group (–N = N–) linked to methine or aromatic sp^2 hybridized C atoms (Zainal et al. 2009). In the present work, Methyl Orange is the model dye compound used for determining the photocatalytic degradation of TiO_2/MgO nanocomposite.

2 Materials and Methods

2.1 *Synthesis of TiO_2/MgO Nanocomposite*

TiO_2/MgO nanocomposite was synthesised by the following method. TiO_2 was prepared from $TiCl_3$. The $TiCl_3$ solution was mixed with deionized water and the pH was maintained between 4.5 and 6.5 using ammonium hydroxide solution. The blue violet coloured solution thus obtained is kept in an oven for 24 h, resulting in a white coloured suspension. The white coloured suspension is washed with distilled water in order to remove the chloride ions and the white coloured precipitate is filtered out. The white coloured precipitate which is the TiO_2 nanoparticle, is coated with MgO by dissolving TiO_2 in distilled water and 0.1 ml HNO_3 and about 3 % (w/w) of MgO nanoparticle (Rizwan Wahab et al. 2007). The solution was diluted with distilled water and stirred for half an hour, which was then evaporated to dryness. This dry powder was used for the photocatalytic treatment of the organic azo dye in a photocatalytic reactor. Characterisation of TiO_2/MgO nanocomposite was done by TEM.

2.2 *Photocatalytic Study*

The photocatalytic study of TiO_2/MgO nanocomposite was done in a photocatalytic reactor set up, having a 500 ml capacity reactor flask, 125 W UV lamp, 0–800 rpm magnetic stirrer, an outer jacket for water circulation which cools the UV lamp and an exhaust fan. The TiO_2/MgO nanocomposite was dispersed in the organic dye solution inside the reactor flask within the photocatalytic reactor. Methyl Orange was taken as the model organic dye compound and the stock solution of 5 ppm of the dye being made by dissolving requisite amount of the dye in distilled water. After certain intervals of time, treated solution was taken out for analysis.

By measuring the following characteristics of the initial dye solution and the final dye solution, the photocatalytic activity of the nanocomposite can be determined

(1) Absorbance (Concentration can be determined)
(2) FTIR (Fourier Transform Infrared Spectroscopy) analysis

2.2.1 Absorbance

The absorbance measurement is done by using Evolution 201 UV Visible Spectrophotometer. Decolourisation efficiency is measured from an equation relating the absorbance values before and after treatment in the photoreactor, where the absorbance values can be known from the UV Spectra at λ_{max} (maximum wavelength) of the respective dye. The expression is as in Eq. (1)

$$\text{Decolourisation}\,(\%) = \frac{A_0 - A_t}{A_0} \times 100 \quad (1)$$

where, A_0 is the initial value of absorbance of the dye solution and A_t is the value of absorbance after treatment with the photocatalyst at time t.

2.2.2 FTIR Analysis

FTIR analysis is done to confirm the degradation of methyl orange dye. The FTIR spectrum is a graph showing absorbance or transmittance versus frequency or wavelength. Bruker Alpha Spectrophotometer were used for the measurements.

3 Results and Discussion

3.1 Characterisation of the Photocatalyst TiO_2/MgO Nanocomposite

Figure 1 shows that the particle size of TiO_2/MgO synthesised is approximately 10–20 nm. The figure shows large, irregular and highly agglomerated particles.

Photocatalytic Study

The rate of photocatalytic activity was calculated from the absorbance values and IR spectrum of the initial and final methyl orange dye solution.

Fig. 1 TEM image of TiO_2/MgO nanocomposite

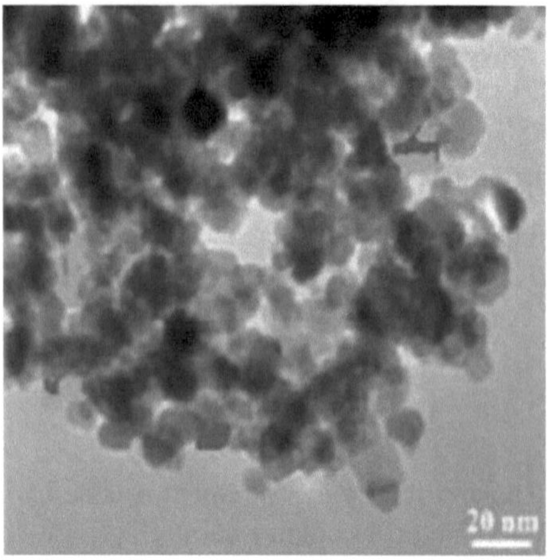

3.2 Absorbance

The absorbance was measured using UV spectrophotometer. Absorbance was found to decrease with time and the absorbance peak disappeared at 90 min, which indicates that 100 % decolourisation of the dye solution occured after 1½ h. Figure 2 is the calibration curve of concentration versus absorbance of methyl orange. From this calibration curve, the unknown concentrations are calculated from the absorbance values.

Figure 3 shows the increase in degradation efficiency with time of methyl orange dye solution.

3.3 FTIR Analysis

The FTIR spectra of Methyl orange solution (Fig. 4), displays peak at 3688 cm^{-1} indicates the presence of free NH group, 3211 cm^{-1} represents the O–H vibration,

Fig. 2 Calibration curve of methyl orange

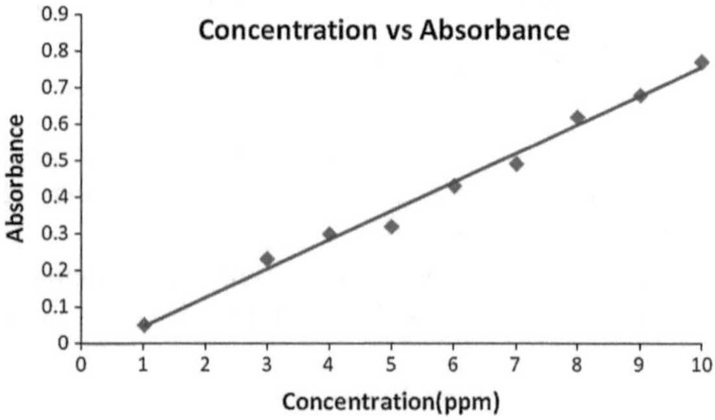

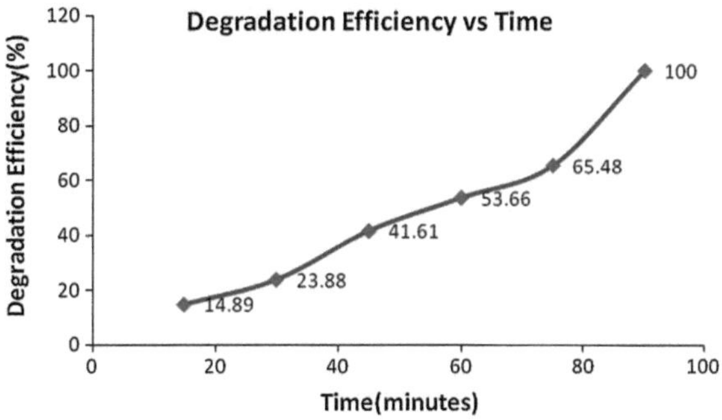

Fig. 3 Degradation efficiency of methyl orange

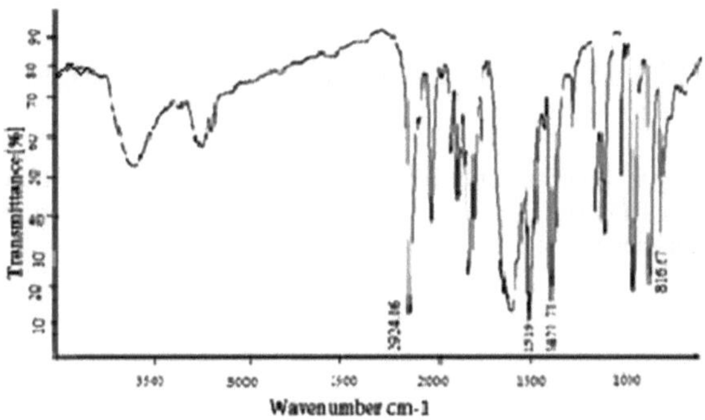

Fig. 4 FTIR spectra of methyl orange

2924.06 cm^{-1} representing the alkyl C–H stretch, 1421.71 cm^{-1} represents the stretching vibrations of S = O, 1040 cm^{-1} indicates the stretching vibration for C–N, 1007 cm^{-1} represents the C–O stretch, 816 cm^{-1} for benzene disubstituted compounds 1519.78 cm^{-1} represents –N = N– stretching of azo compounds. In the FTIR spectrum of the treated solution of methyl orange (Fig. 5) transmittance peaks at 790–840 cm^{-1} is absent, which indicated para disubstituted aromatic compound and the peaks which indicates the presence of –N = N– and –S = O– compounds are also not present, which indicates that methyl orange have been degraded.

3.4 *Comparison of TiO_2 and TiO_2/MgO Nanocomposite*

Figure 6 shows the photocatalytic degradation efficiency of TiO_2 and TiO_2/MgO nanocomposite. TiO_2 alone took 150 min for complete degradation whereas TiO_2/MgO nanocomposite took only 90 min for the complete degradation of methyl orange dye solution.

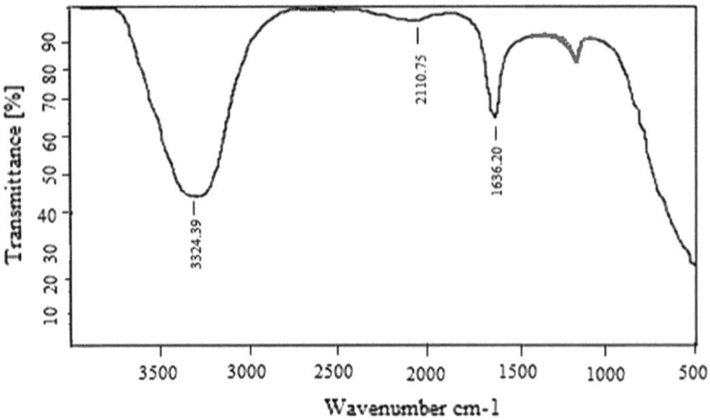

Fig. 5 FTIR Spectra of treated solution of methyl orange

Fig. 6 Comparison of TiO_2 and TiO_2/MgO nanocomposite

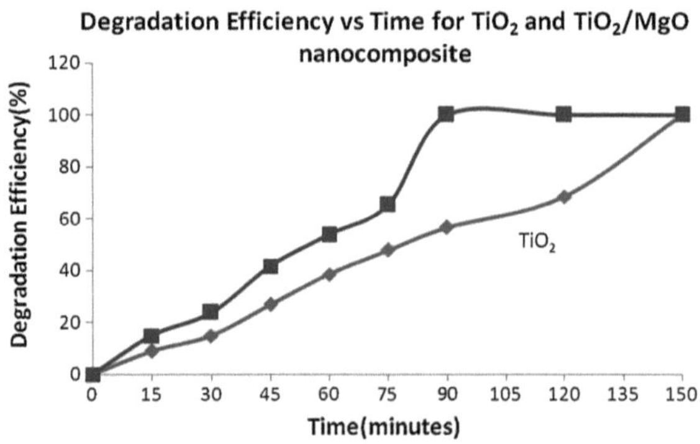

4 Conclusion

The TiO_2/MgO nanocomposite enhances the photocatalytic activity for the degradation of the dye solution. Up to 100 % decolourisation is obtained within 1½ h by TiO_2/MgO nanocomposite and this can be successfully used for breaking down many constituents of the dye solution.

References

Ambrus, Z., Balazs, N., Alapi, T., Wittmann, G., Sipos, P., Dombi, A., Mogyorosi, K.: Synthesis, structure and photocatalytic properties of Fe(III)—doped TiO_2 prepared from $TiCl_3$. Appl. Catal. B **81**, 27–373 (2007)

Bayal, N., Jeevanandam, P.: Synthesis of TiO_2/MgO mixed metal oxide nanoparticles via a sol gel method and studies on their optical properties. Ceram. Int. **40**, 15463–15477 (2014)

Bandara, J., Kuruppu, S.S., Pradeep, U.W.: The promoting effect of MgO layer in sensitized photodegradation of colorants on TiO_2/MgO composite oxide. Colloids Surf. A: Physicochem. Eng. Asp. **276**, 197–202 (2006)

Chen, D., Caruso, R.A.: Recent progress in the synthesis of spherical titania nano structures and their applications. Adv. Func. Mater. **23**, 1356–1374 (2013)

Dhanavel, S., Nivethaa, E.A.K., Narayanan, V., Stephen, A.: Photocatalytic activity of Chitosan/ZnO nanocomposites for degrading methylene blue. Int. J. ChemTech Res. **6**, 1880–1882 (2014)

Farzana, M.H., Meenakshi, S.: Synergistic effect of chitosan and titanium dioxide on the removal of toxic dyes by the photodegradation technique. Ind. Eng. Chem. Res. **53**, 55–63 (2014)

Fujishima, M., Takatori, H., Tada, H.: Interfacial chemical bonding effect on the photocatalytic activity of TiO_2/SiO_2 nanocoupling systems. J. Colloids Interface Sci. **361**, 28–631 (2011)

Gao, D.Z., Watkins, M.B., Shluger, A.L.: Transient mobility mechanism of deposited metal atoms on insulating surfaces: Pd on MgO. J. Phys. Chem. **116**, 14471–14479 (2012)

Hauas, A.A., Lachheb, H.A., Mohamed, K.A., Elaloui, E.A., Guillard, C., Hermann, J.M.: Photocatalytic degradation pathway of Methylene blue in water. Appl. Catal. B **31**, 145–157 (2001)

Lezner, M., Grabowska, E., Zaleska, A.: Preparation and photocatalytic activity of iron modified titanium dioxide photocatalyst. Physicochemical Probl. Miner. Process. **48**, 193–200 (2012)

Wahab, R., Ansari, S.G., Dar, M.A., Kim, Y.S., Shin, H.S.: Synthesis of magnesium oxide nanoparticles by sol-gel process. Mater. Sci. Forum **558–559**, 983–986 (2007)

Zainal, Z., Hui, L.K., Hussein, M.Z., Abdullah, A.H., Hamadneh, I.R.: Characterisation of TiO_2-Chitosan/Glass photocatalyst for the removal of a monoazo dye via photodegradation-adsorption process. J. Hazard. Mater. **164**, 138–145 (2009)

Visible Light Induced $ZnTiO_3$ Photocatalyst Synthesized by Co-Precipitation Process

P. Sirajudheen, K.B. Sanoop and Muhammed Rashid

1 Introduction

Degradation of environmental pollutants is one of the many plausible advantages of photocatalytic reactions. The surface of natural-occurring ceramics or inorganic particles is considered interface where photocatalytic reactions occur, emitting electrons upon ultraviolet irradiation to accelerate degradation of pollutants. The readily changed atoms from the ground state to the excited state upon ultraviolet irradiation are the most distinguished feature of photocayalytic inorganic materials. Excited electrons and electron holes generated during excitation are involved in chemical reactions to increase reaction rates (Ho et al. 2009).

Zinc oxide and titanium oxide are examples of inorganic materials having photocatalytic activity. Of them, zinc oxide has been widely used because it is extensively accessible and one of the inevitable nutrients for human. Using titanium dioxide (TiO_2) as a photocatalyst, the elimination of volatile organic compounds (VOC) through photocatalysis from water and environment is lately established, but it needs UV light to produce the hydroxyl radicals which then consequently mineralizes VOC. The UV induced TiO_2 can be transferred to the range of visible light by doping it with C, N and S particles, which facilitate the red shift incorporation rim (Web-1 2014, Xin et al. 2008, Khan et al. 2002, Yamashita et al. 2002 and Zhou et al. 2001). Another process to stimulate the TiO_2 in the range of visible light is made by incorporating of transition metal ions into the TiO_2 particles (Web-2 2013, Chang and Shen 2006, Wu and Chen 2004 and Web-3 2011). Photocatalytic activity of nickel titanate (Beenakumari 2013), bismuth titanate and strontium titanate Photocatalysts co-doped with antimony and chromium (Chang and Shen 2006 and Kat and Kundo 2002) were already reported in the literature.

P. Sirajudheen (✉) · K.B. Sanoop · M. Rashid
Department of Chemistry, WMO Imam Gazzali Arts and Science College, Koolivayal 670721, Kerala, India
e-mail: sirajpalliyalil@gmail.com

I. Regupathi et al. (eds.), *Recent Advances in Chemical Engineering*,
DOI 10.1007/978-981-10-1633-2_25

Large $ZnTiO_3$ particles with too little compositional homogeneity and unrestrained structural morphologies can be produced through the general solid-state process for the synthesis of $ZnTiO_3$. However, owing to its stoichiometric control, homogeneity in the phase of reactant, high purity and unfussiness of processing, the co-precipitation peroxide method was considered a potential way to prepare powder for the ferroelectric materials. In this work, $ZnTiO_3$ powder was prepared by the co-precipitation method. Through the X-ray diffraction (XRD), the phase transformation of the synthesized titanate was studied. By using TGA, the Fourier Transform Infrared Spectra (FTIR) was taken for confirming the bonding characteristics of SnO and $TiO_{2,}$ and Decomposition temperature was analyzed. The Scherer equation was used to measure the particle size of the zinc titanate. The photocatalytic efficiency was measured on the basis of the deprivation of methyl orange in aqueous solution.

2 Materials and Methods

2.1 Materials

Zinc chloride and titanium tetrachloride are used as preliminary materials. Oxalic acid, hydrogen peroxide, ammonia solutions and urea from Merck are also used without further purification. X-ray diffraction was conducted on $ZnTiO_3$ powders by means of CuKα radiation (Phillips Analytical) and the particle size was evaluated by using the Scherer equation. Bruker FTIR apparatus was used for the IR spectral analysis of $ZnTiO_3$. Using TGA Versa thermo gravimetric analyzer, Thermal studies were conducted. The photo-bleaching of methyl orange was analyzed by means of UV-Visible Spectrophotometer of Shimadzu at 465 nm by incorporating the dye solution with zinc titanate, exposed to daylight for dissimilar irradiation time, using UV-Visible Spectrophotometer. The initial concentration of dye was 10.0 ppm. For the period of the irradiation, the solution was gurgled with air.

2.2 Preparation of the Sample

Titanium tetrachloride and zinc chloride are employed as the preliminary materials for the preparation of $ZnTiO_3$. $TiCl_{4,}$ 0.372 g in 100 ml (0.1 mol) and $ZCl_{2,}$ 0.273 g in 100 ml (0.1 mol) are dissolved in distilled water respectively. An aqueous solution containing ions of zinc and titanium in an equi-molar ratio are prepared by mixing 100 ml of $ZnCl_2$ solution and 100 ml $TiCl_4$ solution. 0.1 mol of oxalic acid solution (50 ml) was added to this mixed solution of titanium tetrachloride and zinc chloride solution. A fresh 200 ml blend solution was prepared by mixing 30 ml of 30 % H_2O_2, 15 ml of 20 % aqueous ammonia, 20 ml of 10 % urea solutions respectively and rest of distilled water by stirring to form a precipitate of complex peroxide (Beenakumari 2013) corresponding to zinc titanate. The subsequent mix up solution had being adding together in drops to the aqueous solution having ions of zinc and titanium in an equimolar ratio. The urea and oxalic acid which are present in the second solution will act as nuclei for the precipitation accomplishment, which is in rule of the particle size. The precipitate was wash down, desiccated and calcined essentially at 800 °C to get nano crystals of $ZnTiO_3$.

2.3 *Photocatalytic Activity*

The photocatalytic efficiency of the prepared powders was evaluated by measuring the absorbance of light at 465 nm by using UV-Visible Spectrophotometer of Shimadzu, The degradation of a model aqueous solution of methyl orange was measured. The primary concentration of methyl orange aqueous solution was 10 ppm exposed under visble light irradiation. During analysis varying amount of zinc titanate sample was added to methyl orange solution and it is exposed under visible light, sun light irradiation. The solution was bubbled with air during the irradiation. The photo-degradation of the dye, methyl orange was sturdily influenced by the existence of zinc titanate present in the system. The degradation of dye increases with increase in exposure time.

3 Result and Discussion

3.1 *Structural and Thermal Analyses*

The X-ray diffraction analysis of the prepared zinc titanate ($ZnTiO_3$) calcinated by different temperatures are shown in Fig. 1.

Fig. 1 XRD of $ZnTiO_3$

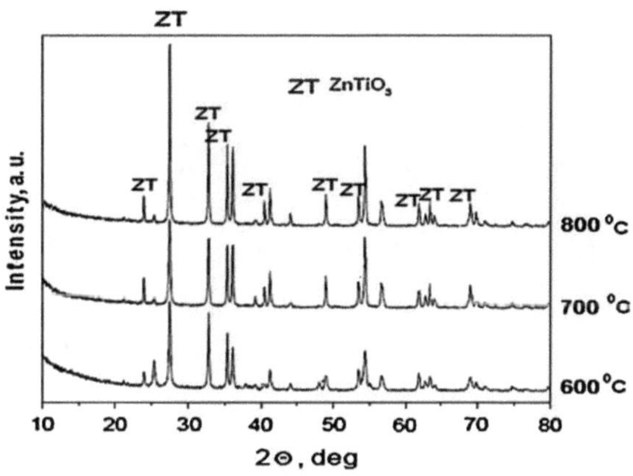

The sample was amorphous at 600 °C and the strongest peaks of cubic $ZnTiO_3$ were observed at 700 °C. According to the XRD analysis (Hosono et al. 2004) occurred the same phase at below 600 °C using titanium tetraisopropoxide and zinc acetylacetonate as main precursors by sol-gel method was similar. Even though, the cubic crystals are still dominant when prepared the cubic $ZnTiO_3$ at 500 °C and their results are revealed by (Andres-Verges and Martinez-Gailego 1992). At 800 °C, number of peaks are associated with the hexagonal $ZnTiO_3$ are appeared in which the cubic peaks were dominant, when using the identical precursors (Wang et al. 2003) the effect was the same. As per (Golovchansky et al. 1998), the hexagonal $ZnTiO_3$ prepared by zinc nitrate hexahydrate and titanium tetraisopropoxide, in which single-phase at 500 °C succeeded by cubic phase and rutile has shown above 600 °C which are closer to those acquired by means of (Aubert et al. 2010). In Sherrer's equation, the average crystalline size as seemed at 800 °C $ZnTiO_3$ particles calculated from the expansion of the diffraction lines. The sample was exemplified by a strong affinity to agglomerate with the average particle size of the total about 1 μm, according to SEM interpretation (Fig. 2a). The microprobe investigation confirmed the presence of $ZnTiO_3$ only (Fig. 2b) with molar ratio of Zn and Ti was found to be 1:1, which is perfectly matched with the XRD results.

The Differential thermal analysis curve of the formerly heat treated at 400 °C sample (Fig. 3) explicates two endothermic peaks at 120 and 380 °C and two exothermic peaks at 450 and 540 °C.

The exothermic peaks may well be related to the flaming of the organic compounds and crystallization of the amorphous compound with formation of $ZnTiO_3$ phase (540 °C) and the endothermic peaks belonging to the dehydratation of the sample. Both exothermic peaks (at 450 and 540 °C) have not been observed in the DTA graph of the formerly heated sample at 500 °C. This reveals that the last exothermic outcome is connected to the crystallization course, which is in good unity with the X-ray diffraction results (Fig. 1). The results are compatible to those given by Wang et al. 2003).

In Fig. 4, the Fourier Transform Infrared Spectra (FTIR) of the prepared zinc titanate ($ZnTiO_3$) samples is shown.

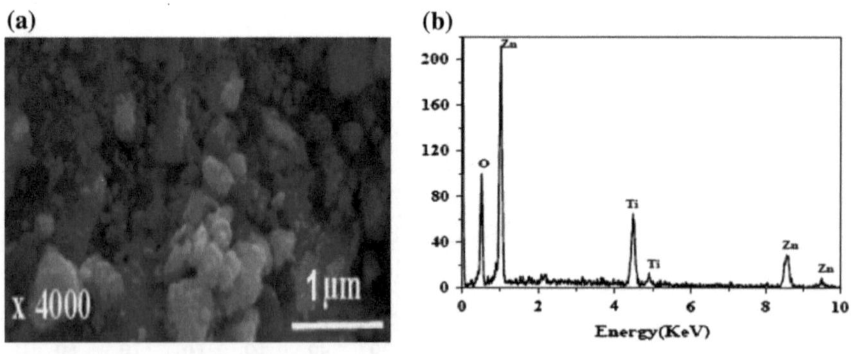

Fig. 2 **a** SEM micrographs of $ZnTiO_3$ **b** EDS spectrum of $ZnTiO_3$

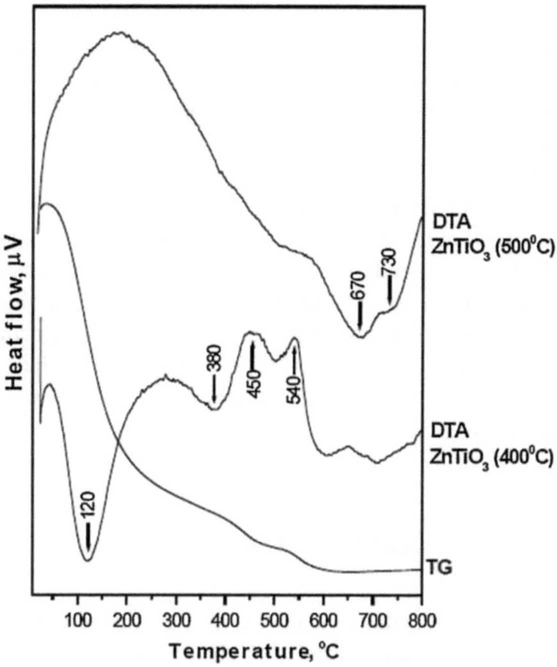

Fig. 3 DTA curve of $ZnTiO_3$

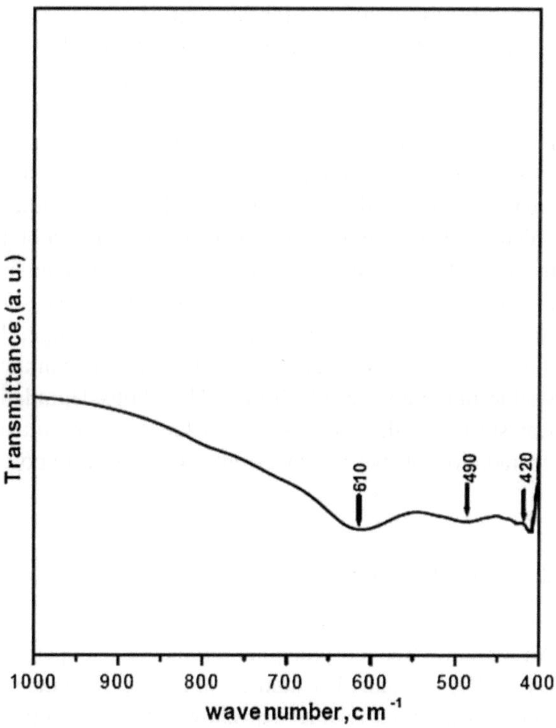

Fig. 4 IR spectra of $ZnTiO_3$

There are some Strong bands at 610, 490 cm^{-1} and a weak band at 420 cm^{-1} are observed in the spectra of $ZnTiO_3$ and the weak band was related to the vibrations of TiO_6 units in $ZnTiO_3$ (Yamaguchi et al. 1987; Shabalin 1982) and could be related to that bands in the incorporation range 700–400 $cm^{-1.}$ According to the X-ray diffraction analysis for this sample, merely one type of ($ZnTiO_3$) was observed. It is also well-known that the peaks analogous to ZnO_n polyhedra are in the similar absorption range (Mancheva et al. 2011; Verges and Gailego 1992). The disparity of the IR spectrum of zinc titanate is associated to the increased strength of the band centered on 420 cm^{-1}. At the same time, as it is well known, this band was characteristic for the vibrations of TiO_6 units building up the rutile adaptation and its existence is in concurrence with the X-ray diffraction outcome (Yurchenko 1981; Murashkevich et al. 2008). The weak band at the vicinity of 450 cm^{-1} might be associated with the stretching vibrations of Ti-O in $ZnTiO_3$ (Yamaguchi et al. 1987; Shabalin 1982).

3.2 *Photocatalytic Activity*

The photo-degradation of the dye, methyl orange was done by adding different amount of nano particle of zinc titanate in 10.0 ppm methyl orange solution exposed in the visible range, sun light irradiation,. The decolorization of methyl orange by means of varying amount of zinc titanate exposed to visible light irradiation for different duration could be found in Fig. 3. The photo-degradation of the dye was sturdily having an effect on the subsistence of zinc titanate in the system. The degradation of the dye increases with increase in exposure time owing to the increased photon assimilation by zinc titanate directs to further transport of photoelectrons and photo holes between valance band and conduction band and thereby creating super radical oxygen and hydroxyl radical in the photo catalytic cell (Web-5 2013). The radical ions generated are accountable for the decolorization of the dye. When the total amount of zinc titanate in the methyl orange solution increased, the rate of degradation of the dye was also increased due to the more absorption of photon from the visible light range. The concentration of dye was diminished towards 2.0 ppm when solution containing 0.2 % of $ZnTiO_3$ in the visible range contact of 90 min. The photo-degradation of the dye, methyl orange was sturdily influenced by the existence of zinc titanate present in the system. The degradation of dye increases with increase in exposure time.

4 Conclusions

An organic free co-precipitation method prepares Zinc titanate nano powders. By XRD, IR and TG studies, the zinc titanate synthesized was characterized. Thermo-gravimetric analysis shows that crystalline configuration of $ZnTiO_3$ is

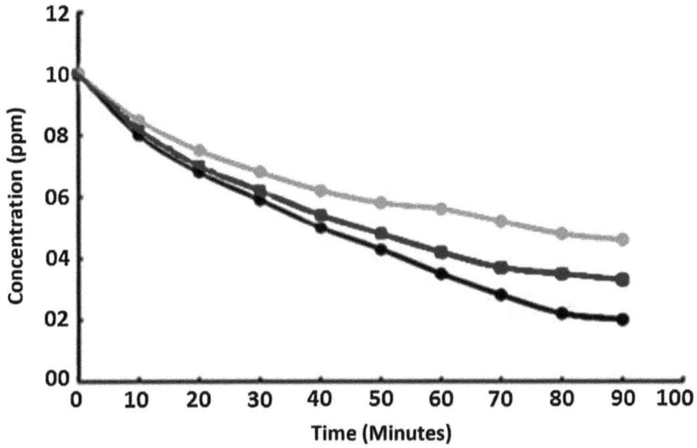

Fig. 5 The Photo-bleaching of Methyl orange solution with $ZnTiO_3$ at varying concentration (*red circle* 0.05, *blue circle* 0.1, *black circle* 0.2 in %)

accomplished over 500 °C. The size of the prepared zinc titanate particles was found to be an average of 100 nm. The methyl orange concentration was reduced from 10.0 to 0.2 ppm by applying 0.2 % zinc titanate exposes to sunlight for 90 min time. The decolorization of the dye in the presence of visible light by $ZnTiO_3$ confirms the photocatalytic efficiency of the prepared compound, zinc titanate in the range of visible light irradiation. The photo-degradation of the dye, methyl orange was sturdily influenced by the existence of zinc titanate present in the system. The degradation of dye increases with increase in exposure time (Fig. 5).

References

Andres-Verges, M., Martinez-Gailego, M.: Spherical and rod-like zinc oxide microcrystals: morphological characterization and microstructural evolution with temperature, Mater. Sci. J., 3756–3762, (1992)

Aubert, T., Grasset, F., Potel, M., Nazabal, V., Cardinal, T., St Pechev, Saito, N.: Synthesis and characterization of Eu3+,Ti4+@ZnO organosols and nanocrystalline c-$ZnTiO_3$ thin films aiming at high transparency and luminescence, Sci. Technol. Adv. Mater., 044401–044409, UK (2010)

Beenakumari, K.S.: Visible Light driven Tin Titanate Photo-Catalyst Prepared by Co-Precipitation Method. Int. J. Mater. Sci. Innov, 174–181, Malasiya (2013)

Chang, C.H., Shen, Y.H.: Synthesis and characterization of chromium doped $SrTiO_3$, Matr. Lett, Elsevier (2006)

Golovchansky, A., Kim, H., Kim, Y.: Zinc Titanates Dielectric Ceramics Prepared by Sol Gel Process, pp. S1167-S1169, J. KoreanPhys. Soc., Korea (1998)

Ho, J., Choy, T.Un., Eue, P.: Zinc oxide having enhanced photocatalytic Activity. IPC8 Class: AC01G902FI, USPC Class: 428402, Washington, US (2009)

Hosono, E., Fujihara, S., Onuki, M., Kimura, T.: Low-temperature synthesis of nanocrystalline zinc titanate materials with high specific surface area, J. Amer. Cer. Soc, 1785–1788, US (2004)

Khan, S.U., Al-Shahry, M., Ingler, W.B.: Efficient photochemical water splitting by a chemically modified Sn-TiO_2, pp. 2243–2245, J. Mater. Sci. USA (2002)

Kato, H., Kundo, A.: Visible light response and photocatalytic activities of TiO_2 and $SrTiO_3$ Photocatalysts Co doped with Antimony and Chromium **106**, J. Phys. Chem. B, Oxford 5029–5034(2002)

Mancheva, M., Iordanova, R., Dimitriev, Y.: Mechanochemical synthesis of nanocrystalline $ZnWO_4$ at room temperature, J. Alloys Compounds, 15–20, US (2011)

Murashkevich, A., Lavitkaya, A., Barannikova, T.: Infrared absorption spectra and of TiO_2-SiO_2 composite, J. Appl. Spectrosc., 730–734, Australia (2008)

Shabalin, B.G.: Synthesis and IR Spectra of Some Rare and New Titanium and Niobium, Mineral. Zh., 54–61 (1982)

Wu, J., Chen, C.H. (2004): Photo-catalytic reduction of carbon dioxide to methane using TiO_2, J. Photochem. Photobiol. A: Chem. Elsevier, 503–508, USA (2004)

Wang, S.F., Gu, F., Lu, M.K.: Preparation and characterization of sol–gel derived $ZnTiO_3$ nanocrystals, Mater. Res. Bullt., 1283–1288, Elsevier (2003)

Wang, S.F., Gu, F., Lu, M.K.: dielectric properties of zinc titanate ceramics, Ceramics, Mater. Res. 1283–1288, (2003)

Xin, S., Jing, H., Dong, C.: Visible-light-induced photocatalyst based on nickel titanate nanoparticles, Ind. Eng. Chem. Res, 4750–4753, Washington (2008)

Yurchenko, E., Kustovar, G., Bacanov, S.: Vibratioanl spectroscopy of inorganic compounds. Nauka, Russian (1981)

Yamashita, H., Harada, H., Misaka, J.: Degradation of propanol diluted in water under visible light irradiation using metal ion-implanted titanium dioxide photocatalyst. J. Photochem. Photobiol. A: Chem. 257–261. Elsevier (2002)

Yamaguchi, O., Morimi, M., Kawabata, H.: Formation and transformation of ZnTiO3, J Am. Ceram. Soc., C-97-C98, US (1987)

Zhou, Z., Ye, Y., Sayama, K., Arakawa, H.: Direct splitting of water under visible light irradiation with an oxide semiconductor photocatalyst, pp. 625–627, Nature, London (2001)

Web sites

Web-1: Takashi Kamegawa' Yasushi Ishiguro' Ryota Kido' Hiromi Yamashita .: Design of Composite Photocatalyst of TiO_2 and Y-Zeolite for Degradation of 2-Propanol in the Gas Phase under UV and Visible Light Irradiation, pp. 16477–16488, doi:10.3390/molecules191016477. www.mdpi.com/1420-3049/19/10/16477/pdf. (2014)

Web-2: Steven L. suibe.: New and Future Developments in Catalysis: Solar Photocatalysis, pp. 116–118, Elsivier, Washington (2013) ttps://books.google.com.pk/books?isbn=0444538739, (2013)

Web-3: Mingfei Shao, Jingbin Han, Min Wei, David G. Evans, Xue Duan.: The synthesis of hierarchical Zn–Ti layered double hydroxide for efficient, pp. 519– 524, Chemical Engineering Journal, Elsevier, Washington (2011) www.ldh-buct.com/wp-content/uploads/2015/03/2011-23.pdf, (2011)

Web-5: K S Beenakumari.: Visible Light driven Tin Titanate Photo-Catalyst Prepared by Co-Precipitation Method, Acadamic Research Online Publishers Sdn. Bhd., Malasiya (2013) www.aropub.org/…/html-visible-light-driven-tin-titanate-photo-catalyst-, (2013)

Part V
Modelling and Simulation of Chemical Processes

Mathematical Modelling on Particulate Removal in Multistage Dual-flow Sieve Plate Column Wet Scrubber

Swamy Kurella, Pavan Kishan Bhukya and B.C. Meikap

1 Introduction

The increase in population and industrial development demand more energy and coal is the only readily available chief fossil fuel which can produce energy to meet world requirements. Coal gasification and coal burning processes release different air pollutants including fly ash and many acid gases into the environment. The fly ash particles from the products of these processes should be removed before they undergo further processes for the energy production to save the equipment and to control emissions of particulate. Particulate matter can be defined as the particles which are less than 100 μm and the particles less than 10 μm are very dangerous due to their high impact on the health of the human beings and other living creatures (Raj et al. 2008). From the effluent gas streams, the large particles can be effectively removed by cyclones. However, the decrease in particle size decreases the particulate collection efficiency of the cyclone (Yoshida et al. 2005; Wark et al. 1998). To overcome this problem wet scrubbing techniques are important by which

S. Kurella (✉) · P.K. Bhukya · B.C. Meikap
Department of Chemical Engineering, Indian Institute of Technology Kharagpur, Kharagpur, India
e-mail: kurellaswamy@che.iitkgp.ernet.in

P.K. Bhukya
e-mail: bpavankishan1@gmail.com

B.C. Meikap
e-mail: bcmeikap@che.iitkgp.ernet.in

S. Kurella · P.K. Bhukya · B.C. Meikap
Department of Chemical Engineering, School of Chemical Engineering, Howard College Campus, University of Kwazulu-Natal (UKZN), Durban 4041, South Africa

I. Regupathi et al. (eds.), *Recent Advances in Chemical Engineering*,
DOI 10.1007/978-981-10-1633-2_26

very fine particulates can be removed effectively. Normally, conventional wet scrubbing systems encounter problems of fouling and clogging. To avoid these situations a three stage dual-flow sieve plate column wet scrubber has been used for the removal of particulate matter. Plate scrubbers can effectively remove the acid gases and dust particles and they are being used in many chemical process industries to control the emissions of air pollutants. Plate without down comer is called as dual-flow plate and dual-flow sieve plate columns have plates with perforations or punched holes and allow the gas and liquid through these holes alternatively in counter current manner by which the system is self cleaned. As there is no down comer for the liquid flow, bubbling area is high and increases the tray capacity by 20 % than normal sieve tray with down comer (Flavio et al. 2014). In dual-flow systems, the liquid forms a layer on the plates and gas passes through this layer and forms small bubbles, because of this advantage of the dual-flow plate, more interfacial area is generated between gas and liquid phases and the availability of particle capturing surface area increases. Thus, the distance travelled by the particle is shortened before it is captured by the liquid phase. This high interfacial area and low-pressure drop make it more economical within the limited operating conditions of dual-flow plate column and its bubbling zone behaves almost as a bubble column. The bubbles capture the particles when they broke and the particles are concentrated into a small volume (Meikap and Biswas 2004; Wang et al. 2012). The particle collection mechanisms by small bubbles involved in wet scrubbers are majorly diffusion, impaction and interception. An attempt was made by Fuchs (1964), who explained the experimental observations of particle pool scrubbing theoretically for the first time and developed the models for the efficiencies for particle collection mechanisms with the assumption of homogeneous distribution of particles in bubble. Kaldor and Phillips (1976) reported that the diffusion is the key mechanism to collect the particles of submicron-size and Lee and Gieseke (1979) have used Kuwabara's cell model flow field to obtain particle removal efficiencies by interception and Brownian diffusion. Kuwabara's extended flow field of cell model is used by Jung and Lee (1998) to deal with the multiple bubbles or droplets flow field by considering the internal circulation effect within the bubbles or droplets on the particle collection efficiency and outside flow field. Recently, Raj et al. (2008) have reported particle removal characteristics in spray column using Kim's theory and developed correlation for particulate removal efficiency in spray column from hot flue gases. Wang et al. (2012) reported the experimental studies on particle removal in fixed valve column and modelled a correlation to predict the particle collection efficiency. However, there are no theoretical models for particulate removal that are especially applicable to the dual-flow sieve tray columns. Therefore, the aim of the study is analysis of hydrodynamics of bubble, system property and operating variables effect on particle collection. This model would be limited to counter flow dual-flow sieve trays. This work tries to verify the fine particle removal characteristics of a dual-flow sieve plate column.

2 Theoretical Modeling

In plate scrubbers, the gas with particulates passes through a liquid (water) film on the plate by forming more air bubbles of small size and the particles are collected on surface of bubble by different interception, inertial impaction and Brownian diffusion. The major assumptions that are made for the development of model to predict the particle collection of dual-flow sieve plate column scrubber are no loading on bubbles, air obeys ideal gas law, particles and bubbles are spherical in shape, uniform distribution bubble size and there is no significant evaporation and condensation. For a fixed valve column, Wang et al. (2012) have developed a model with mathematical approach using mass balance equation. The mass balance for particle collection in a control volume of dual-flow sieve plate column can be given in Eq. (1).

$$\begin{aligned}&(\text{Particle flux to liquid from gas})\\&\quad=(\text{volume swept by moving bubbles})\times(\text{concentration of particle in bubble})\\&\qquad\times\;(\text{particlesfraction collected in swept bubble volume of moving bubbles})\\&\quad=(\text{swept bubbles number }\times\text{ bubble residence time})\times(\text{volumetric flux})\end{aligned} \tag{1}$$

After substituting the terms with their definitions, simplified model is integrated over froth height (H_F) and the resulted model is given in Eq. (2).

$$\eta_{total} = 1 - \exp\left(-\frac{3}{2}\cdot\frac{U}{v_b}\cdot\frac{H_F}{d_b}\eta_{sb}\right) \tag{2}$$

where η_{total} and η_{sb} are overall efficiency and particles fraction collected in single bubble respectively, c the concentration of particle in a bubble (g m^{-3}), U is liquid and bubbles relative velocity (m s^{-1}), v_b the bubble rise velocity (m s^{-1}), d_b bubble diameter (m) and H_F is forth height(m). The Eq. (2) is further simplified using the condition that there is no cross flow of liquid (i.e., relative velocity is equal to bubble velocity) as there is no down comer in this system and it is given in Eq. (3).

$$\eta_{total} = 1 - \exp\left(\frac{3H_F}{2d_b}\eta_{sb}\right) \tag{3}$$

In the present work, liquid hold up, h_L(m), was measured experimentally and the froth height, H_F was calculated using correlation given in Eq. (4) (Flavio et al. 2014).

$$H_F = \frac{h_L}{(1-(1-\epsilon))} \tag{4}$$

where ϵ is porosity of the gas liquid layer on the tray and given in Eq. (5).

$$\epsilon = 1 - 0.0946\left[\frac{\rho_G U_G^2}{\rho_L g h_L}\right]^{-0.2} \tag{5}$$

where ρ_G density of gas (kg m^{-3}), U_G superficial velocity of gas (m s^{-1}), g is acceleration due to gravity (m s^{-2}) and ρ_L density of liquid (kg m^{-3}). Bubbles were observed with naked eye and the size and the rise velocity of bubble on sieve tray were calculated with the correlations given in Eqs. (6) and (7) respectively (Hofhuis 1980).

$$d_b = 0.615\, g^{-0.2}\, P^{0.8}\left[U_G \frac{A_c}{A_h}\right]^{0.4} \tag{6}$$

$$V_b = 0.71(gd_b)^{0.5} + U_G \tag{7}$$

where P is the pitch (m), A_c is column cross sectional area and A_h tray hole area.

The existing modelled equations from the literature were used to predict the particles fraction collected in a bubble by taking into the consideration of the possible mechanisms as discussed below.

2.1 *Diffusion*

Coefficient of diffusion is inversely proportional to particle size, so for small particles it is very high, due to this property submicron particles are collected by the diffusion mechanism. Fuchs (1964) has developed a model for particles fraction collected due to diffusion in a bubble and it is given in Eq. (8) (Wang et al. 2013).

$$\eta_{diff} = 2.94\sqrt{\frac{kTC}{\Pi \mu\, d_p v_b d_b^3}} \tag{8}$$

where T is the temperature (K), k is the Boltzmann constant, d_p is diameter of particle, μ is air viscosity and C is the Cunningham slip correction factor which is taken as unity from the literature.

2.2 *Interception*

The particles which flow in a streamline will be collected when the particle center travels within the radius of a particle from surface of bubble. A model for the particle fraction collected by interception mechanism in a bubble is derived by Jung and Lee (1998) and given in Eq. (9).

$$\eta_{int} = \left(\frac{1-\varepsilon_g}{J} \cdot \frac{1}{d_b}\right) d_p + \left(\frac{1-\varepsilon_g}{J} \cdot \frac{2}{d_b^2}\right) d_p^2 \tag{9}$$

where ε_g is gas hold up and $J = 1 - \frac{6}{5}\varepsilon_g^{1/3} + \frac{1}{5}\varepsilon_g^2$.

2.3 *Impaction*

Impaction plays key role in particulate collecting situations of scrubbers where the gas velocities are more than 0.3 m s^{-1} or more than 5.0 μm particle diameter (Cheng 1973; Kim et al. 2001). For the estimation of collected particle fraction by impaction mechanism Lee and Gieseke (1979) developed a model which is given in Eq. (10)

$$\eta_{imp} = \left(\frac{Stk}{Stk + 0.35}\right)^2 \tag{10}$$

where Stk is stoke number and calculated using Eq. (11).

$$Stk = \frac{\rho_P d_p^2 v_h}{18 \mu d_b} \tag{11}$$

where ρ_P is particle density (kg m^{-3}) and v_h is hole velocity (m s^{-1}).

So, the single bubble particle collection efficiency (η_{sb}) is sum of the collections of above three mechanisms (Sarkar et al. 2007; Kim et al. 2001). Therefore,

$$\eta_{sb} = \eta_{diff} + \eta_{imp} + \eta_{int} \tag{12}$$

3 Results and Discussion

The experiments of fly-ash scrubbing were conducted in three stage dual-flow sieve plate column—2.6 m height, 0.1524 m diameter, 0.6096 m spacing between plates and 396 holes of 3 mm diameter on each plate and gas and liquid distributor plates —with water as scrubbing liquid by varying the liquid, gas and fly ash loading rates. The experimental set up is shown Fig. 1. The variable values are used for the theoretical estimation of particle fraction collected in a single bubble in overall column efficiency prediction. The model equations were developed mathematically and to calculate the theoretical overall efficiency the simulations were carried out

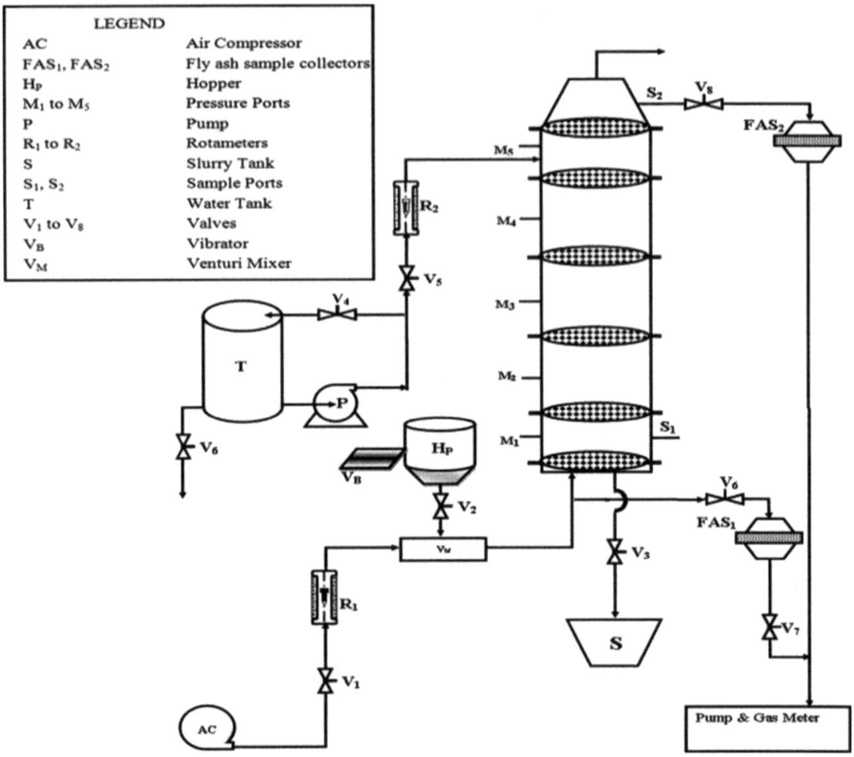

Fig. 1 Experiemntal set-up of dual-flow sieve plate column scrubber for particulate removal

using MATLAB software (Sarkar et al. 2007; Raj et al. 2008). The program was run to observe the particle size, liquid and gas flow rate effect on the particle removal efficiency.

From Fig. 2, it can be seen that the effect of particle size of 1–5 μm and gas flow rate on column overall fly ash collection efficiency at constant 48.183×10^{-6} m^3/s flow rate of liquid and. The efficiency increases steadily with the increase in the particle size and reaches unity (i.e., maximum) for 3.5 μm particle at 19.36×10^{-4} m^3/s gas flow rate and 48.183×10^{-6} m^3/s liquid flow rate. The increase in particle size from 1 to 5 μm increased the overall efficiency where impaction and interception mechanisms play a major role and diffusion shows a minor effect. It can also be observed from the figure that the overall efficiency does not get affected by gas flow rate.

Particle size effect and liquid rate effect on particle removal efficiency at 19.36×10^{-4} m^3/s constant gas flow rate can be seen in Fig. 3. At constant gas flow rate, the theoretical overall efficiency increased with increase in particle size from 1 μm to 5 μm and with flow rate of liquid. As the liquid flow rate increases the forth height on the plate increases which increase the bubble residence time. The liquid flow rate shows significant effect on efficiency when compared to gas flow

Fig. 2 Effect of particle size and flow rate of gas on overall theoretical efficiency at 48.183×10^{-6} m^3/s constant liquid flow rate

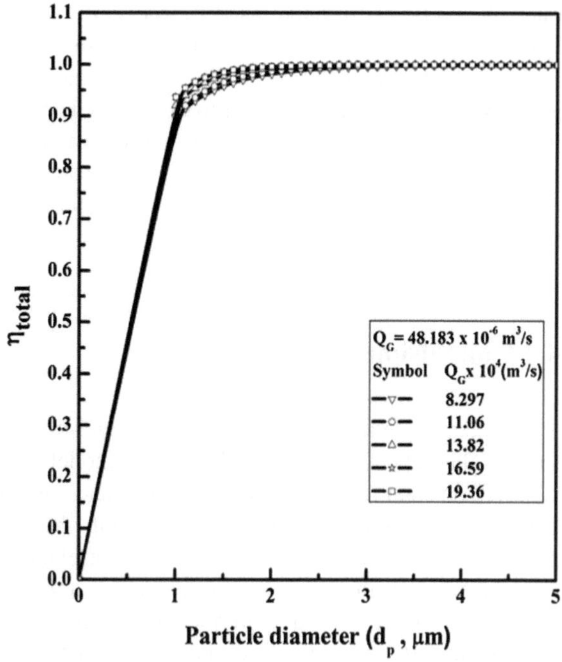

Fig. 3 Effect of particle size and flow rate of liquid on overall theoretical efficiency at 19.36×10^{-4} m^3/s constant gas flow rate

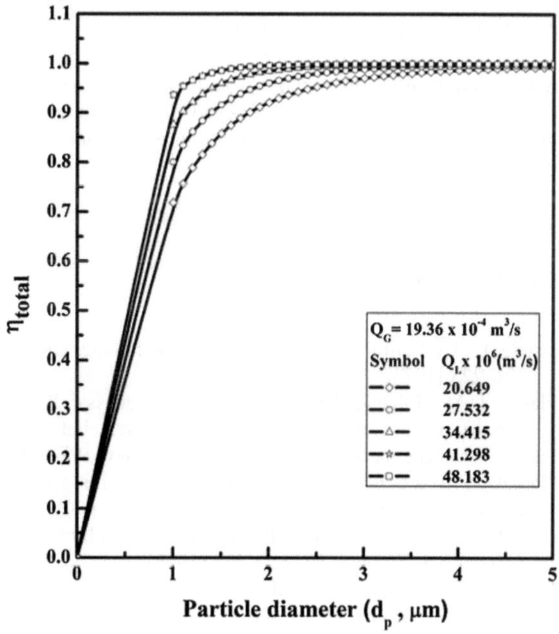

rate. The particle size range used for the study is in the range of 1 to 300 μm and with 10.33 μm surface mean diameter, where the maximum collection efficiency achieved 98.55 % at 48.183 × 10^{-6} m^3/s liquid flow rate and 19.36 × 10^{-4} m^3/s gas flow rate which is supposed to be 100 % based on theoretical prediction from the developed model. This may be due to particles of wide size range of and the some experimental errors or the unknown exact mechanisms involved in the process. Therefore, the idealization of experimental results with predicted results from the developed theoretical model is difficult.

4 Conclusions

The collection efficiency of particle in dual-flow sieve plate column has been studied theoretically and a mathematical model is developed to predict the efficiency. The equations of momentum, mass and continuity are applied over a control volume of a cylindrical element to develop the model to predict single bubble particle collection efficiency. The suitable correlations for different flow parameters were taken from the literature in developing the model. The predicted collection efficiencies are increased with increase in particle size from 1 to 10 μm. The increase in gas flow rate show did not affect the theoretical overall column efficiency where as the flow rate of liquid affects the efficiency notably. The developed model gives the 100 % efficiencies for more than 3.5 μm size particles within the experimental operating conditions.

Acknowledgments The authors would like to acknowledge the Department of Science and Technology (DST), Government of India(GOI) and Indian Institute of Technology (IIT) Kharagpur, India for their financial and technical support (sanction No: DST/INT/SA/P-14/2011) of this research.

References

Cheng, L.: Collection of airborne dust by water sprays. Ind. Eng. Chem. Process Des. Devel. **12**, 221–225 (1973)

Flavio, D.M., Liliana, A.F., Nilson, R.M., Vanessa, B., Ronaldo, H.: Review of hydraulics correlations for sieve trays without downcomers. Ind. Eng. Chem. Res. **53**, 8323–8331 (2014)

Fuchs, N.A.: The mechanics of aerosols. Pergamon Press, New York (1964)

Hofhuis, P.A.M.:. Flow regimes on sieve trays for gas/liquid contacting, Ph.D. thesis, Delft University of Technology, Delft, Netherlands (1980)

Jung, C., Lee, K.: Filtration of fine particles by multiple liquid droplet and gas bubble systems. Aerosol Sci. Technol. **29**, 389–401 (1998)

Kaldor, T.G., Phillips, C.R.: Aerosol scrubbing by foam. Ind. Eng. Chem. Process Design Devel. J. **15**, 199–206 (1976)

Kim, H., Jung, C., Oh, S., Lee, K.: Particle removal efficiency of gravitational wet scrubber considering diffusion, interception, and impaction. Environ. Eng. Sci. **18**, 125–136 (2001)

Lee, K.W., Gieseke, J.A.: Collection of aerosol particles by packed beds. Environ. Sci. Technol. **13**, 466–470 (1979)

Meikap, B., Biswas, M.: Fly-ash removal efficiency in a modified multi-stage bubble column scrubber. Sep. Purif. Technol. **36**, 177–190 (2004)

Raj, M.B., Jain, R.K., Meikap, B.C.: Comprehensive analysis for prediction of dust removalefficiency using twin-fluid atomization in a spray scrubber. Sep. Purif. Technol. **63**, 269–277 (2008)

Sarkar, S., Meikap, B., Chatterjee, S.G.: Modeling of removal of sulfur dioxide from flue gases in a horizontal concurrent gas–liquid scrubber. Chem. Eng. J. **131**, 263–271 (2007)

Wang, Q., Chen, X., Gong, X.: Theoretical and experimental investigation on the characteristics of fly-ash scrubbing in a fixed valve tray column. Am. Inst. Chem. Eng. J. **59**, 2168–2178 (2012)

Wang, Q., Chen, X., Gong, X.: The particle removing characteristics in a fixed valve tray column. Ind. Eng. Chem. Res. **52**, 3441–3452 (2013)

Wark, K., Warner, G.F., Davis, W.T.: Particulate control. In Air pollution: Its Origin and Control 3rd edn. Addison-Wesley

Yoshida, H., Ono, K., Fukui, K.: The effect of a new method of fluid flow control on submicron particle classification in gas-cyclones. Powder Technol. **149**, 139–147 (2005)

Development of a Computational Model of Venturi Scrubber to Simulate Its Performance for Filtered Vented Containment System

Manisha Bal and B.C. Meikap

1 Introduction

Filtered vented containment system is the most effective system to reduce the risk from severe accident occurs due to over pressurization in the Nuclear power plant. Many of the radioactive fission products emitted from the molten core into the containment at the time of an accident. Radioactive iodine-131 is a major constitute that has significant effects in the environment and human health. Iodine can enter the body by ingestion or inhalation. It dissolves in water so it can easily moves from atmospheric into humans and their living organisms. It concentrated in the thyroid gland of human body. Exposure to high levels of radioactive iodine is the chance to occur thyroid cancer in the later life. To avoid this consequences, FVCS is installed in Nuclear power plant. Venturi scrubber and metallic filter are the main parts of the FVCS (Schlueter et al. 1990). Venturi scrubber is most important device used to remove the micron sized dust particles. It is mostly used for its high collection efficiency and low maintenance cost (Lehner et al. 2007). A venturi scrubber has three segments: Convergent segment in which the dust particle is introduced with the air and maximizes the gas velocity at the throat section. Water is injected at the

M. Bal (✉) · B.C. Meikap
Department of Chemical Engineering, Indian Institute of Technology (IIT), Kharagpur, Kharagpur 721302, West Bengal, India
e-mail: chemmanisha1@gmail.com

B.C. Meikap
e-mail: bcmeikap@che.iitkgp.ernet.in

B.C. Meikap
Chemical Engineering, School of Engineering, Howard College, University of Kwazulu-Natal, Durban, South Africa

I. Regupathi et al. (eds.), *Recent Advances in Chemical Engineering*,
DOI 10.1007/978-981-10-1633-2_27

throat section where micron size particle collide with the water droplets mainly through inertial impaction. The third segment, diffuser is used for deceleration of gas to allow recovery of pressure (Guerra et al. 2012). Performance of venturi scrubber has been assessed by the pressure drop and collection efficiency. Computational fluid dynamics has become indispensable tool to develop the design and physical phenomenon in it (Majid et al. 2012).

In the present work, dispersed two phase flow is analysed by Eulerian-lagrangian approach. Gas Flow field inside venturi scrubber is obtained by solving Eulerian K-ε Turbulence model. Trajectory of particles is tracked by the Discrete particle model (DPM) and Influence of the throat gas velocity and liquid mass flow rate on overall collection efficiency of radioactive material are also investigated.

2 Computational Model

2.1 Eulerian—Lagrangian Approach

Gas containing micron sized particle enter at the top of the convergent section of venturi scrubber. Water is injected through the nozzle at the throat section and form droplets. The motion of particle and droplets is analysed by the lagrangian approach which track the each particle. Here gas field is considered as continuum phase and particle is as dispersed phase.

2.1.1 Eulerian Approach

Gas flow field is solved by the Eulerian approach using RNG k-ε turbulence model by considering standard wall function (Ananthanarayanan et al. 1999). Instantaneous Navier-Stokes equations derived the RNG k-ε turbulence model using "renormalization group" (RNG) methods.

Transport equation of k-ε turbulence model

$$\frac{\partial(\rho k)}{\partial t} + \frac{\partial(\rho k u_i)}{\partial x_i} = \frac{\partial}{\partial x_j}\left(\alpha_k \mu_{eff} \frac{\partial k}{\partial x_j}\right) + G_k + G_b - \rho\varepsilon - Y_m + S_k$$

and

$$\frac{\partial(\rho \varepsilon)}{\partial t} + \frac{\partial(\rho \varepsilon u_i)}{\partial x_i} = \frac{\partial}{\partial x_j}\left(\alpha_\varepsilon \mu_{eff} \frac{\partial \varepsilon}{\partial x_j}\right) + c_{1\varepsilon}\frac{\varepsilon}{k}(G_k + C_{3\varepsilon}G_b) - C_{2\varepsilon}\rho\frac{\varepsilon^2}{k} - R_\varepsilon + S_\varepsilon$$

In these equations, S_k and S_ε are source terms that defines the interaction with droplets.

2.1.2 Lagrangian Approach

The force balance of particle in Lagrangian reference frame defines the movement of the particles. The momentum equation for radial-direction (Pak et al. 2006), (LI Shan et al. 2008):

$$\frac{du_d}{dt} = \frac{3}{4}\frac{C_{Df}}{D_d}\frac{\rho}{\rho_d}(u - u_d)|\mathrm{V} - \mathrm{V_d}| + \mathrm{g}$$
$$\frac{du_d}{dt} = \frac{3}{4}\frac{C_{Df}}{D_d}\frac{\rho}{\rho_d}(u - u_d)|\mathrm{V} - \mathrm{V_d}|$$

Droplet trajectories can be evaluated by following equations

$$\frac{dz}{dt} = u_d$$
$$\frac{dr}{dt} = v_d$$

The value for C_{Df} is $C_{Df} = \frac{18.65}{Re_{\mathrm{d}}^{0.84}}$

Droplet Reynolds no $\mathrm{Re_d}$ is $\mathrm{Re_d} = \frac{\rho|\mathrm{V}-\mathrm{V_d}|D_d}{\mu}$

2.2 *Target Efficiency*

Inertial impaction is the main principle behind the collection of particles in a venturi scrubber. Particle captured by a single drop can be obtained as below (Mohebbi et al. 2003):

$$\eta_t = \left(\frac{\psi_c}{\psi_c + 0.7}\right)^2$$
$$\psi_c = \frac{\rho_\mathrm{p}|\mathrm{V} - \mathrm{V_d}|D_p^2}{9\mu\mathrm{D_d}}$$

ψ_c is the inertial impaction parameter.

2.3 Overall Collection Efficiency of Venturi Scrubber

Overall collection efficiency of scrubber is expressed as ratio of the amount of particulate matter collected to the total amount of particulate matter given at the inlet and it follows the below equation (Mohebbi et al. 2003):

$$\eta_{ov} = 1 - \frac{\int_0^R rV(r,z)C_p(r,z)dr}{\int_0^R rV(r,0)C_p(r,0)dr}$$

3 Numerical Solution

Geometry of venturi scrubber is shown in Fig. 1. Mesh of venturi scrubber was created in GAMBIT 2.4.6. 3D Venturi scrubber was discretized into 118,921 cells with faces 246,446 and 24,312 nodes.

CFD ANSYS FLUENT 15 is applied to the solution. This problem is solved by the three dimensional computational domain. Steady state condition is adopted to do the simulation. Velocity inlet is specified at the inlet of the air, as boundary condition and Mass flow rate is applied at water inlet. At the outlet of diffuser section, the pressure outlet boundary condition is applied. No-slip boundary condition is applied for both liquid and gases at all the walls (Ali et al. 2012). Motion of the droplet and dust particles are computed by the DPM model. In the dpm model, particles are injected from the gas inlet surface. Iodine particle with 1 μm is taken. Initial Size of the particle follows uniformly size distribution. Simple algorithm scheme solve the pressure velocity coupling. 2nd order Upwind scheme is selected for momentum equation and turbulent dissipation rate is solved by First order upwind scheme. The convergence criterion chosen for all parameters is RMS residual $<10^{-4}$. Gravitational forces are neglected.

4 Results and Discussion

The particle containing gas field flows downward and scrubbing liquid water was injected at the throat section. Drag force acts on the liquid droplets; as a result, liquid droplets and particle collide by the inertial impaction. Therefore, Particles are captured by the droplets. Droplets are traced by the lagrangian frame of reference. TAB model is used to consider the droplet breakup (Wu et al. 1997). Total flow rate of particle is 10^{-20} kg/s. Discrete Random Walk was considered to predict the turbulent dispersion of particles. Inlet velocity of the gas was limited to 80, 100, 120, 140 m/s and liquid mass flow rate ranged from 0.003 kg/s to 0.075 kg/s.

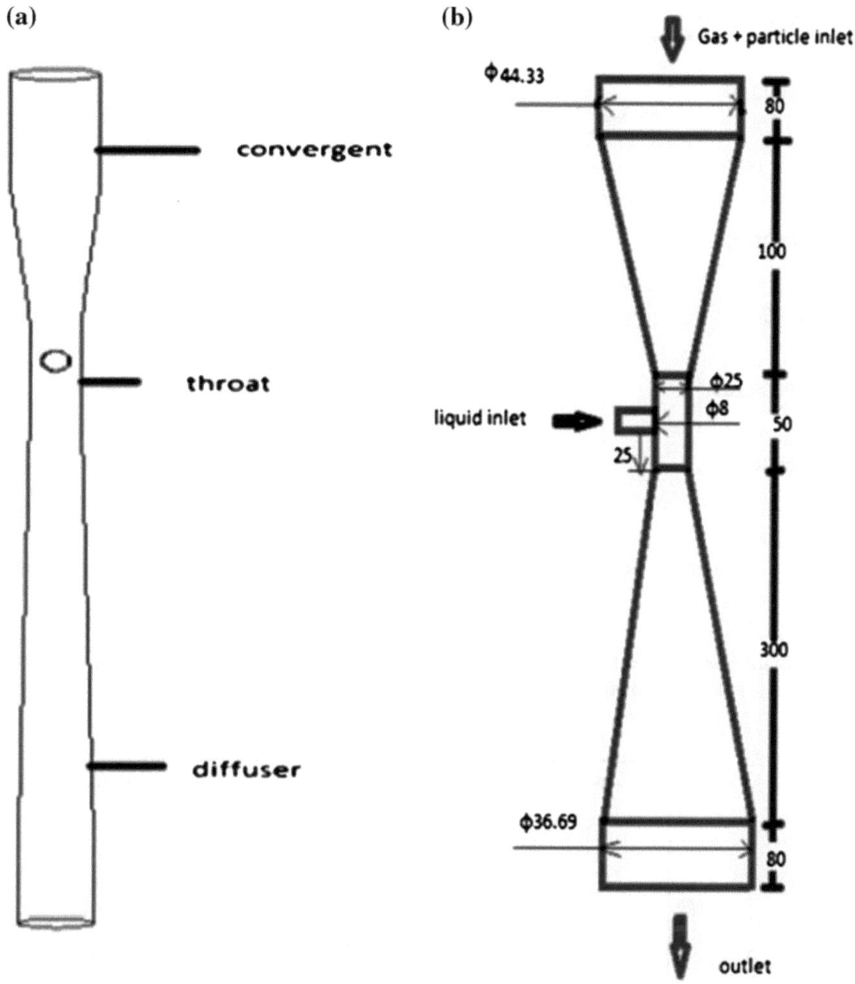

Fig. 1 **a**, **b** Geometry of venturi scrubber (unit: mm)

From the Fig. 2 it shows, gas acceleration occurs in the convergent section due to transformation of potential energy into kinetic energy. Gas reaches the maximum velocity at the throat section, therefore, pressure drop because of the friction of maximum velocity with the wall. In the diffuser, gas recovers some pressure due to loss in velocity. All the kinetic energy obtained at entry of the throat is not return to the diffuser in the form of pressure because of turbulent and viscous dissipations. So, static pressure at the entrance is higher than the exit.

Pressure drop is calculated by the difference between the entry and exit static pressure of venturi scrubber. Figure 3 indicates the dependence of liquid loading on pressure drop. Pressure drop increases with liquid loading at constant gas velocity.

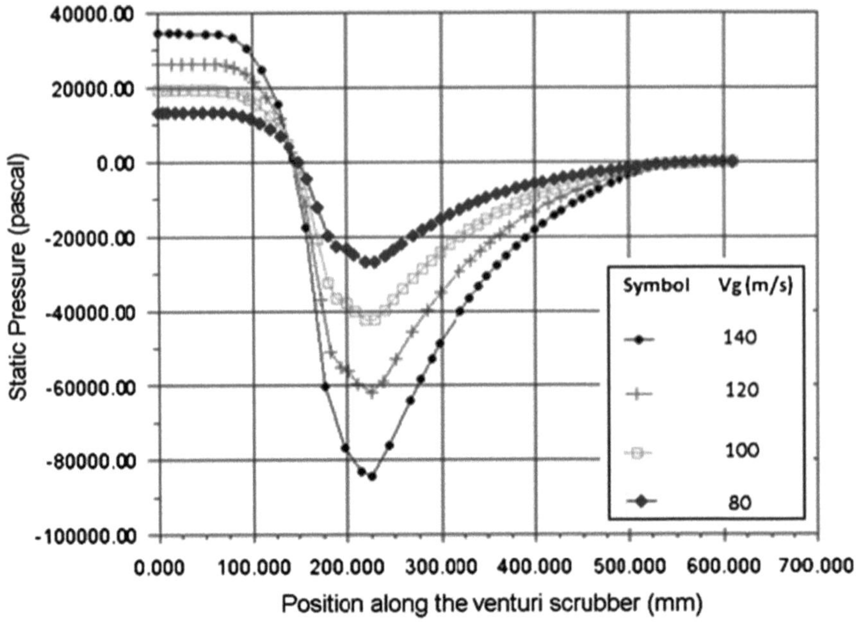

Fig. 2 Static pressure profile along the Venturi scrubber at different gas inlet velocities with constant mass flow rate 0.003 kg/s

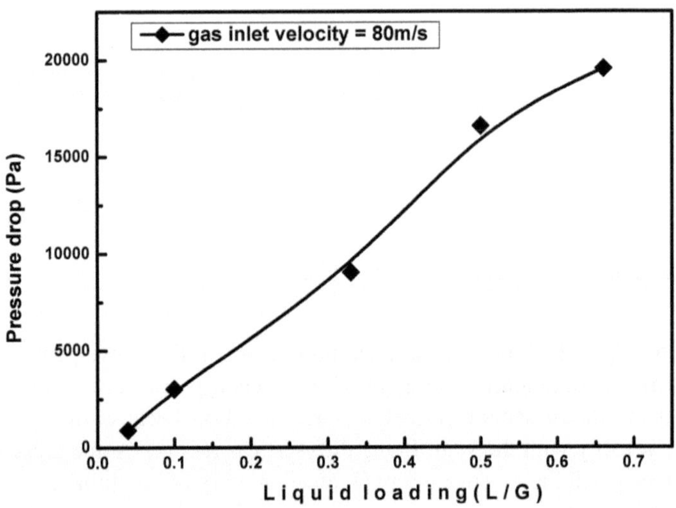

Fig. 3 Effect of liquid loading on the pressure drop at V_g = 80 m/s

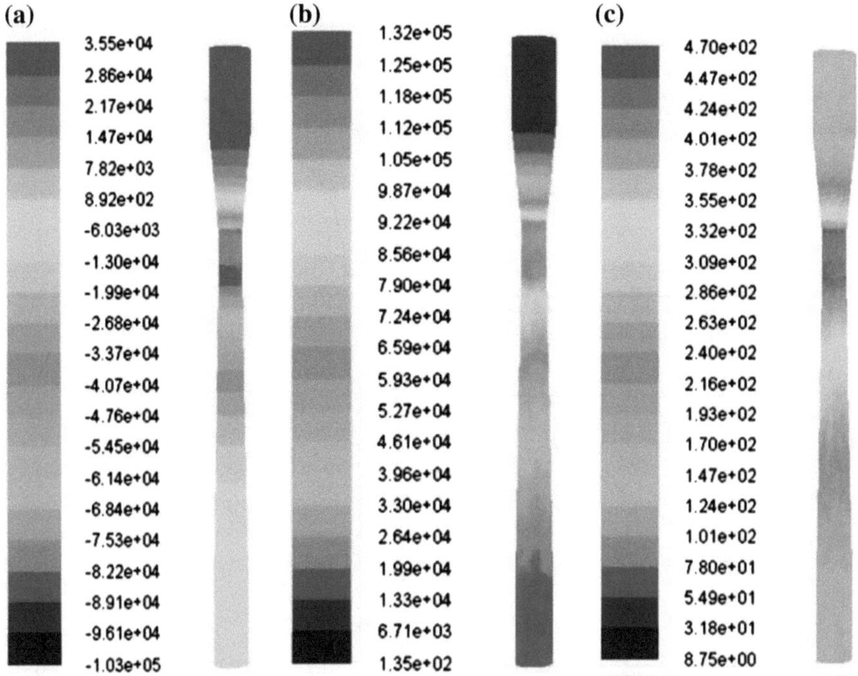

Fig. 4 **a** Contour of static pressure at liquid mass flow rate 0.003 kg/s and gas inlet velocity 140 m/s; **b** Contour of dynamic pressure at liquid mass flow rate 0.003 kg/s and gas inlet velocity 140 m/s; **c** Contour of velocity magnitude at liquid mass flow rate 0.003 kg/s and gas inlet velocity 140 m/s

The contours of static pressure, dynamic pressure, and velocity magnitude are displayed in Fig. 4a–c respectively.

5 Overall Collection Efficiency

In this research, collection efficiency is calculated by the particle tracking. Figure 5a indicates the effect of gas velocity at throat section on the overall collection efficiency that shows the collection efficiency increases with throat gas velocity. This is because increasing impaction with increasing the relative velocity between water droplets and gas. Figure 5b indicates the overall collection efficiency rises with the mass flow rate of liquid due to increasing inertial effect.

From the Fig. 5c shows the influence of the pressure drop on overall collection efficiency. It indicates that high pressure drop across venturi scrubber can provide better collection efficiency. Effect of particle diameter on overall collection efficiency is shown by the Fig. 5d.

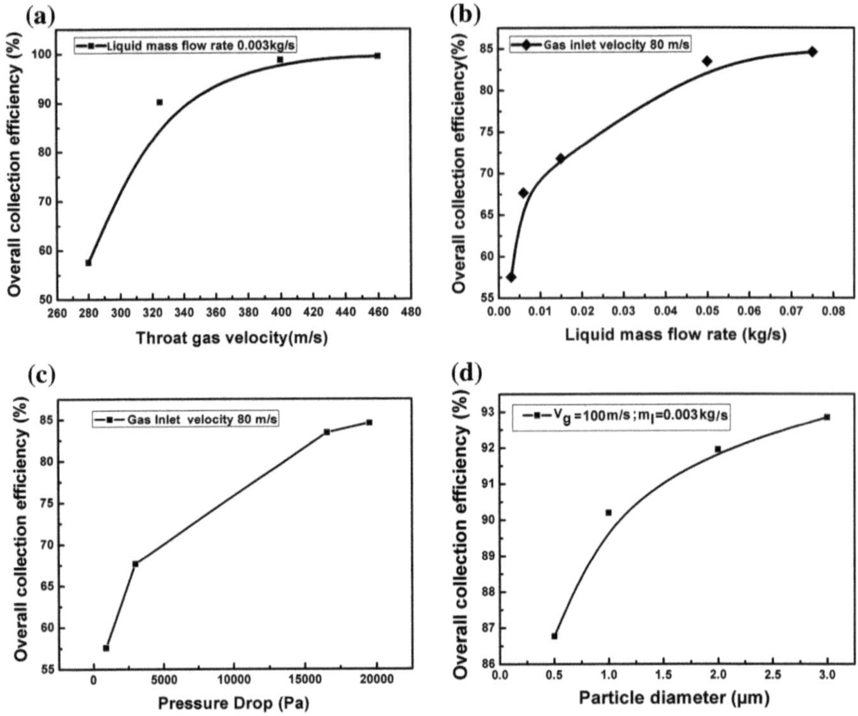

Fig. 5 **a** Effect of gas velocity at throat section on overall collection efficiency; **b** Effect of liquid mass flow rate on overall collection efficiency. **c** Effect of pressure drop on overall collection efficiency; **d** Effect of particle diameter on overall collection efficiency

6 Conclusion

A computational model is developed to estimate the performance of the venturi scrubber using fluent 15. Pressure and velocity distribution inside the venturi scrubber are obtained by using k-ε turbulence model and Discrete phase model. Effect of pressure drop, liquid mass flow rate and throat gas velocity on collection efficiency were also investigated. At liquid mass flow rate 0.2 kg/s and gas inlet velocity 80 m/s get the collection efficiency 99.99 %. Overall collection efficiency of more than 90 % is achieve for dust particles larger than one micron. It is concluded that overall collection efficiency is higher for higher gas flow rate due to increasing the inertial effect and higher pressure drop is favourable for getting higher overall collection efficiency. As a future work, experimental study is required to validate these simulated results.

References

Ali, M., Changqi, Y., Zhongning, S., Jianjun, W., HaiFeng, G.: Analysis of liquid fraction in venturi scrubber by E-E method using CFX. Res. J. Appl. Sci. Eng. Technol. **4**(19), 3830–3833 (2012)

Ananthanarayanan, N.V., Viswanathan, S.: Effect of nozzle arrangement on venturi scrubber performance. Ind. Eng. Chem. Res. **38**, 4889–4900 (1999)

Guerra, V.G., Béttega, R., Gonçalves, J.A.S., José, R.: Coury. pressure drop and liquid distribution in a venturi scrubber: experimental data and CFD simulation. Ind. Eng. Chem. Res. **51**, 8049–8060 (2012)

Lehner, M.: Aerosol separation efficiency of a venturi scrubber working in self-priming mode, pp. 390–402. Aerosol Science and Technology (2007)

Mohebbi, A., Taheri, M., Fathikaljahi, J., Talaie, M.R.: Simulation of an orifice scrubber performance based on Eulerian/Lagrangian method, J Hazard. Mater. **A100**, 13–25 (2003)

Pak, S.I., Chang, K.S.: Performance estimation of a Venturi scrubber using a computational model for capturing dust particles with liquid spray. J. Hazard. Mater. **B13**, 560–573 (2006)

Schlueter, R.O., Schmitz, R.P.: filteredvented containment. Nucl. Eng. Des. **120**, 93–103 (1990)

Shan, L., Cai-ting, L., Guang-ming, Z., Si-min, L., Fei, W., Da-yong, W.: CFD simulation on performance of new type umbrella plate scrubber. Trans. Nonferrous Met. Soc. China **18**, 488–492 (2008)

Wu, P.K., Kirkendall, K.A., Fuller, R.P., Nejad, A.S.: Breakup processes of liquid jets in subsonic crossflows. J. Propul. Power **13**(1), 64–73 (1997)

Mathematical Modelling of Bitumen and Oil Shale Extraction Using Supercritical Fluid

Probal Guha, Tanmay Netke and Sampatrao D. Manjare

1 Introduction

Bitumen is a mixture of hydrocarbons of natural or pyrogenous origin, or combinations of both. These are frequently accompanied by their non-metallic derivatives, which may be in gaseous, liquid, semi-solid or solid form, and are completely soluble in carbon disulphide (CS_2) (Subramanian et al. 1996). The complete solubility in CS_2 is a significant aspect to be noted. Thus, the purity of bitumen can be assessed by its degree of solubility in CS_2 (Meyer and Witt Meyer and Witt 1990). Upon extraction, these hydrocarbons can be upgraded and used for commercial purposes. The difficulty in the current extraction of bitumen is faced due to its high viscosity and density, which makes pumping out bitumen highly energy intensive. Currently, mining is the most widely used process for extraction of bitumen. The basic concept is that, bitumen is mined and then dissolved either in a solvent or mixed with some fluid that effectively reduces the viscosity and density of the bitumen mixture. It is subsequently pumped out for further up gradation and processing to the oil refineries. The current forms of mining are open pit mining, in situ drilling and freeze wall technology (Spright 2004; Mochinaga et al. 2004; Hartzell and Guigard 2012; Read and Whiteoak 2003). The major issue with these current techniques is that they affect the environment immensely.

In view of above, this paper presents the detailed methodology on SCFE technology to extract bitumen. In this work, the mass transfer of bitumen in CO_2 and the sparger used to bubble the fluid has been modeled. Bitumen column here refers to the bitumen bed present in the underground reserve which has been modeled in the form of a column (Hansen et al. 2013; Bredthauer and Williamson 1976).

P. Guha · T. Netke · S.D. Manjare (✉)
Department of Chemical Engineering, BITS-Pilani K.K. Birla Goa Campus, Goa 403726, India
e-mail: manjare@goa.bits-pilani.ac.in

I. Regupathi et al. (eds.), *Recent Advances in Chemical Engineering*,
DOI 10.1007/978-981-10-1633-2_28

2 Methodology

The system here involves drilling holes till the bitumen bed. One set of holes would have spargers installed in them that would feed the supercritical CO_2 into the bitumen bed. The other set of holes would have pipes attached that would feed the gas containing the extracted bitumen into a flash column where the heavy hydrocarbon (bitumen) would be separated from CO_2, which would then be recycled. A heat exchanger and a compressor system would be set up to take the fresh and recycled CO_2 to the supercritical state. While modeling this system, the bitumen bed has been assumed to be analogous to a bubble column, where the liquid column is the bitumen bed and the gas being bubbled into the column is supercritical CO_2. The objective is to extract bitumen by using supercritical CO_2. As bitumen is a highly viscous fluid, the approach here is to sparge the supercritical fluid through the bitumen column. This system can be modeled as bitumen stripping using the supercritical fluid (SCF) as a solvent in a bubble column.

2.1 Bubble Column Introduction

Bubble columns are equipments where the gas is fed into a liquid column in the form of bubbles. The key here is to control the residence time of the supercritical fluid (SCF) within the bed to allow sufficient extraction of bitumen.

2.2 Gas Distribution

The gas is dispersed throughout the cross-section of the entire column in the form of small bubbles to improve the intensity of mass transfer. The gas bubbles for this case are generated by the sparger. The sparger in general is a pipe with holes of desired dimension and spacing. This controls the size and velocity of the bubbles being fed through the bed. Spargers can be of multiple shapes but the most effective design is a ring sparger as this allows an equitable distribution of the SCF mass through the bed. The walls of the sparger are subject to failure as they have to sustain extreme temperature conditions of the supercritical fluid and so choosing the material of construction becomes critical. The design of the sparger has been developed in this paper.

2.3 Bubble Size

Along the height of the column, the bubble size varies greatly due to breakup and coalescence of the bubbles rising up the column. This implies that the size at sparger feed would be different from the size away from the column. The mass transfer of bitumen would occur at a location far away from the sparger, it is this bubble size that is important to predict. This means that the entrance effects of the gas feed can be neglected while modeling the system. Any real estimate of bubble size can only be made by assuming a homogenous flow of gas through the column. The turbulence effects are tough to incorporate in the overall scheme. Based on Kolmogorov's theory of isotropic turbulence, the bubble diameter in a region of high turbulence can be predicted by (Deen 2010):

$$d_{Bs} = \frac{2}{e_m^{0.4}}\left(\frac{\sigma}{\rho_L}\right)^{0.6} \varepsilon_g \left(\frac{\eta_g}{\eta_L}\right)^{0.6} \tag{1}$$

Larger bubbles are more likely to become unstable and breaking up. It has been observed that, for real distributions, the Sauter Diameter is between 40 and 60 % of the largest stable bubble diameter (Zehner 2000).

2.4 Dispersion of the Gas Phase

As described above, the rising gas bubbles create turbulence along the column and causes circulation flow. In addition to the bulk flow, the formation of large and small bubbles, breakup and coalescence results in additional dispersion in the gas phase. The gas-phase dispersion coefficient depends more strongly on gas velocity and column diameter. For this reason, the degree of axial gas mixing is especially relevant when the gas phase is expected to show strong concentration variations. While applying a similar expression as that is being used for the liquid phase, the mass flux is given by (Camy and Condoret 2001):

$$J_G = -D_G \frac{dC}{dz} \tag{2}$$

The diffusivity D_G is taken from literature. Molecular simulations for bitumen based on four-component united-atom model carried out by Hansen, Lemarchand, Nielsen, Dyre and Schroder gives an estimate of the diffusivity. They have assumed the composition of bitumen as 10 asphaltene (A), 10 resin (R), 10 resinous oil (RO) and 82 docosane (D) molecules. Based on the simulations, they identified $D_D > D_{RO} > D_R > D_A$ as the relation in forms of diffusivity. Basing the model on the worst case diffusivity, $D_A = 2.5 \times 10^{-10}$ m^2/s (Sonntag et al. 2002). Hence, assuming:

$$D_G = D_A = 2.5 \times 10^{-10}\ \text{m}^2/\text{s} \tag{3}$$

3 Development of Mathematical Model

In this work we have considered the dynamic system in which the bitumen bed mass and temperature is varying with time. The enthalpy is being introduced into the system by the supercritical CO_2. This problem involves coupling of heat and mass transfer effects. The assumption here is that there is no accumulation of supercritical CO_2 in the bed (Nagy 2010; Pereda et al. 2006; Treybal 1980).

3.1 *Mass Balance*

$$\dot{m}_{in,\mathrm{co}_2} - \left(\dot{m}_{out,\mathrm{co}_2} + \dot{m}_{out,B}\right) = \frac{dm_b}{dt} \tag{4}$$

$$\dot{m}_{in,\mathrm{co}_2} = \dot{m}_{out,\mathrm{co}_2} \geq -\dot{m}_{outB} = \frac{dm_b}{dt} \tag{5}$$

3.2 *Energy Balance*

$$\dot{m}_{in,\mathrm{co}_2} h_{in,\mathrm{co}_2} - \left(\dot{m}_{out,\mathrm{co}_2} h_{out,\mathrm{co}_2} + \dot{m}_{out,B} \dot{h}_{out,B}\right) = \frac{d(m_b h_b)}{dt} \tag{6}$$

As,

$$h_b = c_{p,b}(T - T_i) + \lambda_b \tag{7}$$

Putting in (Eq. 6)

$$\dot{m}_{in,\mathrm{co}_2} h_{in,\mathrm{co}_2} - \left(\dot{m}_{out,\mathrm{co}_2} h_{out,\mathrm{co}_2} + \dot{m}_{out,B} h_{out,B}\right) = \frac{d(m_b(c_{p,b}(T - T_i) + \lambda_b))}{dt} \tag{8}$$

Using (Eq. 5) and rearranging;

$$\begin{aligned} &\dot{m}_{in,\mathrm{co}_2}\left(h_{in,\mathrm{co}_2} - h_{out,\mathrm{co}_2}\right) - \left(\dot{m}_{out,B} h_{out,B}\right) \\ &= (c_{p,b}(T - T_i) + \lambda_b)(-\dot{m}_{out,B}) + m_b c_{p,b}\left(\frac{dT}{dt}\right)_{m_b} \end{aligned} \tag{9}$$

Dividing by T, and rearranging and integrating from $t \in (0, t)\, and\, T \in (T_i, T)$, we get expression for T:

$$T = \frac{T_{in,\text{co}_2}}{\left(\frac{P_{in,\text{co}_2}}{P_{out,\text{co}_2}}\right)^{\frac{R}{c_{p,\text{co}_2}}}} * e^{\left[\left(\ln\left(\frac{T_i}{T_{in,\text{co}_2}}\right) + \frac{R}{c_{p,\text{co}_2}}\ln\left(\frac{P_{in,\text{co}_2}}{P_{out,\text{co}_2}}\right)\right) * e^{\frac{\dot{m}_{in,\text{co}_2} * c_{p,\text{co}_2}}{m_b c_{p,b}} t}\right]} \tag{10}$$

Using Mayer's Formula;

$$c_p - c_v = R \text{ and } \frac{c_p}{c_v} = \gamma \geq \frac{R}{c_{p,\text{co}_2}} = \frac{\gamma_{\text{co}_2} - 1}{\gamma_{\text{co}_2}} \tag{11}$$

$$T = \frac{T_{in,\text{co}_2}}{\left(\frac{P_{in,\text{co}_2}}{P_{out,\text{co}_2}}\right)^{\frac{\gamma_{\text{co}_2}-1}{\gamma_{\text{co}_2}}}} * e^{\left[\left(\ln\left(\frac{T_i}{T_{in,\text{co}_2}}\right) + \frac{\gamma_{\text{co}_2}-1}{\gamma_{\text{co}_2}}\ln\left(\frac{P_{in,\text{co}_2}}{P_{out,\text{co}_2}}\right)\right) * e^{\frac{\dot{m}_{in,\text{co}_2} * c_{p,\text{co}_2}}{m_b c_{p,b}} t}\right]} \tag{12}$$

Equation 12 represents the relation of bed temperature with respect to time as supercritical CO_2 is being introduced into the system. Taking the first derivative with respect to time;

$$\left(\frac{dln(T)}{dt}\right)_{m_b} = \left(\ln\left(\frac{T_i}{T_{in,\text{co}_2}}\right) + \frac{\gamma_{\text{co}_2} - 1}{\gamma_{\text{co}_2}}\ln\left(\frac{P_{in,\text{co}_2}}{P_{out,\text{co}_2}}\right)\right)^2 * \left(\frac{\dot{m}_{in,\text{co}_2} * c_{p,\text{co}_2}}{m_b c_{p,b}}\right)^2 t * e^{\frac{\dot{m}_{in,\text{co}_2} * c_{p,\text{co}_2}}{m_b c_{p,b}} 2t} \tag{13}$$

Substituting Eqs. (13) and (12) into (Eq. 9); this is done to get the relation of mass of bitumen in the bed with respect to time. This helps in determining the duration of the supercritical CO_2 is needed to be fed into the column to strip out the bitumen out.

For simplification, assuming no pressure change in the supercritical fluid at the inlet and outlet condition,

$$\dot{m}_{in,\text{co}_2}\left(c_{p,\text{co}_2} * \left(\ln\left(\frac{T_i}{T_{in,\text{co}_2}}\right)^{e^{\frac{\dot{m}_{in,\text{co}_2} * c_{p,\text{co}_2}}{m_b c_{p,b}} t}}\right)\right) = m_b c_{p,b} * \left(\ln\left(\frac{T_i}{T_{in,\text{co}_2}}\right)\right)^2 * \left(\frac{\dot{m}_{in,\text{co}_2} * c_{p,\text{co}_2}}{m_b c_{p,b}}\right)^2 t * e^{\frac{\dot{m}_{in,\text{co}_2} * c_{p,\text{co}_2}}{m_b c_{p,b}} 2t}$$

$$m_b = ln\left(\frac{T_i}{T_{in,\text{co}_2}}\right) * \frac{\dot{m}_{in,\text{co}_2} * c_{p,\text{co}_2}}{c_{p,b}} t * e^{\frac{\dot{m}_{in,\text{co}_2} * c_{p,\text{co}_2}}{m_b c_{p,b}} t} \tag{14}$$

The above equation is an implicit relationship of the bed mass with respect to time.

Where,

$\dot{m}_{in,\text{co}_2}$ Mass flow rate super-critical CO_2 being fed into the bed (Kg/h)

c_{p,co_2} Specific heat capacity at constant pressure of super-critical CO_2 (kJ/Kg K)
T_{in,co_2} Critical temperature of CO_2 that is being fed into the bitumen bed (K)
T_i Initial bed temperature of bitumen (K)
$c_{p,b}$ Average specific heat capacity of the bitumen bed; this would be based on taking the mass average of specific heat capacity of all the components present in the bed (kJ/Kg K)
m_B Instantaneous mass of bitumen extracted from the bed (Kg/h)
$\mathrm{m_b}$ Mass of bed (kg)
$\mathrm{P_{pipe}}$ It is the output pressure of the vacuum pump being used to fed the gas
$\mathrm{P_{hole}}$ It is the total pressure of the liquid column the gas is sparged through
$\mathrm{V_{pipe}}$ It is dependent on the mass flow rate of the gas

3.3 Interfacial Surface Area

Volumetric flow rate:

$$\dot{V} = \frac{\dot{m}_{in,\text{co}_2}}{\rho_{in,\text{co}_2}} \tag{15}$$

As assumed above the bubble diameter has been estimated by the Sauter diameter, $d_b = d_{b,s}$. Number bubbles fed at any instant is given by

$$\dot{n} = \frac{\dot{V}}{\frac{\pi}{4} d_{b,s}^3} \geq \dot{n} = \frac{\frac{\dot{m}_{in,\text{co}_2}}{\rho_{in,\text{co}_2}}}{\frac{\pi}{6} d_{b,s}^3} \tag{16}$$

Total interfacial area fed into the system,

$$\dot{SA} = \dot{n} * \pi d_{b,s}^2 \tag{17}$$

$$SA = \frac{\frac{\dot{m}_{in,\text{co}_2}}{\rho_{in,\text{co}_2}}}{\frac{\pi}{6} d_{b,s}^3} * \pi \left(\frac{2}{e_m^{0.4}} \left(\frac{\sigma}{\rho_L} \right)^{0.6} \varepsilon_g \left(\frac{\eta_g}{\eta_L} \right)^{0.6} \right)^2 t \tag{18}$$

3.4 Concentration of Bitumen in the Bed

$$C_b = \frac{\frac{m_b}{M_b}}{\frac{(m_{i,b} - m_b)}{\rho_b}} \tag{19}$$

Using the mass flux equation,

$$\frac{dC_b}{dt} = -SA * D_G * \frac{dC_b}{dr} \tag{20}$$

Substituting the equation for C_b, and from (Eq. 18) $m_b = m_b(t)$ has been estimated, therefore

$$\frac{d(m_b)}{dr} = -\frac{\dot{m}_b}{\frac{\frac{\dot{m}_{in,co_2}}{\rho_{in,co_2}}}{\frac{\pi}{6}d_{b,s}^3} * \pi\left(\frac{2}{e_m^{0.4}}\left(\frac{\sigma}{\rho_L}\right)^{0.6}\varepsilon_g\left(\frac{\eta_g}{\eta_L}\right)^{0.6}\right)^2 t * D_G} \tag{21}$$

The above equation gives the mass profile in the radial direction within the bubble (Report 2004).

3.5 *Sparger Design*

The sparger here is used to feed the supercritical CO_2 into a bitumen column with high velocity ensuring an equitable distribution of the gas feed throughout the column. The hole diameter also needs to be designed in such a way that the bubbles formed have low diameter which improves the surface-area of contact thereby improving the mass transfer further. This is based on the fact that stripping occurs at the gas-liquid phase boundary. Higher the phase boundary per unit feed of the supercritical fluid, higher would be the yield of extraction (stripping in this case) (Kantak et al. 1995).

The ring sparger has a ring shape with variable diameter holes to feed the gas through the liquid column. The design objective is to maintain the mass flux through each hole so as to ensure equitable distribution.

While designing, the inlet mass flow rate of the supercritical CO_2 feed needs to be specified. If the Mach number <0.3, the gas is assumed to be incompressible and hence, Bernoulli's equation is valid for the system. For the gas exiting the sparger, the exit pressure is assumed to be the equal to the total pressure of the bitumen column. As the mass is being stripped out of the bed, this pressure would reduce with time (Bird et al. 2002). To calculate the pressure at any instant, the bed density is assumed to be constant, as the cross-sectional area is constant, the only change would come in the height of the bed, which can be predicted from the equation where mass of the bed is estimated with respect to time. Utilizing these assumptions, the momentum, mass and energy balance equations are applied and solved for the exit velocity and sparger hole area. Additionally the pressure head loss due to flow through each hole needs to be considered. As per thumb rule, the holes closer to the gas feed are made smaller in area so as to minimize the pressure drop across each hole.

3.6 Design

The system here is the sparger pipe, the walls of the pipe are assumed to be insulated therefore preventing any loss of energy, as no work is being done by the system. So from the 1st law, there is no change in internal energy and hence it is an isothermal system. The requirement of the design is to have a constant mass flow rate of supercritical CO_2 at each sparger outlet (hole). This would ensure equitable distribution of the supercritical fluid throughout the bed and would help extract the maximum bitumen. As there is no holdup of gas (no accumulation i.e. at steady state) within the sparger, the mass of gas being fed to sparger would be equal to the mass being sparged into the bitumen bed. Therefore, the mass flow rate at each hole can be given by Schnelle (2001):

Mass flow rate at hole = Total Mass flow rate/Number of holes

Assumptions:

Steady state, Inviscid flow, Incompressible, Isothermal flow, Flow rate at each hole is constant, the hole diameter is assumed to be an arithmetic progression with the scheme: $d_h = a + (i - 1) * d$.

Equations used:

Mass Balance: For steady state and incompressible flow, continuity equation can be used

- Total flow rate = $A_{pipe} * V_{pipe} = \sum A_i * V_i = Q$; $1 \leq i \leq n$, $n \geq$ Number of Holes
- Using assumption: $A_i * V_i = \frac{Q}{n}$

  Energy balance: For steady state, inviscid, incompressible and isothermal flow, Bernoulli's Law is valid
- Applying Bernoulli's at the discharge point of the vacuum pump and at the hole inlet of the sparger

$$\frac{1}{n}\sum \frac{4}{(\pi * (a+(i-1) * d)^2)^2} = \frac{\left[2\frac{(Ppipe-Phole)}{\rho} + \left(V_{pipe}\right)^2\right]}{Q};$$

  The terms on the right are specified by: $P_{hole} = Po + \rho_{\text{Bitumen bed}} * g * h_{\text{Bitumen bed}}$, P_{pipe} = the output pressure of the vacuum pump being used to fed the gas, $V_{pipe} = \frac{Q}{A_{pipe}}$.
- The unknowns here are n, a and d;
- Assuming values for n and d, and iterating till the velocity is within the specified range
- Taking Taylor series expansion of the area term till O(2)

$$\left(\frac{4}{(a\text{^}2 * \pi * n)}\right) * \sum\left(1-\left(2(i-1)\frac{d}{a}\right)\right) = \frac{\left[2\frac{(Ppipe-Phole)}{\rho} + \left(V_{pipe}\right)^2\right]}{Q}$$

Solving the summation term:

$$\left(\frac{4}{(a^\wedge 2 * \pi)}\right) * \left(1 - \frac{d}{a}(n-1)\right) = \frac{\left[2\frac{(Ppipe - Phole)}{\rho} + \left(V_{pipe}\right)^2\right]}{Q} \quad (22)$$

Design considerations: ‘d’ must be small to ensure low variations of the V_{hole} because then the hole would have similar area. The number of holes should be less as each the hole would introduce some amount of energy loss in the real system that is being neglected during the design calculations. So, a value based on the above guidelines is assumed for both ‘n’ and ‘d’.

Initial Guess: d is assumed as 0.01 mm, n is assumed as 10 holes, and a is estimated,

V_{hole} is calculated for each hole and is then compared against the set range. If the velocity is beyond the range, another guess is taken by changing the value of ‘d’ and ‘n’ till the velocity is within the required range. The above problem has been resolved using the solver tool in Ms-excel.

In the above design, the equipment is made to fit the requirements of the system. The operating conditions would depend completely upon process considerations such as the rate of gas needed to be sparged, the pressure and temperature of the flow. The above design would give an error within 10 % which is sufficient as a preliminary design. While carrying out detailed design, the turbulence within the pipe and at the hole needs to be considered. Major and minor losses due to roughness of the pipe need to be considered as well.

4 Process Design and Development

Based on the exhaustive literature survey and detailed analysis of the developed model equations, the authors have proposed the flow diagram of the process design and development as shown in the Fig. 1.

5 Results and Conclusion

In this paper the authors have proposed the mathematical model for extraction of bitumen using supercritical CO_2. The relations have been derived for studying variation of bitumen bed mass with time and the mass profile of bitumen in the radial direction within the CO_2 bubble. Further the sparger has been designed in such a way that the bubbles formed have low diameter which improves the surface-area of contact for efficient mass transfer. Subsequently flow diagram for process design and development has been proposed for the system under study and

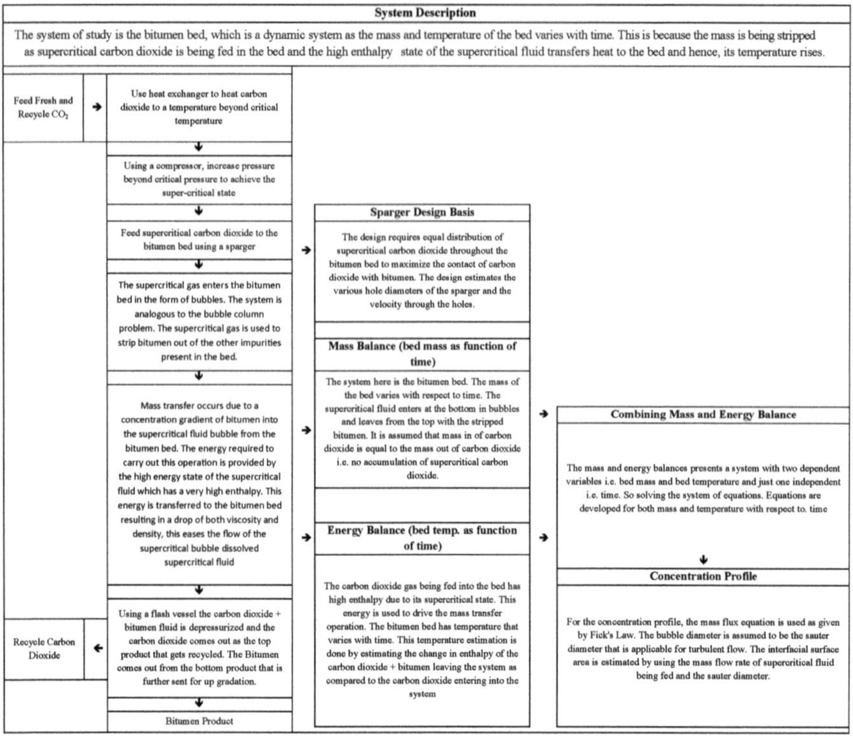

Fig. 1 Proposed flow diagram of process design and development

is presented in Fig. 1. Based on the implicit relation for mass of bitumen extracted, a plot for the bed mass v/s time has been developed for different mass flow rates of supercritical carbon dioxide feed as shown in Fig. 2.

In Fig. 1 the mass flow rates of carbon dioxide are taken in the ratio of 1:2:5. A comparative study is being carried out by varying the mass flow rates and its relation with the extraction mass. As expected, irrespective of the mass flow rate, finally they achieve a saturated level where no further extraction is possible. However, higher solvent flow rate achieves a saturated value lower than that of the lower mass flow rate. This is also expected intuitively as the higher solvent flow rate would have a higher capacity to accommodate bitumen. From the plot it can be seen that by increasing the mass flow rate of supercritical carbon dioxide, the yield of bitumen improves. It can also be seen that the slope decreases as the time increases. This implies that the rate of mass transfer decreases with time. The reason is that, as the mass present in the bed decreases, the driving force i.e. concentration gradient reduces. The graph also shows that the slope is much steeper for $5M_{CO_2,in}$ as compared to $M_{CO_2,in}$ and this can be traced to the fact that the higher the mass flow rate of the solvent, the higher will the gradient for mass transfer. For all the cases i.e. mass flow rates of the solvent, it reaches a saturated value, beyond which

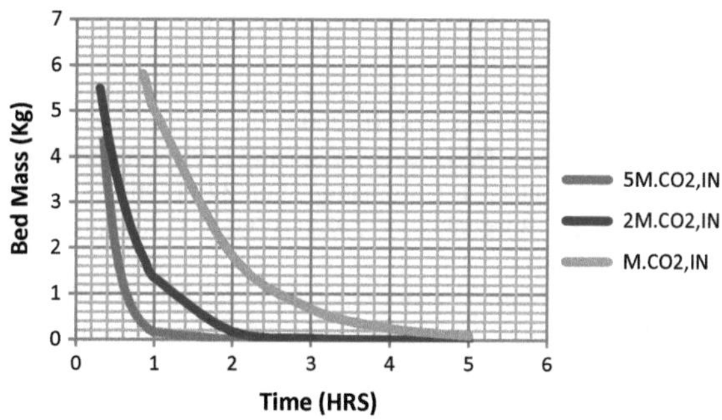

Fig. 2 Variation in bitumen bed mass at different mass flow rates of supercritical CO_2

the concentration gradient would not be sufficient for the extraction to take place. The rate of change of the slope is highest for the maximum flow rate as expected. This is because it would attain a saturation point much before the lower flow rates. The model here gives an accurate picture of the trend that should be followed for the extraction process (Meyer et al. 2012). The developed model equations predict the effectiveness of bitumen removal. Supercritical technology could be a better option for the extraction of bitumen.

References

Bird, R.B., Stewart, W.E., Lighfoot, E.N.: Transport Phenomena, 2nd edn. Wiley (2002)

Bredthauer, R.O., Williamson, T.N.: Mechanical Potential for Oil Shale Mining System for Deep, Thick oil Shale Deposits (1976)

Camy, S. Condoret, J.S.: Dynamic modeling of a fractionating process for a liquid mixture using supercritical carbon dioxide (2001)

Deen N.G.: "Bubble Columns", Ullmann s Encyclopedia of Industrial Chemistry, 15 January 2010

Hansen, J.S., Lemarchand, C.A., Nielsen, E. Dyre, J.C.: Thomas Schroder; Four Component united-atom model of bitumen (2013)

Hartzell, K., Guigard, S.E.: The Use of Supercritical Carbon Dioxide and Modifiers for Enhanced Separation of Bitumen from Athabasca Oil Sand. University of Alberta (2012)

Kantak, M.V., Hesketh, R.P., Kelkar, B.G.: Effect of gas and liquid properties on gas phase dispersion in bubble columns. Chem. Eng. J. & Biochem. Eng. J. (1995)

Meyer, F., Jaeger, P., Eggers, R., Stamenic, M., Milovanovic, S., Zizovic, I.: Effect of CO_2 pre-treatment on sc CO_2 extraction of natural material. Chem. Eng. Process. (2012)

Meyer, R.F., de Witt, W. Jr.: Definiton and world resources of natural bitumens. U.S. Geological Survey (1990)

Mochinaga, H., Onozuka, S.: Properties of Oil sands and Bitumen in Athabasca; CSPG-CSEG-CWLS convention (2004)

Nagy, B.: Modelling of Supercritical Fluid Extraction, Ph.D. Thesis, 2010

Pereda, S., Bottini, S.B., Brignole, E.A.: Fundamentals of Supercritical Fluid Technology. University Piloto de Ingenieriea Quimica (2006)

Read, J., Whiteoak, D.: The Shell Bitumen Handbook (2003)

Report Application foe and Oil Sands Mine, Bitumen Extraction Plant, and Bitumen Upgrading Plant in the Fort McMurray Area; Canadian Natural Resources Limited (2004)

Schnelle: Process Design and the Strategy of Process Design. Handbook Series for Mechanical Engineering (2001)
Spright J.G.: Natural Bitumen (Tar Sands) and Heavy Oil; Coal, Oil Shale, Natural Bitumen, Heavy Oil and Peat Vol. II (2004)
Subramanian, M., Deo, M.D., Hanson, F.V.: Composition analysis of bitumen and bitumen derived products. J. Chromatogr. Sci. **34** (1996)
Sonntag, R.E., Borgnakke, C., Van Wylen, G.J.: Fundamentals of Thermodynamics, 6th edn. Wiley (2002)
Treybal, R.E.: Mass Transfer Operation, 3rd edn. (1980)
Zehner, P.: "Bubble Columns", Ullmann s Encyclopedia of Industrial Chemistry, 15 June 2000

Methanol and Water Crossover in a Passive Direct Methanol Fuel Cell: Mathematical Model

Naveen Shrivastava, Rajkumar Chadge, Sanjeev Bankar and Anil Bamnote

1 Introduction

Fuel cell is an electrochemical device, which converts the chemical energy of fuel directly into the electrical energy (Barelli et al. 2011). Hydrogen is the most preferred fuel for the fuel cells (Beck et al. 2005; Chao et al. 2010) But, it is difficult to store and requires bulky storage medium and hence not suited for the portable applications. Direct liquid fuel cell (DLFC) which utilizes liquid fuel viz. methanol, ethanol, formic acid, dimethyl ether and hydrazine offers high energy density and compact size (Daniela et al. 2010). In this category, direct methanol fuel cell (DMFC) of passive type, which does not require any pump or blower for supply or removal of reactant and product, emerges as potential alternative of batteries for empowering the portable electronic devices (Saarinen et al. 2007).

The methanol crossover and water crossover through polymer electrolyte membrane (PEM) are the major technological problems of passive DMFC (Saarinen et al. 2007). Although, significant research work has been done in the recent past to encounter these problems, MCO and WCO are still not fully understood (Weimin et al. 2006; Young-Chul et al. 2011). In order to control methanol crossover (MCO) and water crossover (WCO), fundamental understanding about these is essential. In this study, a mathematical model that incorporates the heat transfer effect is developed. The model considers methanol transport, water transport, oxygen transport phenomena and the electrochemical reactions. The experimental data is used to validate the model. The model is used to analyze the MCO and WCO through membrane and their variation with the different operating and physical conditions of the cell.

N. Shrivastava · R. Chadge (✉) · S. Bankar · A. Bamnote
Yeshwantrao Chavan College of Engineering, Wanadongri, Hingna Road, Nagpur, India
e-mail: rbchadge@rediffmail.com

I. Regupathi et al. (eds.), *Recent Advances in Chemical Engineering*,
DOI 10.1007/978-981-10-1633-2_29

2 Passive DMFC

Basic components of a passive DMFC are shown in Fig. 1. This includes methanol reservoir to store methanol water solution, anode and cathode current collector (ACC, CCC) for collecting the electrons and supporting membrane electrode assembly (MEA), diffusion layer on the anode and cathode side (ADL, CDL) for uniform distribution of reactant on the anode and cathode catalyst layers (ACL, CCL) for effective electrochemical reactions:

$$CH_3OH + H_2O \rightarrow CO_2 + 6H^+ + 6e^- + heat \quad Anode\ reaction$$

and

$$3/2\, O_2 + 6H^+ + 6e^- \rightarrow 3H_2O + heat \quad Cathode\ reaction$$

Methanol and water, which are not consumed at ACL, crosses the membrane and reaches to cathode side. This is due to combined effect of diffusion and electro-osmotic drag and known as MCO and WCO. MCO not only results in wastage of fuel, but also causes overpotential and leads to decreased cell performance. On the other hand, WCO results in water loss from anode and water flooding at cathode side.

3 Model Assumptions

The passive direct methanol fuel cell is a complex system involving simultaneous mass, charge and energy transfer. Following assumptions are introduced in order to simplify the processes:

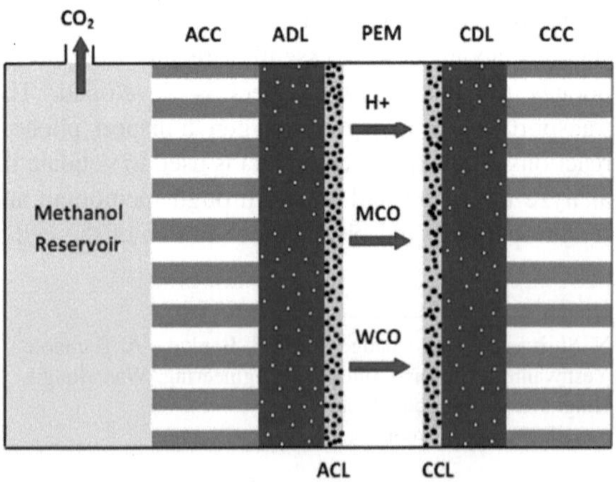

Fig. 1 Change in methanol crossover with current density (ambient temperature: 296 K)

- The steady-state operation of the fuel cell is considered.
- The methanol concentration in reservoir remains same during the operation.
- It is assumed that the methanol crossover is completely oxidized at CCL and therefore the methanol concentration is neglected at CCL.
- Water in cathode side in considered only in gas phase.
- The catalyst layers are treated as an interface and temperature gradient and species concentration gradient are ignored in theses layers.

4 Mathematical Model and Solution Methodology

The mathematical model used to describe the mass transfer, heat transfer, electrochemical reactions and cell voltage for the passive DMFC are summarized in Table 1. An analytical solution of the given mathematical model has been obtained. A MATLAB code has been developed for the simulation.

5 Result and Discussion

To ensure reliability of the model, it has been validated with the available experimental result of Chao et al. (2010), as shown in Fig. 2. For this, Current–voltage (I-V) and current–power (I-P) characterization curves has been generated by the model for 2 and 3 M methanol feed concentration. The operating conditions were considered as same as that of the referred experimental work (Chao et al. 2010). The Fig. 2 indicates that model results are in good agreement with the available experimental data. Hence, the model can be used for the more detailed study.

The Fig. 3 shows the variation of MCO flux with cell current density for 1, 2, 3 and 4 M methanol feed concentrations. It can be observed that at relatively lower methanol concentration (1 and 2 M) the MCO decreases on increasing the cell current density. Whereas, at relatively higher methanol concentrations (3 and M) MCO increases on increasing the current density. On increasing the cell current density, diffusion component of MCO decreases and electro-osmotic drag (EOD) component increases. At low methanol feed concentration the MCO due to diffusion has more significant value that that due to EOD. Whereas, MCO due to EOD plays a dominating role at relatively higher methanol concentration. MCO is not only the fuel wastage but also results in overpotential and decreases the fuel cell performance. Hence, the cell must be operated at relatively lower methanol concentration and at relatively higher current density for achieving low methanol crossover.

Variation in WCO flux with cell current density for variable methanol feed concentration is shown in Fig. 4. The WCO is increasing linearly with current density for every methanol feed concentration. This is due to the combined effect of

Table 1 Model Equations

Methanol transport equations			
Through ADL	$N_{MeOH}^{ADL} = -D_{MeOH,ADL}^{eff}\left(\frac{dC}{dx}\right)_{MeOH,ADL}$	(1)	Diffusion
Through PEM	$N_{MeOH}^{PEM} = -D_{MeOH,PEM}^{eff}\left(\frac{dC}{dx}\right)_{MeOH,PEM} + n_{d,MeOH}^{eff}\left(\frac{I_{cell}}{F}\right)$	(2)	Diffusion and electro osmotic drag
	where, $n_{d,MeOH}^{eff} = n_{d,H_2O}x_{MeOH}^{ACL}$	(3)	
	and, $x_{MeOH}^{ACL} = C_{MeOH}^{ACL}/\left(C_{MeOH}^{ACL} + C_{H_2O}^{ACL}\right)$	(4)	
Mass balance at ACL	$N_{MeOH}^{ADL} = \frac{I_{cell}}{6F} + N_{MeOH}^{PEM}$	(5)	Methanol consumed in MOR, remaining crossovers PEM
Water transport equations			
Through ADL	$N_{H_2O}^{ADL} = -D_{H_2O,ADL}^{eff}\left(\frac{dC}{dx}\right)_{H_2O,ADL}$	(6)	Diffusion (liquid phase)
Through PEM	$N_{H_2O}^{PEM} = -D_{H_2O,PEM}^{eff}\left(\frac{dC}{dx}\right)_{H_2O,PEM} + n_{d,H_2O}^{eff}\left(\frac{I_{cell}}{F}\right)$	(7)	Diffusion and electro osmotic drag
	$n_{d,H_2O}^{eff} = n_{d,H_2O}x_{H_2O}^{ACL}$	(8)	
	$x_{H_2O}^{ACL} = C_{H_2O}^{ACL}\Big/\left(C_{H_2O}^{ACL} + C_{MeOH}^{ACL}\right)$	(9)	
Through CDL	$N_{H_2O}^{CDL} = -D_{H_2O,CDL}^{eff}\left(\frac{dC}{dx}\right)_{H_2O,CDL}$	(10)	Diffusion (gas phase)
Mass balance at ACL	$N_{H_2O}^{ADL} = \frac{I_{cell}}{6F} + N_{H_2O}^{PEM}$	(11)	Water consumed in MOR and remaining Crossover PEM
Mass balance at CCL	$N_{H_2O}^{CDL} = N_{H_2O}^{PEM} + \frac{I_{cell}}{2F} + 2N_{MeOH}^{PEM}$	(12)	Crossovered water reaches at CCL, consumed in oxidation of crossovered methanol and generated in ORR
Oxygen Transport equations			
Through CDL	$N_{O_2}^{CDL} = -D_{O_2,CDL}^{eff}\left(\frac{dC}{dx}\right)_{O_2,CDL}$	(13)	Diffusion
Mass balance at CCL	$N_{O_2}^{CDL} = \frac{I_{cell}}{4F} + \frac{3}{2}N_{MeOH}^{PEM}$	(14)	Oxygen consumed in ORR and oxidation of crossovered methanol
Heat transport equations			
Through ACC, ADL, PEM, CDL, CCC	$q_{component} = -K_{component}^{eff}\left(\frac{dT}{dx}\right)_{component}$	(15)	Conduction

(continued)

Table 1 (continued)

ACC to fuel reservoir	$q_{conv,anode} = h_{anode}(T_{ACC} - T_{MeOH,Rvoir})$	(16)	convection
	$Nu = \frac{h_{anode} L}{K} = 0.54(Gr.\Pr)^{0.25}$	(17)	Correlation for horizontal surface facing up
CCC to ambient	$q_{conv,cathode} = h_{cathode}(T_{CCC} - T_{amb})$	(18)	convection
	$Nu = \frac{h_{cathode} L}{K} = 0.27(Gr.\Pr)^{0.25}$	(19)	Correlation for horizontal surface facing down
Heat Generation at ACL	$q_{ACL} = I_{cell}\eta_a - I_{cell}\left(\frac{\Delta H_a - \Delta G_a}{6F}\right)$	(20)	Anode overpotential causes the heat generation and the heat is consumed in anode electrochemical reaction
Heat Generation at CCL	$q_{CCL} = (I_{cell} + I_{MeOH})\eta_c - (I_{cell} + I_{MeOH})\left(\frac{\Delta H_c - \Delta G_c}{4F}\right) - I_{MeOH}\left(\frac{\Delta H_a - \Delta G_a}{6F}\right) - h_{v,H_2O}N_{H_2O}^{CDL}$	(21)	Heat is generated due to cathode overpotential, oxidation of permeated methanol and cathode electrochemical reaction.
Heat balance at ACL	$q_{ACL} = q_1 + q_2$	(22)	
Heat balance at CCL	$q_{CCL} = q_3 - q_2$	(23)	
Electrochemical Kinetics equations			
Anode	$I_{cell} = I_{ref}^{MeOH} \frac{C_{MeOH}^{ACL}}{C_{ref}^{MeOH}} \exp\left(\frac{\alpha_a F}{RT_{ACL}}\eta_a\right)$	(24)	Tafel equation
Cathode	$I_{cell} + I_{MeOH} = I_{ref}^{O_2} \frac{C_{O_2}^{CCL}}{C_{ref}^{O_2}} \exp\left(\frac{\alpha_c F}{RT_{CCL}}\eta_c\right)$	(25)	Tafel equation
Cell voltage equations			
	$V_{cell} = E_{cell} - \eta_a - \eta_c - I_{cell}R_{cell} - I_{cell}R_{contact}$	(26)	
	$E_{cell} = E_{cell}^0 + (T - T_0)(\partial E_{cell}/\partial T)$	(27)	
	$R_{cell} = \delta_{PEM}/\sigma_{PEM}$	(28)	

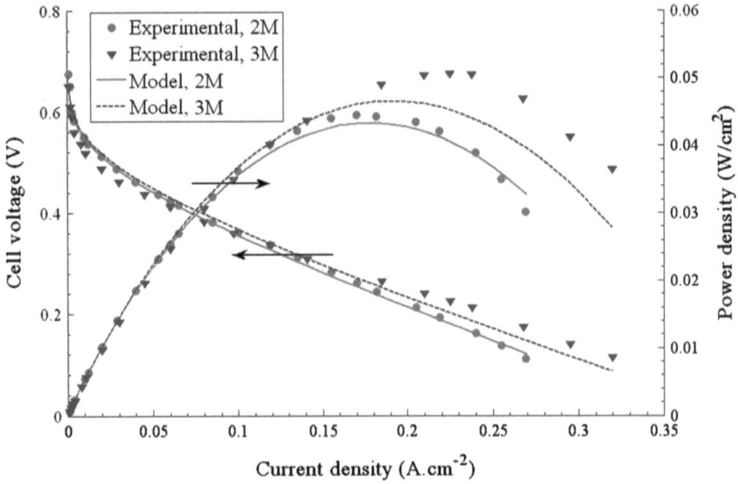

Fig. 2 Change in water crossover with current density (ambient temperature: 296 K)

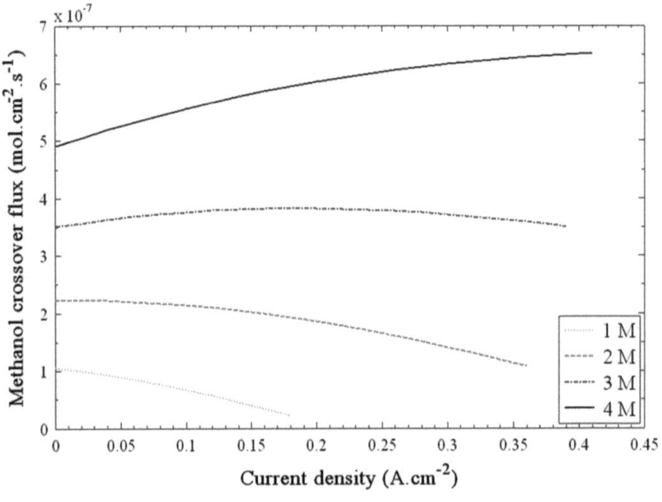

Fig. 3 Change in methanol crossover with current density (ambient temperature: 296 K)

diffusion and EOD of water molecule by the proton. Thus, at higher current density water flooding at cathode will be more severe. This will results in disrupted transport of oxygen and reduce the cell performance. Hence, to make sufficient availability of oxygen for cathode electrochemical reaction, cell must be operated at relatively lower current densities. It is also clear from the Fig. 4 that methanol feed concentration does not make significant effect on WCO.

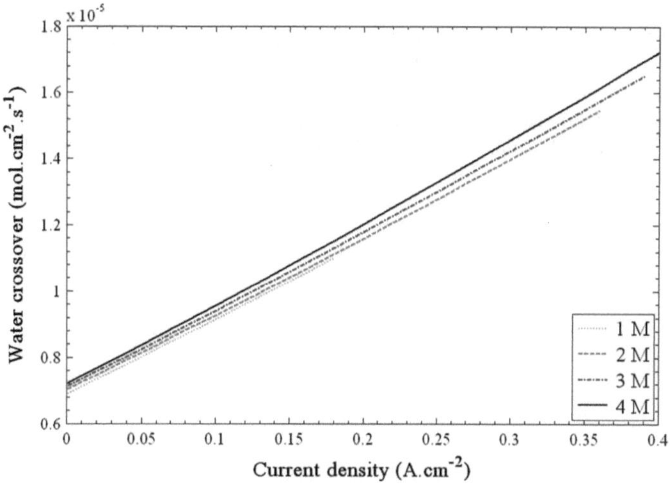

Fig. 4 Change in water crossover with cell current density (ambient temperature: 296 K)

6 Conclusion

- A non-isothermal mathematical model incorporating species transport equations and electrochemical reactions for a passive DMFC have been developed.
- The mathematical model is successfully validated with the experimental data. The model is used to analyze the MCO and WCO in the DMFC.
- For achieving the low MCO, the cell must be operated at relatively lower methanol concentration and at relatively higher current density.
- Also, the WCO can be controlled by the operating the cell at low current density.
- Although, methanol feed concentration does not make any significant affect on WCO. Hence it can be concluded that low current density is a desired operating condition for controlling both, the MCO and WCO.

References

Barelli, L., Bidini, G., Gallorini, F., Ottaviano, A.: Analysis of the operating conditions influence on PEM fuel cell performances by means of a novel semi-empirical model. Int. J. Hydrogen Energy, 10434–10442 (2011)

Beck, K., Byungchan, B., Scibioh M. A., Jaeyoung, L., Heung, Y. H.: On the consequences of methanol crossover in passive air-brea thing direct methanol fuel cells, J. Power Sources 142, 50–55 (2005)

Chao, X., Amir, F., Xianglin, L, Travis, W.: Methanol and water crossover in a passive liquid-feed direct methanol fuel cell. Int J. Hydrogen Energy 1769–1777 (2010)

Daniela, C.F.G., Benjamin, B., David, B., Abdellatif, M.: Methanol fuel processor and PEM fuel cell modeling for mobile application. Int. J. Hydrogen Energy 6S63–6S7 (2010)

Saarinen, V., Himanen, O., Kallio, T., Sundholm, G., Kontturi, K.: A 3D model for the free-breathing direct methanol fuel cell: Methanol crossover aspects and validations with current distribution measurements, J. Power Sources 805–815 (2007)

Weimin, Q., David, P. W., Jun, S., Haijiang, W., Jiuiun, Z.: Architecture for portable direct liquid fuel cells. J. Power Sources 202–213 (2006)

Young-Chul, P., Dong-Hyun, P., Sang-Keun, D., Sang- Kyung, K., Seongyop, L., Doo-Hwan, J.: Operating characteristics and performance stability of 5 W class direct methanol fuel cell stacks with different cathode flow patterns. Int. J. Hydrogen Energy 1853–1861 (2011)

Tuning of PID Controllers for First Order Plus Time Delay Unstable Systems

Saxena Nikita and M. Chidambaram

1 Introduction

Over past 50 years, the Proportional-Derivative-Integral (PID) controller remains a central element in the process industries. The Proportional (P) mode adjusts controller output according to the error size, the Integral (I) mode eliminate offset at the steady state whereas the Derivative (D) mode increases the speed of the response. Due to these useful functions along with its lack of complexity, efficient online tuning and superior performance approximately 97 % of process industries use PID controllers. Efficient rules and automatic methods for tuning the PID controllers are desired. Many methods, such as gain margin/phase margin method, pole placement technique, optimization technique (minimizing IAE/ISE/ITAE), direct synthesis method, internal model control method, equating coefficient method and robust loop shaping have been reviewed by Vilanova and Visioli (2012); Yu (1999). One of the methods of tuning the PID controllers based on stability analysis is the Ziegler-Nichols method (1942). In the method, the system is brought to the condition of marginal instability. At the point, the system exhibits a sustained oscillation commonly known as limit cycles in the response. The ultimate gain and the ultimate frequency (referred as critical points) determines the PID controller settings. It is a analytical method of designing the PID controller. The method is also automated by using relay auto-tuning method which is a single step method. The PID tuning rules proposed were based on the simulation of a large number of processes. As the key criterion is a quarter decay ratio, a large overshoot and an oscillatory response is obtained. Many researchers proposed methods based on the critical points and proposed different setting to improve the method. Astrom and Hagglund (1988, 1995, 2006), Tyreus-Luyben (1992), Yu (1999) and Smith et al.

S. Nikita · M. Chidambaram (✉)
Department of Chemical Engineering, Indian Institute of Technology Madras, Chennai 600036, India
e-mail: chidam@iitm.ac.in

I. Regupathi et al. (eds.), *Recent Advances in Chemical Engineering*,
DOI 10.1007/978-981-10-1633-2_30

(2003) proposed tuning formulae to improve the ZN method. Many such controller tuning rules are reviewed by O'Dwyer (2009).

But the scope of tuning PID controllers for unstable system based on critical points is restricted. The following study includes description on the general controller design procedure along with the performance and the robustness measures in Sect. 1.2. Section 1.3 discusses the fundamentals of improved method for determining the PID settings for unstable FOPTD systems for varying time delay to time constant ratios, followed by Sect. 1.4, where the proposed method is applied on first order non-linear bioreactor model and is compared with the different reported methods. Section 1.5 summarises the study.

2 Controller Design Method

In this paper, the single loop feedback controller structure is used, where G_p is the process transfer function, G_c is the PID controller transfer function of the form given by

$$G_c(s) = K_c + \frac{K_I}{s} + K_D s \tag{1}$$

where, $K_I = \frac{k_c}{\tau_{I,}}; K_D = k_c * \tau_D$

The proposed method is implemented on unstable processes. It should be recalled in the context of unstable system that they have a peculiar property of exhibiting the maximum value of controller gain $(K_{c,\max})$ as well as the minimum value of controller gain $(K_{c,\min})$.

The tuning rule proposed by Ziegler and Nichols (1942) for the PID controllers are based on the systems critical values. In the present work, simple equations for calculating PID settings for unstable FOPTD are proposed based on the reported work of Nikita and Chidambaram (2016). The amplitude ratio (A_r) and phase angle (φ) for the FOPTD system are realized by solving the magnitude criterion and the phase angle criterion. The controller gain is set equal to $1/A_r$ so as to make the gain unity of the system at π radian phase angle. At crossover frequency, the closed loop system will attain the condition of marginal stability and will give limit cycle with frequency equal to $\omega_{\max}$. To eliminate offset and to expedite the response, the integral mode and the derivative mode is added to the controller.

It is observed that once the integral and derivative modes become active, the controller gain is affected. For stable systems based on the bode plots, Douglas (1972) proposed a trial and error procedure to ascertain the updated values of controller constants. The integral and derivative times are considered respectively as $5/\omega_c$ and $1/\omega_c$. Enhanced results are obtained when the controller is tuned around these values.

To compare the performance of the system based on proposed method over the other methods present in literature, Integral of the Time weighted Absolute Error (ITAE), maximum sensitivity as defined by Vilanova and Visioli (2012), gain margin and phase margin in controller output values are considered for unit step change in the input and in the disturbance.

3 Proposed Method

The main objective of the control system design is the fast set point tracking while also ensuring the system is insensitive to the uncertainties in the model. The FOPTD system is given as:

$$G_p(s) = \frac{K_p \exp(-\varepsilon s)}{s-1} \tag{2}$$

where, ε is the time delay to time constant ratio and $K_p = 1$. The corresponding amplitude ratio and the phase angle are written as:

$$A_r = \frac{1}{\sqrt{\omega^2+1}}; \varphi = -\varepsilon s - \pi + \tan^{-1}(\omega) \tag{3}$$

At the crossover frequency, phase angle is equal to −180. Hence, Eq. (3) can be solved to obtain the minimum frequency and the maximum frequency $(\omega_{c,min}, \omega_{c,max})$, for different values of ε. On substituting the $\omega_{c,min}, \omega_{c,max}$ values in Eq. (3), the minimum and the maximum values of amplitude ratio is calculated, and corresponding controller gain can be determined as by using the relation $K_c = \frac{1}{A_r}$.

The values of reset time (τ_I) and derivative time (τ_D) constant are calculated using $\omega_{c,max}$ as: $P_{u,max} = \frac{2\pi}{\omega_{c,max}}; \tau_I = \frac{P_{u,max}}{2}; \tau_D - \frac{P_{u,max}}{8}$

The controller is designed based on the Zeigler-Nichols method. For the proposed method,τ_I and τ_D constant are calculated using the equation:

$$\tau_I' = \frac{5}{\omega_{c,max}}; \tau_D' = \frac{0.8}{\omega_{c,max}} \tag{4}$$

The τ_I' and τ'_D values obtained in Eq. (4) are corresponding to the series PID controller. The corresponding values for the parallel PID form (Yu, 1999) are computed by:

$$\tau_{Ip} = \tau_I' + \tau_D'; \quad \tau_{Dp} = \frac{\tau_I' * \tau_D'}{\tau_I' + \tau_D'} \tag{5}$$

Once controller is designed and integral plus derivative action becomes effective, the value of K_c changes. It needs to be updated again. For that purpose, the transfer function of PID controller is added to the FOPTD system as:

$$G_p(s) = \left(1 + \frac{1}{\tau_I s} + \tau_D s\right)\left(\frac{\exp(-\varepsilon s)}{s-1}\right) \tag{6}$$

The amplitude ratio and phase angle of the new system are written:

$$\varphi = \tan^{-1}\left(\tau_D\omega - \frac{1}{\tau_I\omega}\right) - \varepsilon\omega - \pi + \tan^{-1}(\omega) \tag{7}$$

$$A_r = \frac{\sqrt{\left(\tau_D\omega - \frac{1}{\tau_I\omega}\right)^2 + 1}}{\sqrt{\omega^2 + 1}} \tag{8}$$

At the crossover frequency, Eq. (7) is solved for the updated $\omega_{c,min}$ and $\omega_{c,max}$ corresponding to which the updated $K_{c,min}$ and $K_{c,max}$ for different values of ε are obtained (Table 1). The $K_{c,des}$ is taken to be the average of $K_{c,min}$ and $K_{c,max}$. The following correlations are proposed for the unstable FOPTD system:

$$K_c K_p = 1.045\varepsilon^{-0.96} \text{ when } 0.01 < \varepsilon < 0.1 \tag{9a}$$

$$K_c K_p = 1.041\varepsilon^{-0.94} \text{ when } 0.1 < \varepsilon < 0.8 \tag{9b}$$

$$\frac{\tau_I}{\tau} = 8.059\varepsilon^3 - 4.511\varepsilon^2 + 3.620\varepsilon - 0.036 \tag{9c}$$

$$\frac{\tau_D}{\tau} = 1.414\varepsilon^3 - 0.791\varepsilon^2 + 0.635\varepsilon - 0.006 \tag{9d}$$

The controller performance is evaluated for three different value of ε (0.05, 0.5 and 0.8) given in Table 1 and the performance is evaluated in the closed loop system for a unit step change in the set point and unit change in load as shown in Fig. 1. The proposed method gives enhanced results (Table 2). The maximum sensitivity (MS), gain margin (G_m) and phase margin (φ_m) for different values of ε are given in Table 3. It can be clearly seen that the proposed method is more robust than the ZN method indicating the closed loop system is less sensitive to the variations in the process dynamics.

Table 1 Controller parameters for FOPTD system for different values of ε

	ZN method			Proposed method		
ε	$K_{c,des}$	τ_I	τ_D	$K_{c,des}$	τ_I	τ_D
0.05	15.8911	0.1021	0.0255	22.6045	0.1850	0.0224
0.5	1.6183	1.3477	0.3369	2.0562	2.4882	0.2959
0.8	1.8935	3.3100	0.8275	1.2167	6.1111	0.7266

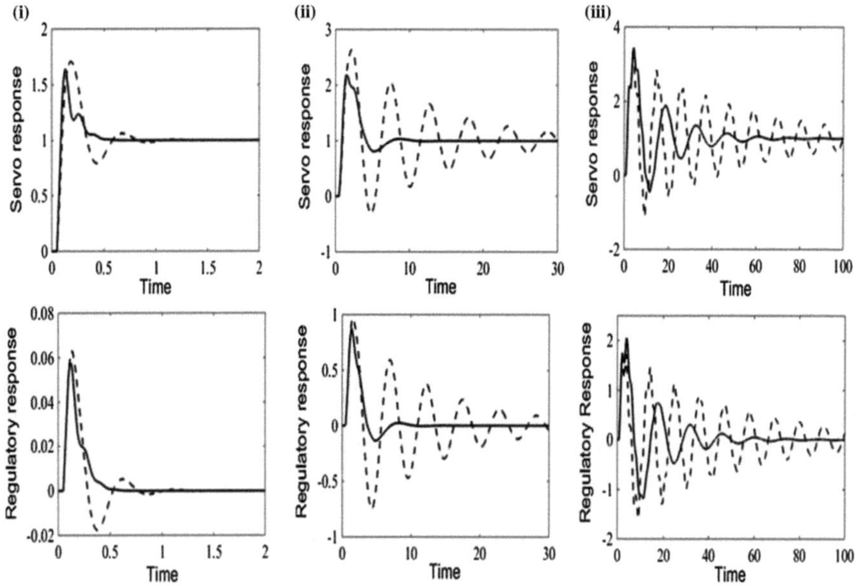

Fig. 1 Process response for step change **i** ε = 0.05; **ii** ε = 0.5; **iii** ε = 0.8 (*Dotted* ZN method; *Solid* Proposed method)

Table 2 ITAE value for different cases

	Servo response			Regulatory response		
Method	ZN	Proposed	Imp* (%)	ZN	Proposed	Imp* (%)
ε = 0.05	0.0493	0.0187	61	0.0031	0.0015	52
ε = 0.5	158.2	8.035	95	85.92	4.487	94
ε = 0.8	2202	440.3	80	1656	358	78

*Imp denotes percentage reduction in the values

Table 3 Maximum sensitivity, gain margin and phase margin for different values of ε

	ε = 0.05		ε = 0.5		ε = 0.8	
Method	ZN	Proposed	ZN	Proposed	ZN	Proposed
MS	1.984	3.017	10.267	2.173	13.052	8.671
G_m	2.106	1.495	1.1079	1.851	1.083	1.130
Φ_m	60.532	38.713	11.178	54.775	8.788	13.244

4 Case Study: Unstable Non-linear Bioreactor

The proposed method is applied to a Non-linear continuous bioreactor that exhibits output multiplicity. The model given by Agarwal and Lim (1984) is considered. The system is assumed to be at unstable steady state condition. At t = 0,

Table 4 Controller parameter for different methods and corresponding ITAE values

Parameters	Proposed method	Majhi and Atherton method	Thyagarajan and Yu method	Tyreus and Luyben method
K_c	−0.92915	−1.1855	−0.9439	−0.8157
τ_I	3.9874	4.7255	5.5363	9.5031
τ_D	0.4741	0.4377	0.5235	0.6857
ITAE (servo)	3.271	13.16	4.522	9.462
ITAE (S_f decreased)	0.7871	0.8173	1.284	3.315
ITAE (S_f increased)	0.7289	0.7515	1.185	3.002

X = 0.9951; S = 1.5122 and D = 0.3 l/h. The method is applied to the local linearised model with the parameters: $K_p = -0.5898$, $\tau = 5.6125$ and $\theta = 1.0$. The controller parameters are calculated for the Non-Linear system using the same method as discussed in Sect. 3. The method is compared with the methods present in literature. The controller parameters obtained for Majhi and Atherton method (2000); Thyagarajan and Yu method (2003) and Tyreus-Luyben method (1992) along with the proposed method are shown in Table 4. Appreciable reduction in ITAE values in servo response is obtained (step change in X from 0.9951 to 1.294). Also when disturbance is added to the substrate feed (±10 %), the proposed method gives enhanced performance as seen in Table 4.

Although the proposed method presented here is based on the transfer function model, it can be automated using relay auto-tuning method where identification of transfer function model is not necessary for controller design. The work can be further extended for application in industrial auto-tuners.

5 Conclusions

An improved ZN method is proposed for tuning the PID controllers. The simulation results of unstable FOPTD model with different ε values are given. The correlation for the unstable FOPTD system based on the simulation studies are proposed. The method is implemented to a Non-Linear bio reactor model and compared with three other methods present in literature. Significant improvement in performances along with the robust behaviour is seen for the proposed method.

References

Agarwal, P. Lim, H.C.: Analysis of various control schemes for continuous bioreactors. Adv. Biochem. Eng./Biotechnol **30**, 61–90 (1984)

Astrom, K.J., Hagglund, T.: Automatic Tuning of PID controllers. Instument Society of America, North Carolina (1988)

Astrom, K.J., Hagglund, T.: PID Controllers: Theory, Design and Tuning, 2nd edn. Instument Society ofAmerica, North Carolina (1995)

Astrom, K.J., Hagglund, T.: Advanced PID control. Instument Society of America, North Carolina (2006)

Douglas, J.M.: Process Dynamics and Control, 2 Control System Synthesis. Prentice Hall, Englewood Cliffs, NJ (1972)

Majhi, S., Atherton, D.P.: Online tuning of controllers for an unstable FOPTD process. IEEE Proc. CTA **146**, 415–425 (2000)

Nikita, S. Chidambaram, M.: Improved Continuous Cycling Method of Tuning PID Controllers for Unstable Systems. Indian Chemical Engineer (2016) in press

O'Dwyer, A.: Handbook of PI/PID Controller Tuning Rules, 3rd Edn. Imperial College Press, (2009)

Smith, C.L.: Intelligently Tune PID Controllers, Chemical Engineering, January, pp. 56–62, (2003)

Thyagarajan, T., Yu, C.C.: Improved auto-tuning using shape factor from relay feedback, Ind. Eng. Chem. Res. **42**, 4425–4440 (2003)

Tyreus, B.D., Luyben, W.L.: Tuning PI controllers for integrator/dead-time processes. Ind. Eng. Chem. Res. **31**, 2625 (1992)

Vilanova, R., Visioli, A.: PID Control in Third Millennium. Springer verlag Ltd., London (2012)

Yu, C.C.: Auto Tuning of PID Controllers. Springer, Berlin (1999)

Ziegler, J.G., Nichols, N.B.: Optimum settings for automatic controllers. Trans. ASME **64**, 759–765 (1942)

Modelling of Interactive Multivariable Systems for Control

Chandra Shekar Besta and Manickam Chidambaram

1 Introduction

In general, the control of multivariable processes is complex in comparison with the scalar systems. The complexity is because of the presence of interactions among the loops. Figure 1 shows a schematic view of a TITO system with two inputs and two outputs. To measure the degree of interaction, steady state Relative Gain Array (RGA) proposed by Bristol (1966) is used. The RGA method deals with the steady-state gains of the system and it proposes a suitable pairing for the control structure. The input-output pairing having the relative gain element close to 1 is preferred. The pairing having negative RGA element should be avoided. Recent methods to decide the suitable pairing are given by McAvoy et al. (2003), Xiong et al. (2005), He et al. (2009) and Monshizadh-Naini et al. (2009). The performance of the decentralized controllers is appreciable in case of the less interacting system. For the design of decentralized controllers, the diagonal element should be approximated by a First order plus time delay models and the well-known method of PI designing controllers can be applied.

Recently for control of SISO systems, the PID controller designed based on the critically damped SOPTD model gives much improved performance than that of FOPTD model. Both the models have the three parameters (Ram and Chidambaram 2015; Simi and Chidambaram 2015). The focus of present work is to evaluate the performance of the centralized PID controllers for multivariable systems which are approximated by these two simple models.

C.S. Besta · M. Chidambaram (✉)
Chemical Engineering, Indian Institute of Technology Madras, Madras 600036, India
e-mail: chidam@iitm.ac.in

I. Regupathi et al. (eds.), *Recent Advances in Chemical Engineering*,
DOI 10.1007/978-981-10-1633-2_31

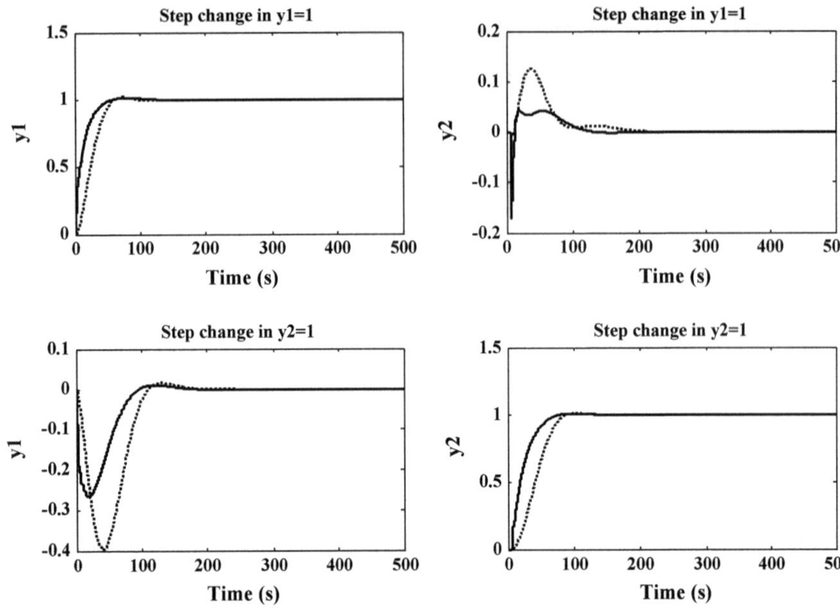

Fig. 1 Servo responses (*solid* CSOPTD-centralized; *dash* FOPTD-centralized)

2 Multi Variable System

To analyse the performance of the proposed scheme, a two-input two-output (TITO) multivariable system with time delay is considered. $G_p(s)$ and $G_c(s)$ represent the plant transfer function and the centralized controller respectively. Controller output and the plant output are represented by u_i and y_i.

$$G_P(s) = \begin{bmatrix} g_{P,11}(s) & g_{P,12}(s) \\ g_{P,21}(s) & g_{P,22}(s) \end{bmatrix} \tag{1}$$

$$K_c(s) = \begin{bmatrix} g_{c,11}(s) & g_{c,12}(s) \\ g_{c,21}(s) & g_{c,22}(s) \end{bmatrix} \tag{2}$$

Each element of the process transfer function matrix is considered as a critical damped SOPDT stable model or by a FOPTD model

$$g_{p-f}(s) = \frac{k_p e^{-\theta_1 s}}{(\tau_1 s + 1)} \quad g_{p-s}(s) = \frac{k_p e^{-\theta_2 s}}{(\tau_2 s + 1)^2} \tag{3}$$

3 Model Order Reduction

Critical damped SOPTD identification The model parameters are computed by using the open loop time response of the higher order model by Rangaiah and Krishnaswamy (1994) method. The critical damped SOPTD parameters (k_p, θ,τ) are determined using two points from the step response namely t_1 and t_2, where t_1 and t_2 are the time when the fractional responses are 14 and 55 % of the final value respectively. The model parameters are calculated from the following simplified relations:

$$k_p = \frac{\Delta y_\infty}{\Delta u_\infty} \quad \frac{t_2 - t_1}{\tau} = 1.18911 \quad \frac{t_2 - \theta}{\tau} = 1.843205 \tag{4}$$

FOPTD identification For FOPTD parameters, the method proposed by Sundaresan and Krishnaswamy (1978) is used. By this method, the FOPTD model parameters (k_p, θ, τ) are calculated from the open loop time response of the higher order model. To identify the FOPTD model, only two values from the step response curve are required namely t_1 and t_2, where t_1 and t_2 are the time when the fractional responses are 35.3 and 85.3 % of the final value respectively. The following simplified relations obtained from (Sundaresan and Krishnaswamy 1978) method are used:

$$k_p = \frac{\Delta y_\infty}{\Delta u_\infty} \quad \tau = 0.64(t_2 - t_1) \quad \theta = 1.3t_1 - 0.29t_2 \tag{5}$$

4 Control System Design

The TITO system is defined by $Y(s) = G_P(s)U(s)$. The present study considers each individual transfer function $g_{p,ij}(s)$ as a higher order system. For SISO systems, a controller design scheme based on the internal model control (IMC) principle is reviewed by Rivera et al. (1986) and Morari and Zafiriou (1989). In the proposed method, the proportional gain (K_C), and the integral time (τ_I) and derivative time (τ_D) for all individual transfer function of the process $g_{c,ii}(s)$ are calculated by the SISO method. For the present case study, the IMC method is applied to calculate the controller settings.

For scalar FOPTD model the setting used are

$$\mathrm{k_{c-F}} = \frac{1}{k_p}\left[\frac{\tau}{\tau_c + \theta}\right] \quad \tau_{I-F} = \tau \tag{6}$$

For scalar C-SOPTD model, the Controller settings are calculated from:

$$k_{c-S} = \frac{1}{k_p}\left[\frac{2\tau}{\tau_c + \theta}\right] \quad \tau_{I-S} = 2\tau \quad \tau_{D-S} = \frac{\tau}{2} \tag{7}$$

where $\tau_c = \alpha\tau$, α is a constant and it is assumed as 1 for the present studies.

After obtaining the individual PI/PID controller settings [Eq. (6–7)], the centralized controller K_c, is obtained by the method proposed by Tanttu and Lieslehto (1991):

$$K_{c-TL} = \begin{bmatrix} 1/k_{c,11} & 1/k_{c,12} \\ 1/k_{c,21} & 1/k_{c,22} \end{bmatrix}^{-1} \tag{8}$$

K_{I-TL} and K_{D-TL} can be obtained in the similar manner. The centralized controllers are assumed to be in the form of Eqs. (8), where K_C, τ_I and τ_D are matrices of size $n \times n$, $K_C = [k_{C,ij}]$, $\tau_I = [\tau_{I,ij}]$ and $\tau_D = [\tau_{D,ij}]$, $(i, j = 1, 2)$ ($k_{C,ij}$, $\tau_{I,ij}$ and $\tau_{D,ij}$ are the scalar controller settings for individual transfer function). Here, $K_{I,ij} = K_{C,ij}/\tau_{I,ij}$ is the integral gain for the loop ij. To analyse the performance of the control systems, ISE, ITAE, IAE and TV indices are considered for the study purpose. The robustness analysis is carried out by using the inverse maximum singular value of the input and output multiplicative uncertainties.

5 Simulation Studies

Example: Consider the critically damped SOPTD model:

$$G_P(s) = \begin{bmatrix} \frac{12.8e^{-s}}{(16.7s+1)^2} & \frac{-18.9e^{-3s}}{(21s+1)^2} \\ \frac{6.6e^{-7s}}{(10.9s+1)^2} & \frac{-19.4e^{-6s}}{(14.4s+1)^2} \end{bmatrix} \Lambda = \begin{bmatrix} 2.0094 & -1.0094 \\ -1.0094 & 2.0094 \end{bmatrix} \tag{9}$$

It should be kept in mind that for the above case study the interactions are greater than 1 ($\lambda_{11} > 1$). The Relative Gain Array (Bristol 1966) (Λ) can be calculated [Eq. (9)]. The details of the calculations of the identified model parameters of the C-SOPTD and FOPTD models are given in Table 1. PID controllers for the identified critically damped SOPTD models were designed by the IMC method.

Table 1 Parameters of the identified model

S. No.		11	12	21	22
FOPTD	t_1	21.76	29.11	20.55	23.89
	t_2	57.68	74.38	44.04	54.9
	TF	$\frac{12.8e^{-11.5608s}}{24.12s+1}$	$\frac{-18.9e^{-16.2728s}}{30.3309s+1}$	$\frac{6.6e^{-13.9434s}}{15.7383s+1}$	$\frac{-19.4e^{-15.136s}}{20.7767s+1}$

The calculated centralized PID settings are given for C-SOPTD by Eq. (10). PI controllers for the identified FOPTD models are designed by the IMC method. The calculated centralized PI settings are given for FOPTD systems by Eq. (11).

CSOPTD-PID Controller:

$$k_{c-S} = \begin{bmatrix} 0.3962 & -0.3114 \\ 0.1562 & -0.1956 \end{bmatrix}; \quad \tau_{I-S} = \begin{bmatrix} 36.1139 & 24.7638 \\ 47.7102 & 31.1402 \end{bmatrix}$$
$$\tau_{D-S} = \begin{bmatrix} 9.1293 & 6.2601 \\ 12.0608 & 31.1402 \end{bmatrix} \tag{10}$$

FOPTD-PI Controller:

$$k_{c-F} = \begin{bmatrix} 0.1226 & -0.1062 \\ 0.0455 & -0.0692 \end{bmatrix}; \quad \tau_{I-F} = \begin{bmatrix} 25.6326 & 17.5584 \\ 33.8385 & 22.0796 \end{bmatrix} \tag{11}$$

Figure 1 depicts the closed-loop servo responses and interactions for a unit step change in y_{r1} along with the response and interaction for the unit step change in y_{r2}. The closed-loop regulatory responses for a step change in d_1 or in d_2 are shown in

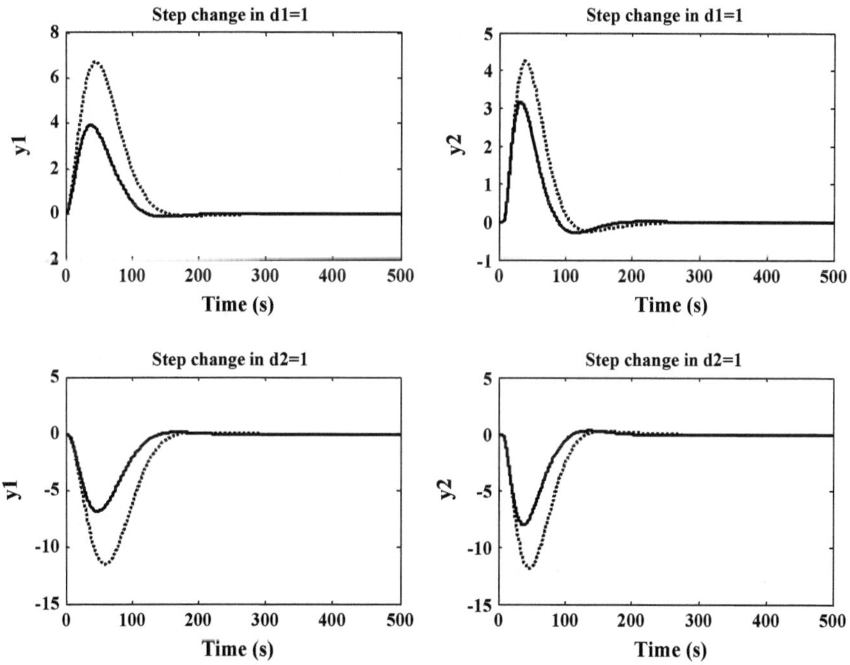

Fig. 2 Regulatory responses (*solid* CSOPTD-centralized; *dash* FOPTD-centralized)

Fig. 2. Figures 1 and 2 suggests that the proposed C-SOPTD centralized PID controllers give improved performance when compared to that of the FOPTD based centralized PI controllers. For the proposed centralised control scheme, the interactions are less.

The stability bound for the above case study is given in Fig. 3. The curve divides the entire region in two parts; region below the curve portrays the stability region while the region above the curve portrays instability region. From Fig. 3, we can see that the area under the curve for the proposed control schemes is high indicating the more robust behaviour. As quoted above, the performance of the proposed centralized control system is superior to that of the proposed decentralized control systems.

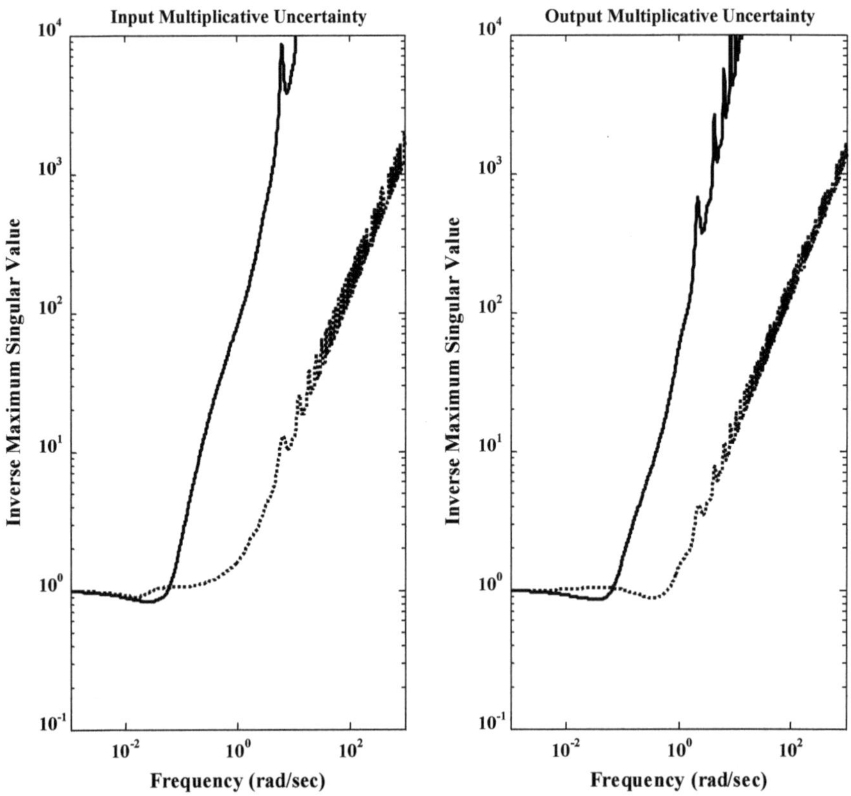

Fig. 3 Stability regions of input and output uncertainties (*Solid* CSOPTD-centralized; *dash* FOPTD-centralized)

6 Conclusions

The centralized SOPTD-PID and centralized FOPTD PI and PID controllers are designed by TL method using the tuning parameters same in the IMC method. A TITO system exhibiting high interactions is simulated to show the effectiveness of the proposed scheme. The performance of the proposed centralized SOPTD-PID controllers is enhanced in comparison with the centralised FOPTD-PI/PID controllers for both servo and regulatory problems. The robustness analysis is studied for both the controllers structure using the Inverse Maximum Singular Value versus Frequency plot.

References

Bristol, E.H.: On a new measure of interaction for multivariable process. IEEE Trans. Autom. Control 133–134 (1966)

He, M.-J., Cai, W.-J., Ni, W., Xie, L.-H.: RNGA based control system configuration for multivariable processes. J. Process Control **19**(6), 1036–1042 (2009)

Mc Avoy, T., Arkun, Y., Chen, R., Robinson, D., Schnelle, P.D.: A new approach to defining a dynamic relative gain. Control Eng. Pract. **11**(8), 907–914 (2003)

Monshizadh-Naini, N., Fatehi, A., Khaki-Sedigh, A.: Input—Output pairing using effective relative energy array. Indian Eng. Chem. Res. **48**, 7137–7144 (2009)

Morari, M., Zafiriou, E.: Robust Process Control. Prentice-Hall, Englewood Cliffs, New Jersey (1989)

Ram, V.D., Chidambaram, M.: On-line controller tuning for critically damped SOPTD systems. Chem. Eng. Commun. **202**, 48–58 (2015)

Rangaiah, G.P., Krishnaswamy, P.R.: Estimating second order plus dead time model parameters. Indian Eng. Chem. Res. **33**(4), 1867–1871 (1994)

Rivera, D.E., Morarl, M., Skogestad, S.: Internal model control. 4. P I D controller design. Indian Eng. Chem. Res. **25**, 252–265 (1986)

Simi, S., Chidambaram, M.: Tuning of PID controllers for critically damped second order plus time delay systems. Indian Chem. Eng. **57**(1), 32–51 (2015)

Sundaresan, K.R., Krishnaswamy, P.R.: Estimating of time delay constant parameters in time, frequency, and Laplace domains. Can. J. Chem. Eng. **56**, 257–260 (1978)

Tanttu, J.T., Lieslehto, J.: A comparative study of some multivariable PI controller tuning methods. In: Devanathan, R. (ed.) Intelligent Tuning and Adaptive Control, pp. 357–362. IFAC, Singapore (1991)

Xiong, Q., Cai, W.-J., He, M.-J.: A practical loop pairing criterion for multivariable processes. J. Process Control **15**(7), 741–747 (2005)

Bayesian Inference Approach to Estimate Robin Coefficient Using Metropolis Hastings Algorithm

Sai Krishna Dammalapati, Vishal Murmu and Gnanasekaran Nagarajan

1 Introduction

The applications of Inverse Heat Transfer Phenomenon are found in various branches of engineering and science including in chemical engineering. The boundary condition's determination, thermal properties, unknown heat source and heat transfer coefficients are important for efficient thermal systems. It is evitable that the majority of the Inverse Heat Transfer problems (IHTP) are unstable, ill-posed, and unsolvable when compared to the direct problems.

There have been several numerical methods proposed to estimate the thermo physical properties. A comprehensive parameter and function estimation process has been discussed in (Ozisik and Orlande 2000). Boundary element method has been used for parametric estimation (Onyango et al. 2009). Bayesian Inference approach has been used in many works of estimation. Bayesian inference approach has been used in parameter estimation for multi-mode heat transfer problems (Parthasarathy and Balaji 2008). Bayesian inference approach has been used for estimating physical and transport properties (Reddy et al. 2012). Simultaneous estimation of heat transfer coefficient and thermal conductivity was done by Bayesian approach (Gnanasekaran and Balaji 2011). Estimation of the two dimensional heat transfer coefficient was done by applying the iterative regularization method (Masson et al. 2004).

S.K. Dammalapati (✉) · V. Murmu · G. Nagarajan (✉)
NITK Surathkal, Srinivasa Nagar, Mangalore, India
e-mail: saikrishnadammalapati@gmail.com

G. Nagarajan
e-mail: gnanasekaran@nitk.edu.in

V. Murmu
e-mail: vishalmurmu@yahoo.com

I. Regupathi et al. (eds.), *Recent Advances in Chemical Engineering*,
DOI 10.1007/978-981-10-1633-2_32

Based on the literature review, the authors found that Robin coefficient is very important while designing any thermal systems and often referred as corrosion damage. But very few works have been attempted to investigate the Robin coefficient based on stochastic method. Conjugate gradient method has been worked on identification of Robin coefficient (Yang et al. 2009). The Bayesian inference is a stochastic method and prior information about the unknown parameter can easily be incorporated in the Bayesian framework. So, a parabolic heat conduction problem is solved for convection boundary conditions and an attempt is made to estimate the Robin coefficient with reasonable accuracy using Markov Chain Monte Carlo (MCMC) algorithm. This algorithm has been used for determining thermal diffusivity in heat transfer experiments (Gnanasekaran and Balaji 2013).

In the future work section, the efficacy of Hybrid Monte Carlo algorithm over Monte Carlo algorithm is discussed.

2 Mathematical Model of the Problem

To determine the Robin coefficient (ρ), the mathematical model of the heat transfer problem is given by the following parabolic equation:

$$\frac{\partial \theta}{\partial t}(x,t) = \frac{\partial^2 u}{\partial x^2}(x,t), \quad (x,t)\epsilon Q := (0,1) \times \left(0, t_f\right] \tag{1}$$

$$\theta(x,0) = g(x), \quad x\epsilon[0,1] \tag{2}$$

$$\frac{\partial \theta}{\partial x}(0,t) + \rho(t)\theta(0,t) = h_0(t) \quad t\epsilon\left[0,t_f\right] \tag{3}$$

$$\frac{\partial \theta}{\partial x}(1,t) + \rho(t)\theta(1,t) = h_1(t) \quad t\epsilon\left[0,t_f\right] \tag{4}$$

where $t_f > 0$ the final time which is fixed randomly, g is, h_0, h_1 are the given functions, and θ represents the temperature, which is a function of x and t. $\rho(t) \geq 0$ Is the coefficient of heat transfer representing the corrosion damage and is a function of time. In general, it is always interpreted as a Robin coefficient during energy interaction and acts as an interface makes to the overall thermal resistance of the system. In the above equations, when $\rho(t)$ is known, then the problem is a direct, a well posed. The characteristics of this coefficient are unknown in majority of the practical problems and thus require additional information to determine the physical process in a complete sense.

The measurement data is considered as,

$$Y(t) = \theta(1,t), \quad t\epsilon\left[0,t_f\right] \tag{5}$$

Let $\boldsymbol{F}(\rho)$ be the vector of the determined temperatures at $x = 1$. Then by solving the following equation, we obtain the solution to the inverse heat transfer problem.

$$\boldsymbol{F}(\rho) = Y \tag{6}$$

The proposed problem, like the other inverse problems, is an ill-posed and unsolvable problem. Large deviations from the exact solution may also emerge owing to the contamination of the measurement data with random errors. So, to deal with the instability of the problem, in this paper, Bayesian inference approach is used.

2.1 Numerical Implementation of the Direct Problem

The forward-finite difference method (FDM) is used to solve Eq. (6) and x is discretized into M points, uniformly spaced, as shown below.

$$x_i = x_{i-1} + h, \quad x_0 = 0, \quad x_M = 1, \quad i = 1, 2, \ldots M, \quad h = \frac{1}{M} \tag{7}$$

Also, the discretization of time t is given by,

$$t_n = t_{n-1} + \tau, \quad t_0 = 0 \quad t_N = t_f, \quad n = 1, 2 \ldots N \quad \tau = \frac{t_f}{N} \tag{8}$$

where N is any positive integer.

Denoting θ_i^n as the approximate value of (x_i, t_n), the FDM used to discretize Eqs. (1)–(4) is as follows:

$$\frac{\theta_i^{n+1} - \theta_i^n}{\tau} = \frac{1}{2h^2}\left(\theta_{i-1}^n - 2\theta_i^n + \theta_{i+1}^n + \theta_{i-1}^{n+1} - 2\theta_i^{n+1} + \theta_{i+1}^{n+1}\right), \tag{9}$$

$$\theta_i^1 = g(x_i), \tag{10}$$

$$\begin{aligned} \frac{\theta_2^n - \theta_1^n}{h} + \frac{\theta_2^{n+1} - \theta_1^{n+1}}{h} = &\left[\rho(t_n)\theta_1^n + \rho(t_{n+1})\theta_1^{n+1}\right] - \left[h_0(t_n) + h_0(t_{n+1})\right] \\ &+ \frac{h}{\tau}\left(\theta_1^{n+1} - \theta_1^n\right) \end{aligned} \tag{11}$$

$$\begin{aligned} \frac{\theta_M^n - \theta_{M-1}^n}{h} + \frac{\theta_M^{n+1} - \theta_{M-1}^{n+1}}{h} = &-\left[\rho(t_n)\theta_M^n + \rho(t_{n+1})\theta_M^{n+1}\right] + \left[h_1(t_n) + h_1(t_{n+1})\right] \\ &+ \frac{h}{\tau}\left(\theta_M^{n+1} - \theta_M^n\right) \end{aligned} \tag{12}$$

The above set of equations can be represented as a matrix equation as follows,

$$A\theta^{n+1} = B\theta^n + C, \tag{13}$$

$$\theta^1 = G, \tag{14}$$

where the matrices,

$$\theta^n = \left[\theta_1^n, \theta_2^n, \ldots \theta_M^n\right]^T, \tag{15}$$

$$C = \frac{\tau}{h}\left[h_0(t_n) + h_0(t_{n+1}), 0, \ldots 0, h_0(t_n) + h_0(t_{n+1})\right]^T, \tag{16}$$

$$G = \left[g(x_1), g(x_2), g(x_3) \ldots g(x_M)\right]^T, \tag{17}$$

And the matrices A and B are $(M \times M)$ tri-diagonal matrices with

$$A = \begin{matrix} 1+r+\frac{\tau}{h}\rho(t_{n+1}) & -r & 0 & 0 & 0 \\ -\frac{r}{2} & 1+r & -\frac{r}{2} & 0 & 0 \\ 0 & \ddots & \ddots & \ddots & 0 \\ 0 & 0 & -\frac{r}{2} & 1+r & -\frac{r}{2} \\ 0 & 0 & 0 & -r & 1+r+\frac{\tau}{h}\rho(t_{n+1}) \end{matrix} \tag{18}$$

$$B = \begin{matrix} 1-r-\frac{\tau}{h}\rho(t_{n+1}) & r & 0 & 0 & 0 \\ \frac{r}{2} & 1-r & \frac{r}{2} & 0 & 0 \\ 0 & \ddots & \ddots & \ddots & 0 \\ 0 & 0 & \frac{r}{2} & 1-r & \frac{r}{2} \\ 0 & 0 & 0 & r & 1-r-\frac{\tau}{h}\rho(t_{n+1}) \end{matrix} \tag{19}$$

where $r = \frac{\tau}{h^2}$. One can also incorporate noise to the measurement to mimic the actual experimental data. For example,

$$Y = \boldsymbol{F}(\rho) + \omega \tag{20}$$

where ω indicates the presence of random error, which may arise due to the errors in measurement and in calculations. Then the solution to the inverse heat transfer problem can be obtained by estimating the real Robin coefficient $\rho(t)$ such that the determined temperatures $\boldsymbol{F}(\rho)$, with this estimate, can match the measurement data Y.

3 Bayesian Inference Approach

Bayesian inference method is different from the deterministic methods in the way that they utilize all the prior information to form the posterior density function p (x/Y), where x is the quantity to be determined from the measurements Y. To quantify the quality of the result, any statistics about x can be determined. By the recent advancement of MCMC methods like Metropolis Hastings and Hybrid Monte Carlo, the Bayesian inference method has become very much beneficial due to its lesser computational time equivalent to deterministic methods.

Bayesian inference approach uses the Bayes law,

$$\pi(\rho|Y) = \pi(Y|\rho)\pi(\rho), \tag{21}$$

where $\pi(Y|\rho)$ the likelihood is function and $\pi(\rho)$ is the prior. Hereby $\pi(.)$ indicates the probability density function (pdf). Through the prior distribution function $\pi(\rho)$ expert knowledge about the expected solution can be incorporated prior to any measurements. The likelihood function $\pi(Y|\rho)$ combines knowledge about the model and the noise.

In the Bayesian method, the construction of the likelihood function $\pi(Y|\rho)$ is the most direct part. Assuming a simple model, the random errors in Eq. (7) can be considered as they are independent and identically distributed.

Then the likelihood function is given by,

$$\pi(Y|\rho)\alpha(\sigma^2)^{\frac{-y}{2}} \exp\left(-\frac{(\boldsymbol{F}(\rho) - Y)^T(\boldsymbol{F}(\rho) - Y)}{2\sigma^2}\right) \tag{22}$$

where y is the dimension of the matrix Y.

The main step is the construction of the prior density function which is used for prior modeling in Markov random field (MRF). For this work, following MRF is taken

$$\pi(\rho)\alpha \quad \lambda^{\frac{m}{2}} \exp\left(-\frac{1}{2}\lambda\rho^T W\rho\right) \tag{23}$$

With Eqs. (9) and (10) the PPDF can be evaluated as follows,

$$\pi(\rho|Y)\alpha \exp\left(-\frac{(\boldsymbol{F}(\rho) - Y)^T(\boldsymbol{F}(\rho) - Y)}{2\sigma^2}\right) \exp\left(-\frac{1}{2}\lambda\rho^T W\rho\right) \tag{24}$$

Here, it is assumed that λ is the regularization parameter. m, is the dimension of the vector.

4 Metropolis Hastings

The Metropolis Hastings (MH) algorithm is one of the prominent and powerful methods in MCMC methods. The Metropolis Hastings method employed in this work is detailed as follows:

1. Initialize $\rho^{(0)}$.
2. For i = 0: $N_{mcmc} - 1$ Sample ρ^* using the proposal density $q(\rho|\rho^{(i)})$
3. Sample $\theta \sim \theta(0,1)$
4. Compute $A = \min\left(1, \frac{\pi(\rho^{(*)}|Y)}{\pi(\rho^{(i)}|Y)}\right)$ considering q as a symmetric distribution.
5. If $\theta < A \quad \rho^{(i+1)} = \rho^{(*)}$
   Else $\quad \rho^{(i+1)} = \rho^{(i)}$
   As the Markov chain converges, the recorded samples form the target posterior distribution function.

5 Experimental Results

In this section the efficiency and the accuracy of the Bayesian inference approach are illustrated by a numerical example. The following functions and values have been considered, in the numerical experiment.

$$t_f = 1 \quad g(x) = x^2 + 1 \quad h_0(t) = t(2t+1) \quad h_1(t) = 2 + 2t(2t+1)$$

The problem is solved using MATLAB software. The length of the chain or the number of samples is taken to be 10,000. The value of h and τ are considered to be 0.1, by taking M and N as 10. λ, The regularization parameter is taken as 10.

In this example the Robin coefficient is defined as a smooth function $\rho(t) = t$. Figure 1 shows the markov chain process. The initial guess of the Robin coefficient is 0.1 and eventually it reached the solution within 500 samples. The advantage of the Bayesian framework along with Metropolis Hastings algorithm is, irrespective of the initial guess the markov process converges to the solution. Totally 10,000 samples have been considered to obtain better mean and MAP estimates. The posterior probability density function is plotted against the 10,000 samples and it is shown in Fig. 2.

Similarly, or the same functions taken above in the example and all the other parameters being the same except by varying the measurement data of the temperatures Y, the following plots can be obtained.

Plots corresponding to the value of measurement data at Robin coefficient = 0.4 has been shown in Figs. 3 and 4. An initial guess of 0.1 is assumed for the Robin coefficient and it is clearly seen in the Fig. 3 that the solution is reached within few number of samples.

Similarly, plots corresponding to the value of measurement data at Robin coefficient = 0.3 is shown in Figs. 5 and 6.

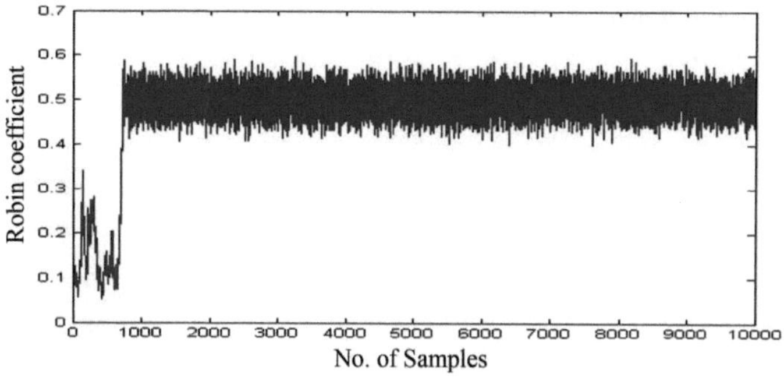

Fig. 1 Markov chain process of Robin coefficient

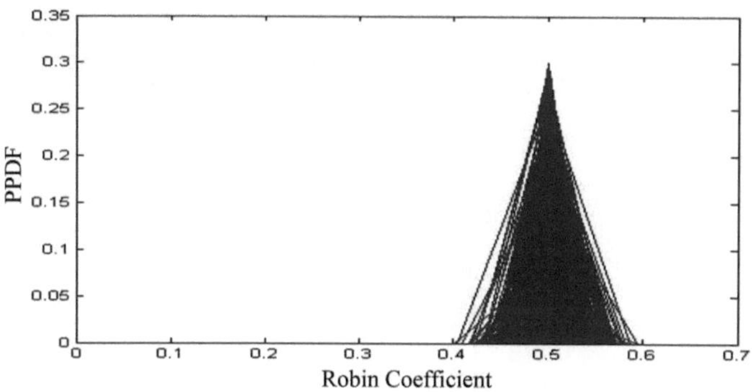

Fig. 2 PPDF of Robin coefficient

Fig. 3 Robin coefficient versus no. of samples

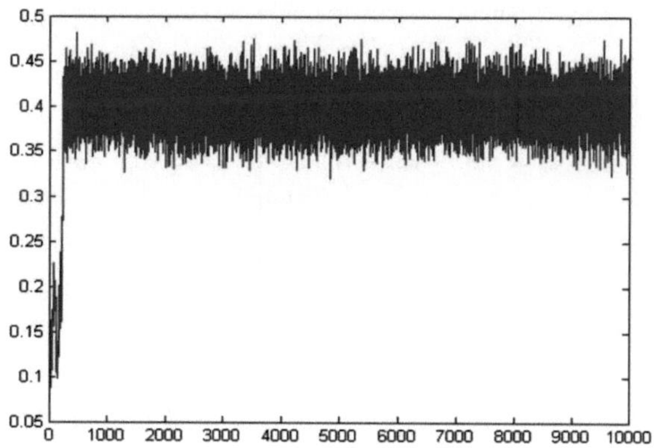

Fig. 4 PPDF versus Robin coefficient

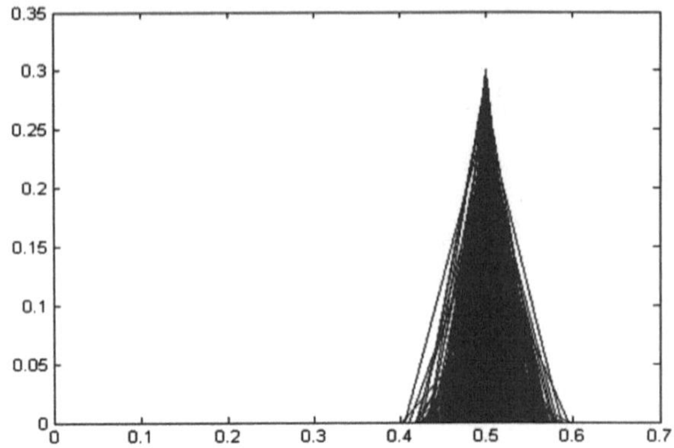

Fig. 5 Robin coefficient versus no. of samples

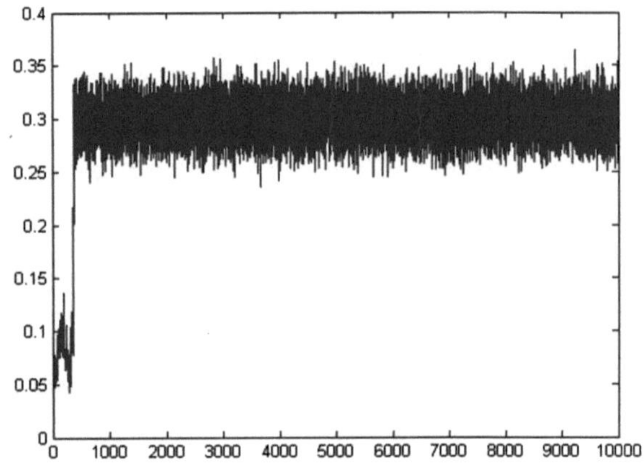

Fig. 6 PPDF versus Robin coefficient

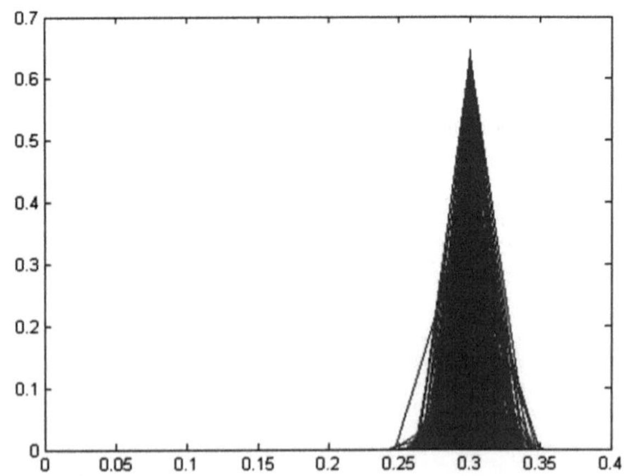

6 Future Work

In this paper the authors reinforced the work done by other researchers in the field of Bayesian inference there by determining the Robin coefficient of an inverse heat conduction problem. Hybrid Monte Carlo is a proven algorithm in many fields of science that it is better than Metropolis Hastings in convergence of the solution and in the acceptance ratio. The future scope of the present work is to employ HMC in a similar fashion for the inverse heat conduction problem and compare its performance with that of Metropolis Hastings.

7 Conclusions

Bayesian inference approach is employed, in this paper, using the Metropolis Hastings algorithm to identify a Robin coefficient in parabolic problems. 1-D transient conduction equation has been considered and convection boundary conditions were applied. Implicit scheme was incorporated to obtain the temperature distribution for the conduction problem. The Robin coefficient was estimated efficiently using Metropolis-Hastings algorithm. The efficiency of the Bayesian framework successfully tested for different initial guess and satisfactory results have been obtained. The results prove that the Bayesian inference approach can be used to obtain an accurate solution to the parametric estimation problems.

References

Gnanasekaran, N., Balaji, C.: A Bayesian approach for the simultaneous estimation of surface heat transfer coefficient and thermal conductivity from steady state experiments on fins. Int. J. Heat and Mass Transf. **54**, 3060–3068 (2011)

Gnanasekaran, N., Balaji, C.: Markov Chain Monte Carlo (MCMC) approach for the determination of thermal diffusivity using transient fin heat transfer experiments. Int. J. Therm. Sci. **63**, 46–54 (2013)

Masson, P.L., Loulou, T., Artioukhine, E.: Estimation of a 2D convection heat transfer coefficient during a test: comparison between two methods and experimental validation. Inverse Prob. Sci. Eng. **12**, 595–617 (2004)

Onyango, T.T.M., Ingham, D.B., Lesnic, D., Slodicka, M.: Determination of a time-dependent heat transfer coefficient from non-standard boundary measurements. Math. Comput. Simul. **79**, 1577–1584 (2009)

Ozisik, M.N., Orlande, H.R.B.: Inverse Heat Transfer: Fundamentals and Applications. Taylor and Francis, New York (2000)

Parthasarathy, S., Balaji, C.: Estimation of parameters in multi-mode heat transfer problems using Bayesian inference-effect of noise and a prior. Int. J. Heat Mass Transf. **51**, 2313–2334 (2008)

Reddy, B.K., Gnanasekaran, N., Balaji, C.: Estimation of thermo-physical and transport properties with Bayesian inference using transient liquid crystal thermography experiments. J. Phys. Conf. Ser. **395**(1), 012082 (2012)

Yang, F., Yan, L., Wei, T.: The identification of a Robin coefficient by a conjugate gradient method. Int. J. Numer. Methods Eng. **78**, 800–816 (2009)

Modeling and Simulation of Non-catalytic Transesterification of Cottonseed Oil

Pradeep Shende and Raju Mankar

List of symbols

CSO	Cottonseed oil
DG	Diglyceride
GL	Glycerol
k_{i0}	Initial rate constant of ith component
k_i	Rate constant of ith component
ME	Methyl esters of fatty acids
$MeOH$	Methanol
MG	Monoglyceride
TG	Triglyceride
C_{TG}	Concentration of triglyceride
C_{DG}	Concentration of diglyceride
C_{MG}	Concentration of monoglyceride
C_{GL}	Concentration of glycerol
C_{MeOH}	Concentration of methanol
C_{ME}	Concentration of methyl ester

1 Introduction

Transesteification is a fundamental reaction in the synthesis of various oleo-chemicals. The reaction involves the transesterification of triglyceride molecules present in the vegetable oil with the alcohol. The various alcohols were used

P. Shende (✉) · R. Mankar
Department of Chemical Engineering, Laxminarayan Institute of Technology,
Rashtrasant Tukadoji Maharaj Nagpur University, Nagpur 440033, Maharashtra, India
e-mail: pgshende@yahoo.com

R. Mankar
e-mail: rajumankar@gmail.com

I. Regupathi et al. (eds.), *Recent Advances in Chemical Engineering*,
DOI 10.1007/978-981-10-1633-2_33

in the synthesis of alcohol esters of fatty acids. The most widely used alcohol is methanol. Now days, the methyl esters of fatty acids are gaining the important substitute of diesel, because of its several advantages over the diesel obtained from the crude oil.

transesterification reaction between the triglycerides and alcohol is usually carried out with or without catalysts. The catalyst used in the transesterification reactions may be acids or bases types. The kinetics of the transesterification reaction largely depends on the type and amount of catalyst, alcohol (methanol, ethanol, propanol, or butanol) to triglyceride molar ratio, oil composition and temperature.

Mechanism of the transesterification of vegetable oil using base type catalyst includes KOH or NaOH are well studied due to distinct advantage of faster reaction rates. But, during the transesterification reaction with the base catalyst, the formation of soap due to saponification reaction and water accumulation disturbs the equilibrium of transesterification reaction and results in less conversion. In addition, difficulties are also faced during the removal of the catalyst from the product. Several studies on the transesterification of various oils using base catalysts were carried out by various researchers, Freedman et al. (1986), Darnoko and Cheryan (2000), Karmee et al. (2006), Doell et al. (2008).

Non-catalytic alcoholysis (Dasari et al. 2003) kinetics of soyabean oil with different alcohols such as methanol, ethanol and isopropanol was studied and reported the effect of temperature on the % conversion. Similar study was reported by Diasakou et al. (1998) on soyabean oil with methanol.

A batch reactor with a two phase liquid-liquid system was used to study kinetics of methanolysis of rapeseed oil by Islam and Bundy (2012) and reported that the various operating parameters such as concentration of initial substance, temperature and stirring speed have significant effect on transesterification. The different reactor system for the better yield of bio-diesel by the transesterification (Stiefel and Dassori 2009) of soyabean and rapeseed oil is also reported. The transesterification, a stepwise and reversible reaction, Noureddini and Zhu (1997) of soyabean oil with methanol was also investigated.

During this study, the kinetic model of transesterification process of cottonseed oil (CSO) was selected for developing the code for solving a set of differential equations involved in the transesterification kinetics using Visual Basic programming. Euler's Method with initial values was used for solving the set of differential equations. An attempt was also made to study the effect of the different parameters on the transesterification reaction and correlate the results with the data published in open literature.

2 Mathematical Model

Usually, the transesterification reaction of vegetable oil is considered as per the reaction scheme mentioned below. The reaction is usually follows in the stepwise manner, where the triglyceride (TG) reacts with methanol to produce diglyceride

(DG) which further reacts with methanol to produce monoglyceride (MG). In the final step, the monoglyceride reacts with methanol to give glycerol (GL) and methyl esters of fatty acid (ME).

$$TG + MeOH \underset{k_2}{\overset{k_1}{\longleftrightarrow}} DG + ME$$

$$DG + MeOH \underset{k_4}{\overset{k_3}{\longleftrightarrow}} MG + ME$$

$$MG + MeOH \underset{k_6}{\overset{k_5}{\longleftrightarrow}} GL + ME$$

The overall reaction is:

$$TG + 3MeOH \longleftrightarrow GL + 3ME$$

These reactions can be either equilibrium reactions or forward reactions depending on the working conditions and type of catalyst (if any is involved) and molar ratio of oil:alcohol.

The proposed mechanism assumes homogenous non-catalytic reactions such that each reaction step is reversible and kinetics is first order with respect to the concentration of the reacting components. The kinetic equations involved in the transesterification process on the basis of above mechanism and model assumptions are as follows:

$$\frac{dC_{TG}}{dt} = -k_1 * C_{TG}(t) * C_{MeOH}(t) + k_2 * C_{DG}(t) * C_{ME}(t) \quad (1)$$

$$\frac{dC_{DG}}{dt} = k_1 * C_{TG}(t) * C_{MeOH}(t) - k_2 * C_{DG}(t) * C_{ME}(t) - k_3 * C_{DG}(t) * C_{MeOH}(t) + k_4 * C_{MG}(t) * C_{BD}(t) \quad (2)$$

$$\frac{dC_{MG}}{dt} = k_3 * C_{DG}(t) * C_{MeOH}(t) - k_4 * C_{MG}(t) * C_{ME}(t) - k_5 * C_{MG}(t) * C_{MeOH}(t) + k_6 * C_{GL}(t) * C_{BD}(t) \quad (3)$$

$$\frac{dC_{GL}}{dt} = k_3 * C_{MG}(t) * C_{MeOH}(t) + k_6 * C_{GL}(t) * C_{ME}(t) \quad (4)$$

$$\frac{dC_{ME}}{dt} = -3\langle\frac{dC_{TG}}{dt}\rangle - 2\langle\frac{dC_{DG}}{dt}\rangle - 1\langle\frac{dC_{MG}}{dt}\rangle \quad (5)$$

$$\frac{dC_{MeOH}}{dt} = -\frac{dC_{ME}}{dt} \tag{6}$$

The rate constants in the forward and reverse reactions are written as follows:

$$k_i = k_{i,o} \cdot e^{\left(\frac{-E_a}{RT}\right)}. \tag{7}$$

These model equations reported in the literature (Wenzel et al. 2006) were used for the study the kinetics of transesterification of cottonseed oil using methanol under non-catalytic condition by varying the operating parameters.

The transformation of system of ordinary differential equations in component concentration is written as a mass ratio i.e. $X_{\text{Component}}$ (kg component/kg ester).

3 Results and Discussion

Kinetics of transesterification of cottonseed oil was studied by developing the computer code based on Euler's initial value method for solving differential equation using Visual Basic programming. The Fig. 1 shows the form used for entering various constants and variables involved during this study. The kinetic rate constant for forward and backward reactions were suggested by the numerous researchers and are available in open literature. In this study, kinetic equations were solved by using the rate constants suggested by the Wenzel et al. (2006). Other

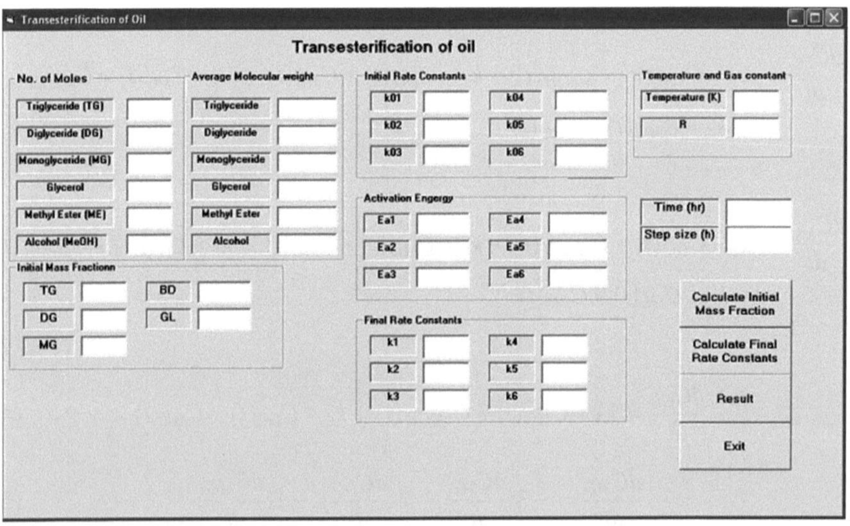

Fig. 1 Visual basic form for entering the data for the study of kinetics of transesterification of CSO

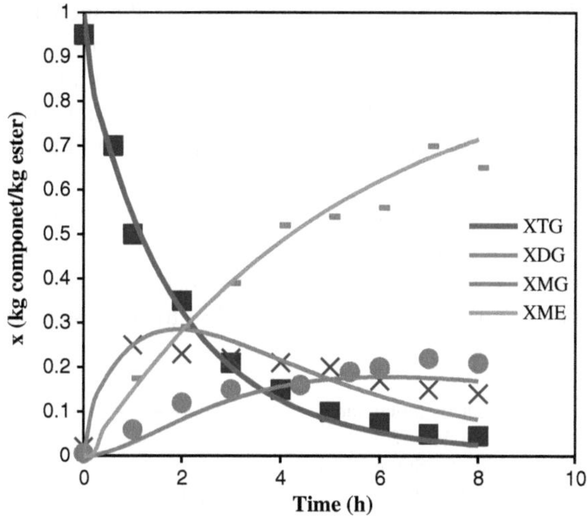

Fig. 2 Simulation results for non-catalytic transesterification of CSO at reaction conditions i.e. temperature: 220 °C and molar ratio: TG:MeOH: 1:27

parameters such as average molecular weight of TG, DG, and MG are taken from (Bailey 2005). Also, the effect of the operating parameters such as temperature and molar ratio of CSO:MeOH on the conversion of TG to ME we studied.

The conversion data was so generated for non-catalytic transesterification. In this study, two different temperatures i.e. 220 and 235 °C under pressure were used. The change in the concentration profile of TG, DG, MG, GL and ME with respect to time for the reaction at 220 and 235 °C respectively at mole ratio TG:MeOH, 1:27 are shown in Figs. 2 and 3.

Simulation results in Figs. 2 and 3, show that there is continuous decrease in the concentration of TG with respect to the time approaching zero, while other

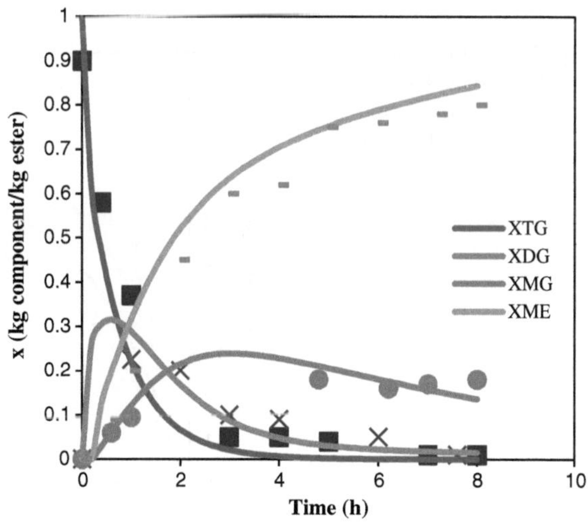

Fig. 3 Simulation results for non-catalytic transesterification of CSO at reaction conditions: temperature 235 °C, molar ratio TG:MeOH: 1:27

glycerides such as DG and MG shows the increase in concentration in the initial stage of the reaction and reaches a maximum value and then decrease slowly up to some appreciable value. Hence, some amount of DG and MG always remain in the final composition. Model simulation shown in Figs. 2 and 3, gives very good agreement between the simulation results and the data reported by Diasakou et al. (1998).

Simulations were also carried out to study the effect of change in the TG to Methanol concentration from 1.06 to 1.27 for temperatures 220 and 235 °C respectively. The results thus obtained are shown in Figs. 4 and 5. As seen from these figures, the increase in yield of methyl ester (ME) from 72 to 97 % at the end

Fig. 4 Effect of molar ratio on the conversion of Triglyceride of CSO to ME at temperature 220 °C

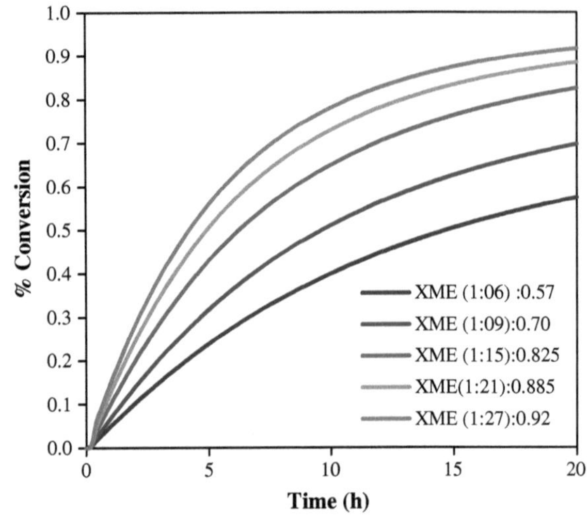

Fig. 5 Effect of molar ratio on the conversion of Triglyceride of CSO to ME ester at temperature 235 °C

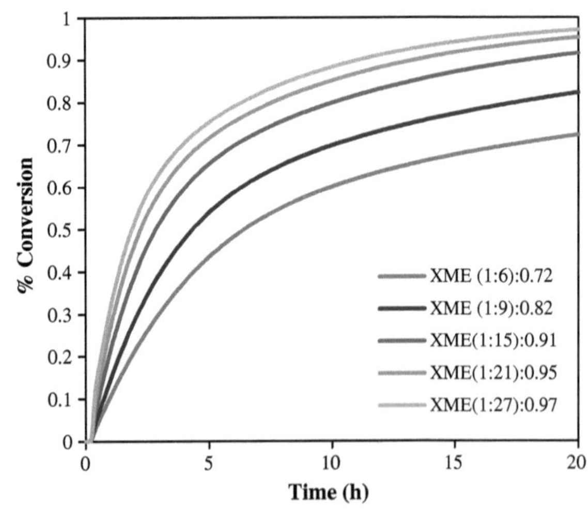

of 20 h for the ratio Triglyceride:Methanol of 1.27 at a temperature of 235 °C, which is in good agreement with experimental data reported by the Diasakou et al. (1998).

4 Conclusions

In this study, a Visual Basic program was developed using the Euler's method to solve the set of differential equations involved in the kinetic model of transesterification of cottonseed oil. The simulation results can be obtained easily for similar transesterification kinetics for various temperatures and different molar ratios of Triglyceride: Methanol. These results can be utilized to obtain optimal conditions of temperature and molar ratios which should be used to achieve maximum possible conversion in minimum amount of time. The parametric sensitivity studies can also be easily carried out using this program for various similar systems with some adaptations.

References

Bailey, A.E.: Bailey's Industrial Oils and Fats Product. Shahidi, F. (Ed.), Vol. 2. Wiley-Interscience, New York (2005)

Darnoko, D., Cheryan, M.: Kinetics of palm oil transesterification in a batch reactor. J. Am. Oil Chem. Soc. **77**, 1263–1266 (2000)

Dasari, M.A., Goff, M.J., Suppes, Gj: Noncatalytic alcoholysis kinetics of soybean oil. J. Am. Oil Chem. Soc. **80**, 189–192 (2003)

Diasakou, M., Louloudi, A., Papayannakos, N.: Kinetics of the non-catalytic transesterification of soyabean oil. Fuel **77**, 1297–1302 (1998)

Doell, R., Konar, K.S., Boocock, D.G.B.: Kinetic parameters of a homogenous transmethylation of Soyabean oil. J. Am. Oil Chem. Soc. **85**, 271–276 (2008)

Freedman, B., Butterfield, R.O., Pryde, E.H.: Transesterification kinetics of SoybeanOil. J Am. Oil Chem. Soc. **63**, 1375–1380 (1986)

Islam, Shahidul, Md, Bundy, C.: Kinetics of rapeseed oil methanolysis in presence of KOH catalyst–studied with gas chromatography. Int. J. Sci. Technol. **2**, 762–768 (2012)

Karmee, S.K., Chandna, D., Ravi, R., Chadha, A.: Kinetics of base-catalyzed transesterification of triglyceride from Pongamia Oil. J. Am. Oil Chem. Soc. **83**, 873–877 (2006)

Noureddini, H., Zhu, D.: Kinetics of transesterification of Soyabean Oil. J. Am. Oil Chem. Soc. **74**, 1457–1463 (1997)

Stiefel, S., Dassori, G.: Simulation of biodiesel production through transesterification of Vegetable Oils. Ind. Eng. Chem. Res. **48**, 1068–1071 (2009)

Wenzel, B., Tait, M., Modenes, A., Krounmu, A.: Modelling chemical kinetics of Soyabean Oil transesterification process for biodiesel production: an analysis of molar ratio between Alcohol and Soyabean Oil temperature changes on the process conversion rate. Bioautomation **5**, 13–22 (2006)

Simulation of Hydroperoxide Process for the Production of Propylene Oxide

V. Vaishali and P.R. Naren

1 Introduction

Propylene Oxide (PO), also known as Epoxypropane is one of the significant organic chemical produced in chemical industries. It is a colourless low boiling liquid with a molecular formula C_3H_6O. It is used as a chemical intermediate. It finds application in the production of polyether polyols which in turn is used to manufacture polyurethane. It also finds use in the production of propylene glycol and propylene glycol ethers. PO can be produced in various methods including Chlorohydrin, Hydroperoxide, Hydrogen peroxide and Cumene processes. Table 1 gives a comparison between various production processes.

2 Selection of Process for Present Work

Production of PO along with styrene as co-product using Ethylbenzene hydroperoxide (POSM) is chosen for the present work. Chlorohydrin process is slowly replaced by other alternative processes because of large disposal problems. Peroxide process involves a large amount of by-product production, either styrene or methyl tert-butyl' ether. So profit depends on the demand of co-product. Owing to the fact that use of

V. Vaishali · P.R. Naren (✉)
School of Chemical & Biotechnology, SASTRA University, Thanjavur 613401, India
e-mail: prnaren@scbt.sastra.edu

V. Vaishali
e-mail: vaishvenkat12@gmail.com

I. Regupathi et al. (eds.), *Recent Advances in Chemical Engineering*,
DOI 10.1007/978-981-10-1633-2_34

Table 1 Comparison of various production processes of PO (Kirk-Othmer 2001)

	Chlorohydrin	Peroxide	H_2O_2	Cumene
Raw materials	Propylene, Chlorine, H_2O, Ca $(OH)_2$	Propylene, ethylbenzene/isobutane, O_2	Propylene, Hydrogen peroxide, methanol	Propylene, cumene, O_2
By-products	$CaCl_2$	Styrene monomer/tert-butanol	Nil	Nil
Catalyst used	Nil	Molybdenum/Ti–Si	Titanium silicate (TS-1)	Ti–Si epoxidation catalyst
Selectivity to PO	87–90 %	95 %	>95 %	99 %

methyl tert-butyl ether as a fuel additive is banned, companies in recent times prefer POSM process (Nijhuis et al. 2006). Hence *POSM process is chosen for the current work.*

3 Process Description

The detailed process flow diagram for Propylene Oxide production is shown in Fig. 1. The first step in the production of PO involves oxidation of Ethylbenzene to Ethylbenzene hydroperoxide (Eq. (1)) in liquid phase. A continuous stirred tank reactor is used for the air oxidation process. Oxidation takes place at 140–150 °C and 206–275 kPa (Kirk-Othmer 2001). Selectivity to by-products include 8–10 %, 5–7 % and <1 % of acetophenone, 1-phenyl ethanol, and organic acids respectively (Kirk-Othmer 2001). The pressure of liquid product from the CSTR is reduced using pressure reducing valve and then sent to a flash column. Unreacted Ethylbenzene is recovered and the peroxide formed is sent to the next reactor. Propylene and Ethylbenzene hydroperoxide (EBHP) reacts in a fixed bed reactor to give Propylene Oxide and 1-Phenylethanol (Eq. (2)). Epoxidation takes place at 95–130 °C and 2500–4000 kPa (Kirk-Othmer 2001). Ti(IV)/SiO_2 heterogeneous catalyst is used for the reaction. A conversion of 95–99 % of EBHP is achieved and selectivity to PO is 92–96 % (Kirk-Othmer 2001). The mixture is sent to a flash drum in which propylene is recovered. This is followed by a distillation column where the top product is Propylene Oxide. Bottom product is sent to the next column in which Ethylbenzene is recovered. The current work is limited to PO production, 1-phenyl ethanol dehydration to styrene is not discussed. The co-product is produced in the range of 2.4:1 styrene/propylene oxide.

Peroxidation

$$\underset{\text{Ethylbenzene}}{C_6H_5CH_2CH_3(1)} + \underset{\text{Oxygen}}{O_2(1)} \rightarrow \underset{\text{Ethylbenzene Hydroperoxide}}{C_6H_5CH(OOH)CH_3(1)} \qquad \Delta H_r^0 = -96.884\ \text{kJ/mol} \tag{1}$$

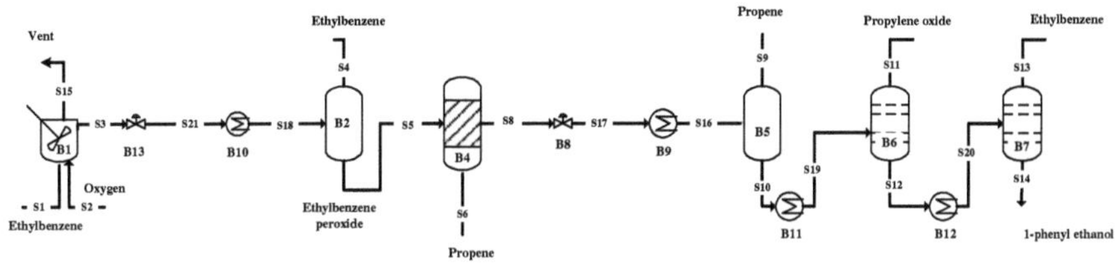

Block No.	Block Name	Block No.	Block Name
B1	CSTR	B7	Separator 4
B2	Separator 1	B8	Pressure reducer 2
B4	Fixed Bed Reactor	B9	Heat exchanger 1
B5	Separator 2	B10	Heat exchanger 2
B6	Separator 3	B11	Heat exchanger 3
B13	Pressure reducer 1	B12	Heat exchanger 4

Fig. 1 Process flow diagram for Propylene Oxide manufacture

Epoxidation

$$\underset{\text{Ethylbenzene}}{C_6H_5CH(OOH)CH_3(g)} + \underset{\text{Propylene}}{CH_3CHCH_2(g)} \xrightarrow{\text{Ti(IV)/SiO}_2}$$

$$\underset{\text{1-Phenylethanol}}{C_6H_5CH(OH)CH_3(g)} + \underset{\text{Propylene Oxide}}{CH_3CHO\ CH_2(g)} \qquad (2)$$

$$\Delta H_r^0 = -198.739\ \text{kJ/mol}$$

4 Material and Energy Balance

Basis chosen for present work is 100 kmol/h of PO. To begin with, by-products were ignored during material and energy balance calculations. List of equipments required were analysed and the thermophysical properties were taken from standard sources (Aspen Plus V8.6, Poling et al. 2001; Yaws 2004; Sinnott 2005). Group contribution method (Joback Group Contribution) was used for compounds (EBHP and 1-Phenylethanol), wherever physical and thermodynamic data were not available. Theoretical yield was calculated based on stoichiometry. Yield for single pass conversion (15 %) was very low (3 %), which indicated that either single CSTR with larger residence time (or larger volume)/series of CSTRs with each 15 % conversion/recycle streams were used in industrial scale in order to increase yield. Residence time for 15 % conversion was available in literature (Kirk-Othmer 2001). Assuming first order kinetics, rate constant at reaction temperature was calculated to be 0.0888 h^{-1}. Using CSTR performance equation, residence time for various conversions was found. Conversion of 80 % (residence time 45 h) was assumed to perform material and energy balance calculations for the entire process. Propylene oxide yield was calculated to be 20 % at 80 % conversion of Ethylbenene. Aspen Plus V8.6 was used to determine operating conditions (temperature and pressure) in flash drum and distillation column which in turn was used to fix separation efficiency of each column. Separation efficiency was used in spreadsheet to determine vapour and liquid flowrate in distillation column and flash drum. From literature (Kirk-Othmer 2001) it was found that propylene should be fed in excess to the fixed bed reactor in order to achieve 99 % conversion. Hence Propylene inlet flowrate was assumed to be 1.5 times that of EBHP. Block wise results for material and energy balance is shown in Table 2.

5 Design of Fixed Bed Reactor and Multicomponent Distillation Column

5.1 *Fixed Bed Reactor*

In the production of Propylene Oxide, Epoxidation reaction is carried out adiabatically in a two phase (gas-solid) fixed bed reactor using Ti(IV)SiO_2 as catalyst. Rate constants for pseudo first order reaction were obtained from literature (Oyama 2008). Plug flow for the gas phase in the fixed bed reactor was assumed. The reaction kinetics was assumed to be first order. Only forward reaction was considered as prior calculation indicated no significant backward reaction. Uniform catalyst hold-up of 0.6 was assumed to compute the reactor volume. *Scilab 5. 5. 1* was used to solve simultaneous ODEs (Eq. (3–4)) (Fogler 2006; Levenspiel 2006). Temperature and conversion versus volume of the reactor are shown in Fig. 2. It is observed that a volume of 4.4 m^3 was required to achieve the desired conversion of 99 % of EBHP.

$$\frac{dX}{dV} = \frac{-r_A}{F_{A0}} \tag{3}$$

$$\frac{dT}{dV} = \frac{-r_A(-\Delta H_r)}{F_{A0}(\sum C_{P_i} + (\Delta C_P X))} \tag{4}$$

5.2 *Multicomponent Distillation Column*

Short cut method for Multicomponent Distillation Design using *Fenske-Underwood-Gilliland Correlation* was used in the present work (Ludwig 1999). Limits were set between two key components (Propylene Oxide and Ethylbenzene) to specify separation between top and bottom products. Propylene and 1-phenylethanol are non-distributed components. Conditions for weeping and entrainment were checked and were found to be appropriate. Results are shown in Table 3.

Table 2 Blockwise material and energy balance for PO production

Block number	Block name	Total mass flow rate (kg/h)		Total molar flow rate (kmol/h)		Total volumetric flow rate (m^3/h)		Enthalpy (kJ/h)	
		In	Out	In	Out	In	Out	In	Out
B1	CSTR	23235.581	23235.581	336.748	202.049	4197.299	856.768	0.000	4517416.701
B2	Separator 1	22157.989	22157.989	168.374	168.374	21.425	2946.181	7613135.211	7613135.211
B4	Fixed bed reactor	23129.264	23129.264	285.585	285.585	3899.780	10643.763	8244521.568	30288349.945
B5	Separator 2	23129.264	23129.264	285.585	285.585	1342.686	1125.345	383042.306	383042.306
B6	Separator 3	21225.584	21225.584	243.899	243.899	595.117	595.117	349972.544	4757565.336
B7	Separator 4	14379.921	14379.921	119.102	119.102	14.184	14.227	4893031.963	5256730.634
B9	Heat exchanger 1	23129.264	23129.264	285.585	285.585	10643.763	1342.686	30288349.945	383042.306
B10	Heat exchanger 2	22157.989	22157.989	168.374	168.374	21.425	21.425	4389467.861	7613135.211
B11	Heat exchanger 3	21225.584	21225.584	243.899	243899	595.117	595.117	349972.544	349972.544
B12	Heat exchanger 4	14379.921	14379.921	119.102	119.102	14.184	14.184	4893031.963	4893031.963

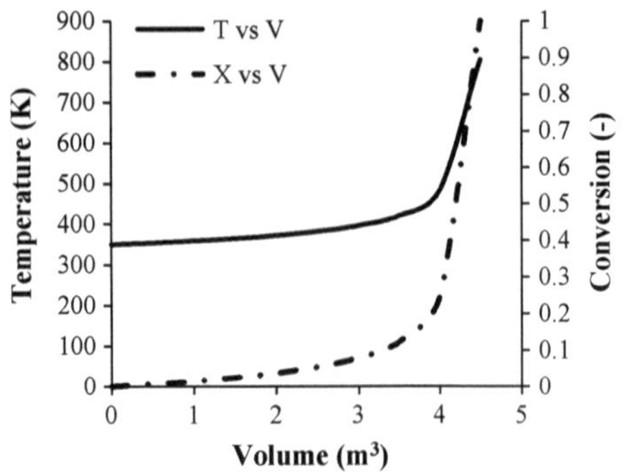

Fig. 2 Reactor temperature and EBHP conversion as function of volume of fixed bed reactor

Table 3 Multicomponent distillation design results

Parameter	Value
Number of ideal stages	9.8229
Number of actual stages	28.52
Height of column	14.76 m
Diameter of column	1.3 m
Net Area	1.168 m^2
Hole area	0.100 m^2
Active area	1.008 m^2
Downcomer area	0.159 m^2
Diameter of hole	5 mm
Number of holes	5137.6

6 Aspen Plus Simulation

The entire process flow sheet for PO production without recycle was simulated using *Aspen Plus V8.6*. Base property method was chosen as Ideal. Since accurate kinetic data were not available in literature for the two main reactions, namely Peroxidation and Epoxidation, to begin with stoichiometric reactors based on known fractional conversion were chosen. Two outlet flash that uses rigorous vapour-liquid equilibrium was chosen for flash operations. Distillation column that incorporates Winn-Underwood-Gilliland method was used for carrying out multi-component distillation. Pressure reducing valve based on adiabatic flash was used as pressure reducing equipment. Although simulation of entire PFD with complete recycle could not be simulated successfully due to convergence errors, simulation of PFD using one recycle stream was attempted successfully. Figure 3 shows the process flow diagram using Aspen Plus V8.6, wherein stoichiometric reactor was replaced by CSTR and stream S8 (vapour outlet from flash), essentially containing Ethylbenzene was recycled back to CSTR. Rate constant and residence time in CSTR were assigned a value of 2.45E−05 s^{-1} and 45.33 h. Table 4 compares results of Aspen Plus simulation with and without recycle.

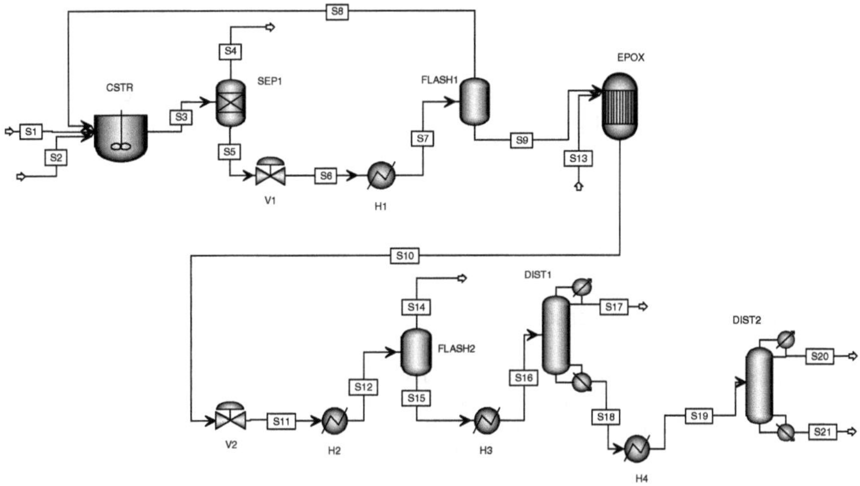

Fig. 3 Process flow diagram using Aspen Plus for PO production with recycle

Table 4 Comparison of Aspen Plus results for PO production with and without recycle

Stream No.	Parameter	With recycle	Without recycle
S7	Temperature (°C)	190.6	225
S17	Molar flowrate of PO (kmol/h)	120.82	100
S10	Mole fraction of propylene (−)	0.101	0.199
S14	Pressure (bar)	1.2	2

7 Effluent Treatment

Effluent treatment is necessary to minimize hazardous emissions into the atmosphere. Gaseous effluents are combusted to CO_2 and water (vapour) before releasing into the atmosphere. Recovery of the effluent stream is necessary since effluent stream chiefly consists of hydrocarbons. The effluent stream containing hydrocarbons can be sent to a fractionating column or flash drum where 99 % of the components are recovered. Solid waste in case of PO production through Hydroperoxide process is the catalyst used (Ti(IV)/SiO_2). Deactivated catalyst is replaced, incinerated to remove residual hydrocarbons, and disposed in a landfill. Liquid effluent consists of PO, Ethylbenzene, EBHP and 1-phenylethanol. The COD should be brought down to 250 mg/l before discharging into water bodies. *Environmental factor* is defined as the ratio of mass of waste generated to mass of desired product. Mass efficiency of a process can be judged by E Factor. The process is environment friendly if the E Factor is low. Values in the range of 0–0.1 is acceptable for bulk chemical industry. E factor with and without recovery was found to be 0.004 and 0.394 respectively for PO production. From the values of E Factor, it was concluded that it was essential to recover major hydrocarbons from the effluent stream.

8 Conclusion

Propylene oxide production using Hydroperoxide process was simulated for a plant capacity of 100 kmol/h of PO. The present work showcased use of spread sheet and Aspen Plus tools to aid process development and design of equipments. Aspen Plus (V 8.6, Aspen Tech., Inc.,) proved to be useful in finding optimum operating conditions for separating equipments. The process was found to be environmental friendly due to low value of E factor. Further work will involve extending the process to convert by-product 1-Phenylethanol to Styrene to further derive value added chemical from the process.

Acknowledgments One of the authors, V. Vaishali wishes to thank Dr. Gautham B Jegadeesan, Senior Assistant Professor, School of Chemical and Biotechnology, SASTRA University for his valuable suggestions on effluent treatment studies during the course of this project work. The simulations were performed on AspenONE University licence (SYS917047/917047) funded by SASTRA University.

References

Fogler, H.S.: Elements of Chemical Reaction Engineering, 4th edn. Pearson Prentice Hall, New Jersey (2006)

Kirk-Othmer, D.F.: Encyclopedia of Chemical Technology, 4th edn., Vol. 20. Wiley Interscience Publication (2001)

Levenspiel. O.: Chemical Reaction Engineering, 3rd edn., John Wiley and Sons, New York (2006)

Ludwig, E.E.: Applied Process Design for Chemical and Petrochemical Plant, 3rd edn., Vol. 1. Butterworth-Heinemann, USA (1999)

Nijhuis, T.A., Makkee, M., Moulijn, J.A., Weckhuysen, B.M.: The production of propylene oxide: catalytic processes and recent developments. Indian Eng. Chem. Res. **45**, 3447–3459 (2006)

Oyama, S.T.: Mechanism in Homogeneous and Heterogeneous Epoxidation Catalysis, pp. 355–386. Elsevier (2008)

Poling, B.E., Prausnitz, J.M., O'Connell, J.P.: The Properties of Gases and Liquids, 5th edn. McGraw-Hill, New York (2001)

Sinnott, R.K.: Coulson and Richardson‘s Chemical Engineering, 4th edn., Vol. 6. Elsevier Butterworth-Heinemann, USA (2005)

Yaws, C.L.: Yaws' Handbook of Thermodynamic and Physical Properties of Chemical Compounds (2004)

Simulation of Natural Convection in Diamond Shaped Cavity Filled with Air or Water

Dorca Polamuri and Sunil Kumar Thamida

1 Introduction

Granular materials have different thermal properties when compared with the properties of individual particles. The presence of interstitial fluid and particles' packing makes the difference. One of the important properties of granular material is effective thermal conductivity (Nguyen et al. 2009). There are a wide number of applications of granular material in thermal systems such as for heat recovery, insulation, and heat accumulation in silos. In many industrial processes such as powder processing, food technology, chemical reactor design, and powder metallurgy heat transfer in granular media occurs (Strumillo 1986). Heat transfer in granular materials occurs due to conduction through the particles, conduction through the fluid, radiation between particles and natural convection in the fluids between the particles. The effective thermal conductivity is also influenced by the contact resistance between the particles. The types of heat supply in general for heating granular material are hot air supply as a fluid in the voids, bottom wall heating and side wall heating. The density gradient of fluid in the voids due to temperature gradient causes natural convection in voids (Mamun et al. 2003; Roy and Basak 2005; Sathiyamoorthy et al. 2007). Void fraction of granular material depends on the type of packing used, whether random or regular. For simulation purpose, the geometry of the void space in granular material depends on the type of assumed packing. Nevertheless a diamond shaped cavity (a squared turned by 45°) in 2D can be taken as a representative void geometry.

D. Polamuri · S.K. Thamida (✉)
Department of Chemical Engineering, National Institute of Technology Warangal, Telangana State 506004, India
e-mail: sunil76@nitw.ac.in

D. Polamuri
e-mail: angelchandrika@gmail.com

I. Regupathi et al. (eds.), *Recent Advances in Chemical Engineering*,
DOI 10.1007/978-981-10-1633-2_35

Side wall heating leads to temperature difference across the voids. It results in natural convection. When bottom wall is heated, it is Rayleigh-Bernard convection. The latter is caused by instability (Chandrasekhar et al. 1981) whereas side wall heating causes natural convection phenomenon. Both types of convection or flow are due to buoyancy effect in the fluid. The buoyancy effect is brought in by assuming that the density of fluid decreases with temperature as per Boussinesq approximation of linear decrease with increase in temperature.

In this paper, a diamond shaped cavity filled with air or water is studied for characterising natural convection with side wall heating. By keeping the side walls temperatures unchanged, only the gravity direction is changed by 90° to study Rayleigh-Bernard convection.

2 Mathematical Modelling

2.1 *Governing Equations*

The schematic diagram of a void or cavity is shown in Fig. 1a, b. There are four possible directions for gravity namely downward, upward towards right and towards left.

The heat transfer and fluid flow have to be solved simultaneously (Holman and Bhattacharyya 2011) using following Eqs. (1)–(3).

Continuity equation:

$$\frac{\partial \rho}{\partial t} + \nabla \cdot (\rho \boldsymbol{u}) = 0 \tag{1}$$

Navier-Stokes equations for Newtonian fluid:

$$\rho \frac{\partial \boldsymbol{u}}{\partial t} + \rho(\boldsymbol{u} \cdot \nabla)\boldsymbol{u} = -\nabla p + \mu \nabla^2 \boldsymbol{u} + \rho \boldsymbol{g} \tag{2}$$

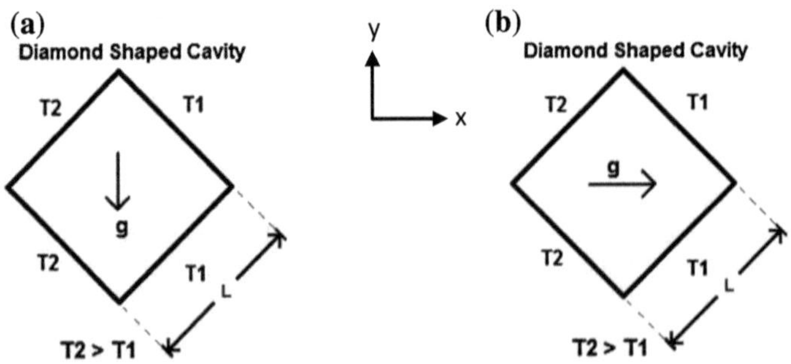

Fig. 1 Schematic diagram of diamond shaped cavity having different direction of gravity, **a** gravity in *vertical* direction **b** gravity in *horizontal* direction

Heat conduction-convection-generation equation

$$\rho c_p\left[\frac{\partial T}{\partial t}+\boldsymbol{u}\cdot\nabla T\right]=k\nabla^2 T+s_0 \tag{3}$$

Here $s_0 = 0$ since there is no volumetric heat generation in the fluid. Equations (1)–(3) are reformulated as in two dimensions as below:

Navier-Stokes equations with Boussinesq approximation at steady state

$$\rho\left[u_x\frac{\partial u_x}{\partial x}+u_y\frac{\partial u_x}{\partial y}\right]=-\frac{\partial p}{\partial x}+\mu\left[\frac{\partial^2 u_x}{\partial x^2}+\frac{\partial^2 u_x}{\partial y^2}\right]+\rho_o(1-\beta(T-T_{avg}))g_x \tag{4}$$

$$\rho\left[u_x\frac{\partial u_y}{\partial x}+u_y\frac{\partial u_y}{\partial y}\right]=-\frac{\partial p}{\partial y}+\mu\left[\frac{\partial^2 u_y}{\partial x^2}+\frac{\partial^2 u_y}{\partial y^2}\right]+\rho_o(1-\beta(T-T_{avg}))g_y \tag{5}$$

Heat conduction-convection equation:

$$\rho c_p\left[u_x\frac{\partial T}{\partial x}+u_y\frac{\partial T}{\partial y}\right]=k\left[\frac{\partial^2 T}{\partial x^2}+\frac{\partial^2 T}{\partial y^2}\right] \tag{6}$$

Here $\boldsymbol{u}$ the velocity vector has components u_x in the x direction and u_y in the y direction. T is the local fluid temperature, ρ is the density, g is the acceleration due to gravity in x or y directions, k is the thermal conductivity, p is the local pressure, μ is the viscosity, and c_p is the specific heat capacity. At steady state, the maximum velocity (V_{max}) magnitude anywhere in the fluid domain is sought for from the solution to the above set of equations. It would indicate the strength of natural convection generated. The important dimensionless number that comes out from Eq. (5) is Grashof number defined as

$$Gr=\frac{g\beta(T_{w1}-T_{w2})L^3}{\vartheta^2} \tag{7}$$

where g is the acceleration due to gravity = 9.8 m/s^2, β is the coefficient of thermal expansion of the fluid $\beta=-\frac{1}{\rho}\frac{\partial\rho}{\partial T}$, $T_{w1} = T_2$ is the left hand side wall temperature and $T_{w2} = T_1$ is the right hand side wall temperature as in Fig. 1, L is the length of the side of the square cavity in m, and ϑ is the kinematic viscosity in m^2/s. The study is to change Gr and to obtain V_{max}, which is achieved by altering the length of the side of the square cavity. The boundary condition for velocity is no slip condition on all the four walls of the cavity. For heat transfer, they are dirichlet boundary conditions since constant wall temperatures are assigned. These boundary conditions are unique as compared to other studies of literature (Roy and Basak 2005; Sathiyamoorthy et al. 2007). Obtaining an analytical answer for T(x, y), u_x(x, y) and u_y(x, y) is difficult hence simulation tool such as COMSOL Multiphysics (www.comsol.com).

COMSOL is used for obtaining the solution to velocity field and temperature field and various post processed results.

2.2 Simulation Procedure

The steps involved in simulating both the heat transfer and fluid flow simultaneously in COMSOL Multiphysics are: creation of the geometry of required dimension (in this study we considered a square rotated by 45° with 'L' as length of the side), adding physics to the geometry, selecting heat transfer and laminar flow with stationary study. The initial temperature for heat transfer is assigned as constant throughout the fluid domain where as in laminar flow the fluid is assumed to be at rest and fluid properties are provided by the user. The boundary conditions used are no slip boundary condition in laminar flow and the temperature of left side wall (T_{w1}) and right side wall (T_{w2}). Here $T_{w1} > T_{w2}$. The type of mesh used is 'fine mesh' so that the simulation approaches accurate solution without much discontinuity. Then computed solutions for temperature and velocity field are post processed to obtain results for analysis. The maximum velocity as a function of time in the fluid is obtained from the solution and tabulated. The solver technique used by COMSOL for solving Navier-Stokes and heat transfer equations is finite element method. The following thermal and flow properties of air and water are used in the simulation (Table 1).

3 Simulation Results

The intensity of natural convection is obtained from a measure of maximum absolute velocity anywhere in the fluid domain. The obtained V_{max} as a function of Grashof number Gr is plotted in Fig. 2a for air as medium and Fig. 2b for water as medium and for all four possible directions of gravity.

Table 1 Thermal Properties used in COMSOL simulation

Properties	Air	Water
Thermal conductivity, W/m K	0.0261	0.6137
Specific heat, J/kg K	1009.65	4182
Density, kg/m^3	1.3904	995.81
Kinematic viscosity, m^2/s	1.819×10^{-5}	0.9334×10^{-6}
Volume force, N/m^3 (T in K)	$9.8 \times ((-0.006 \times T) + 3.195)$	$9.8 \times ((-0.2611 \times T) + 1002.8)$

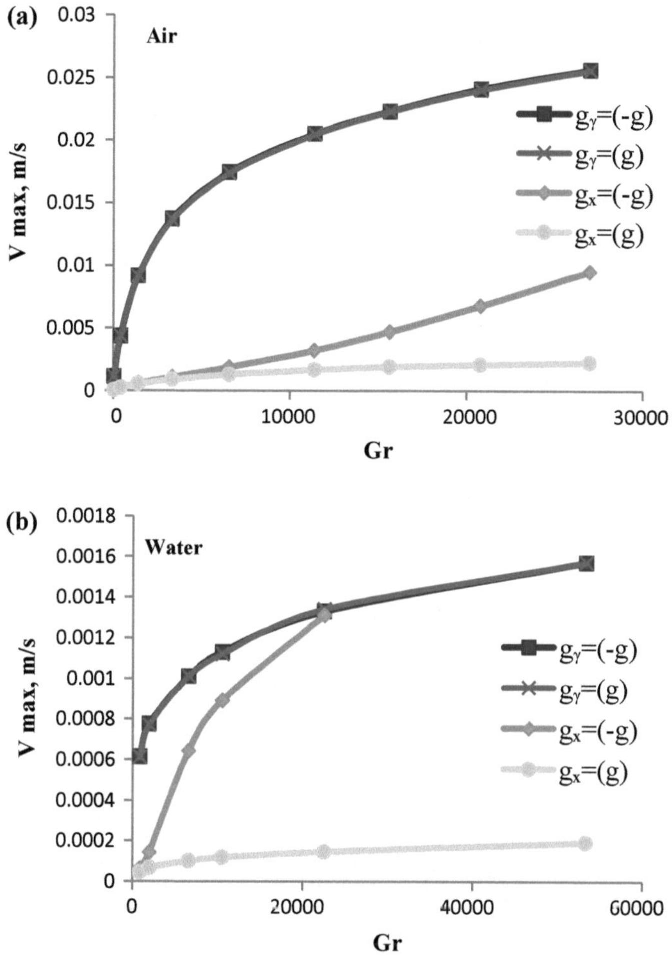

Fig. 2 Natural convection intensity for various media in the cavity: **a** air and **b** water

It can be noticed that for gravity in vertical direction the natural convection intensity is higher compared to gravity in horizontal direction. For gravity in either upward or downward direction the same results is obtained because of mirror effect. The direction of velocity would be reserved but the absolute value will be same. For gravity in horizontal direction, the negative 'x' direction implies Rayleigh-Bernard convection (instability possible) whereas for positive 'x' direction of gravity only a small (negligible) convection occurs, since the buoyancy effect doesn't yield to instability in this situation. The trend is similar for air and water (V_{max} vs. Gr) but the magnitude is higher for air as medium than for water as medium, the reason could be higher kinematic viscosity 10^{-5} m^2/s for air than water 10^{-6} m^2/s. The patterns of the velocity magnitude depicting the circulations caused by natural convection are shown in Fig. 3 for air and Fig. 4 for water for various void sizes and direction of gravity.

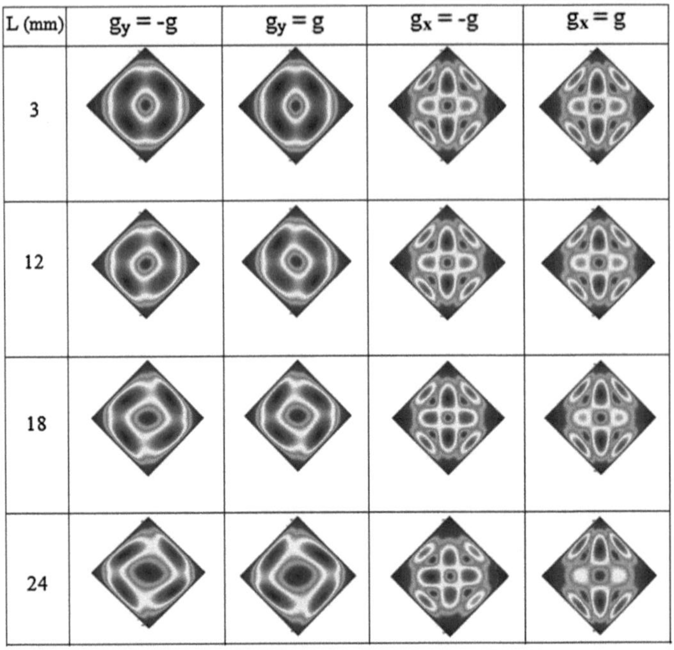

Fig. 3 Patterns of velocity magnitude for air as medium in the voids

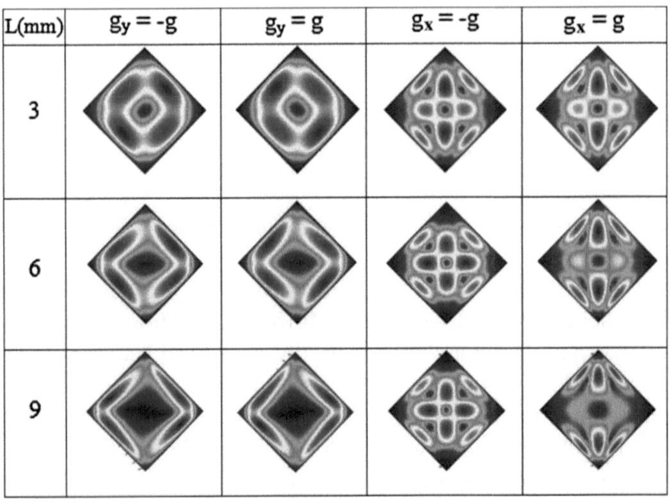

Fig. 4 Patterns of velocity magnitude for water as medium in the voids

4 Conclusion

The natural convection phenomenon is simulated for a diamond shaped cavity with side wall heating (Dirichlet boundary conditions) using COMSOL multiphysics software. It was found from simulation results that the intensity of natural convection is highest for gravity in vertical direction. For gravity in horizontal direction, it is akin to Rayleigh-Bernard convection (Chandrasekhar et al. 1981). Therefore present trend in results as in Fig. 2a, b are in consonance with literature (Roy and Basak 2005; Sathiyamoorthy et al. 2007; Chandrasekhar et al. 1981). But, details of flow and temperature patterns obtained are novel for the diamond shaped cavity. For air, the intensity of natural convection is 10 times more than that of water. These results will find application in studying the heat transfer mechanism in granular material.

References

Chandrasekhar, S.: Hydrodynamic and Hydromagnetic Stability. Dover Publications Inc. (1981)

COMSOL Multiphysics. For product information refer to www.comsol.com

Holman, J.P., Souvik Bhattacharyya.: Heat Transfer, (In SI Units), 10th edn. McGraw Hill (2011)

Mamun, M.A.H., Leong, W.H., Hollands, K.G.T., Johnson, D.A.: Cubical cavity natural convection benchmark experiments: an extension. Int. J. Heat Mass Transf. **46**, 3655–3660 (2003)

Nguyen, V.D., Cogne, C., Guessasma, M., Bellenger, E., Fortin, J.: Discrete modeling of granular flow with thermal transfer: application to the discharge of silos. Appl. Therm. Eng. **29**, 8–9 (2009)

Roy, S., Basak, Tanmay: Finite element analysis of natural convection flows in a square cavity with non-uniformly heated wall(s). Int. J. Eng. Sci. **43**, 668–680 (2005)

Sathiyamoorthy, M., Tanmay Basak, S., Roy, I.Pop: Steady natural convection flows in a square cavity with linearly heated side wall(s). Int. J. Heat Mass Transf. **50**, 766–775 (2007)

Strumillo, C., Kudra, T.: Drying: Principles, Applications and Design. Gordon and Breach Science publishers, Switzerland (1986)

CFD Simulation of Heat Transfer Using Nanofluids in Microchannel Having Dimples and Protrusions

Rahul Kunnath, K. Neelima chowdary and A. Venu Vinod

List of Symbols

$k_{nf,}$ $k_{p,}$ k_b	Thermal conductivity of nanofluid, nano particles and base fluid (W/m K)
$\mu_{nf,}$ μ_b	Viscosity of nanofluid, base fluid (Pa · s)
$\rho_{nf,}$ ρ_b, ρ_p	Density of nanofluid, base fluid, nanoparticles (kg/m^3)
$C_{pnf,}$ $C_{pb,}$ C_{pp}	Specific heat of nanofluid, base fluid, nanoparticles (J/kg K)

1 Introduction

Numerous studies have been reported to improve the process of heat transfer and thus increase the efficiency of thermal systems by employing various active and passive techniques. Introducing nanofluid inside and dimples/protrusion on the wall of a microchannel heat sink can be considered as one such technique. Microchannel heat sinks are highly efficient cooling devices because of less coolant demand and dimensions, due to enhancement in convective heat transfer. But traditional working fluids become the main problem for the improvement of heat transfer performance. Most of the related studies showed that the base fluid with added nanoparticles produced considerable enhancement to the coefficient of heat transfer. Although nanofluid has higher thermal conductivity, their addition to a base fluid will increase the viscosity of that fluid, which can cause increase in pressure drop. Introducing dimples on the surface of the channel becomes significant here as it can improve the heat transfer at low pressure penalty. It was suggested by Lee et al. (2005) that the conventional Navies–Stokes analysis could be used in a rectangular microchannel for predicting heat transfer. Lan et al. (2012) studied the heat transfer performance in microchannel with dimple and protrusion, using water

R. Kunnath · K. Neelima chowdary · A.V. Vinod (✉)
Department of Chemical Engineering, National Institute of Technology Warangal, Warangal, India
e-mail: avv122@yahoo.com

I. Regupathi et al. (eds.), *Recent Advances in Chemical Engineering*,
DOI 10.1007/978-981-10-1633-2_36

Table 1 Properties of Cuo-Water nanofluid at T = 300 K

Nanofluid concentration ϕ%	k (W/mK)	C_p (J/kgK)	ρ (kg/m^3)	μ (Pa · s)
0	0.6	4180	998.20	9.93×10^{-4}
1	0.6178	3956.52	1053.31	1.08×10^{-3}
2	0.6368	3753.46	1108.42	1.19×10^{-3}
3	0.6570	3569.66	1163.53	1.32×10^{-3}
4	0.6784	3402.452	1218.649	1.47×10^{-3}

as the coolant and it was concluded that dimples are effective passive heat transfer augmentation for macroscale channels. But before employing the combination of nanofluid and dimple/protrusion on a thermal system, along with the maximized heat transfer, minimized entropy generation also has to be considered. The present study aims at (i) investigating the convective heat transfer and thermal performance of CuO–water nanofluid in microchannel with dimple and/or protrusion in laminar flow with different nanoparticles volume concentrations and velocity and (ii) compare the heat transfer performance between nanofluid and base fluid (water).

2 Properties of Nanofluids

CuO–H_2O nanofluid was used in the present study. The volume fraction of nanoparticle is varied from 0 to 4 %. A single-phase model is used to represent nanofluid. Nanofluid properties density (Pak et al. 1998), specific heat (Ho et al. 2008), dynamic viscosity (Roy et al. 2004) and thermal conductivity (Bruggeman 1935) were determined using the models reported in literature respectively (Table 1).

3 Physical Model and Numerical Method

3.1 Geometry and Boundary Conditions

The microchannel consists of an inlet, outlet and wall boundaries, and is 50 μm (*W*), and 200 μm (*H*) in cross-section as in Fig. 1. Dimples/protrusions are arranged on the wall having a width of 200 μm and length 300 μm either in staggered or in aligned manner as in Fig. 1. The dimple/protrusion is a sector of a sphere with 100 μm diameter and 20 μm depth, where *S*, *P* are the distances of the adjacent dimple/protrusion in the streamwise and spanwise directions (Table 2).

In this study, CuO–H_2O nanofluid enters microchannel with a bulk temperature of 300 K, providing uniform heat flux of 5×10^5 W/m^2 and no slip boundary condition at the wall by varying volume fraction of nanoparticle from 0 to 4 % and with inlet velocities 1.2, 3.7, 6.2, 8.7 m/s in laminar region.

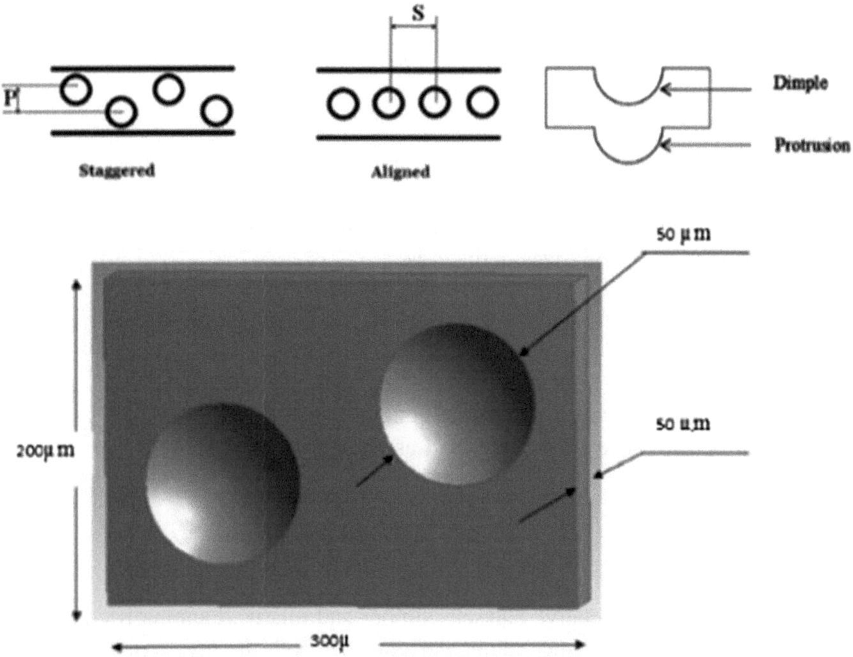

Fig. 1 Geometry of microchannel

Table 2 Types of geometry

Group	D/μm	δ/μm	S/D	P/D	Structure	Type
1	100	20	1.5	0	Dimpled	Aligned
2	100	20	1.5	0.5	Dimpled	Staggered
3	100	20	1.5	0	Dimpled + Protrusioned	Aligned
4	100	20	1.5	0.5	Dimpled + Protrusioned	Staggered

3.2 Data Reduction

In this case, Reynolds number is defined as (Belnap et al. 2002)

$$Re = \frac{\rho_{nf} U D_h}{\mu_{nf}} \quad (1)$$

where U is the average inlet velocity and D_h is hydraulic diameter

$$D_h = 2WH/(W + H)$$

Nusselt number,

$$Nu = \frac{hD_h}{k_{nf}} \tag{2}$$

where k_{nf} is the thermal conductivity of nanofluid and h is heat transfer coefficient defined as, $h = q/\Delta T$, where ΔT is the difference between the mean fluid temperature and mean wall temperature, q is the heat flux applied to the wall.

Fanning friction factor f was defined as,

$$f = \frac{\left[\left(\frac{\Delta P}{L}\right)D_h\right]}{2\rho_{nf}U^2} \tag{3}$$

where Δp is pressure drop, L is channel length

Thermal performance *TP* is described as (Belnap et al. 2002)

$$TP = \left[\left(\frac{Nu}{Nu_o}\right)\left(1/\sqrt[3]{\left(\frac{f}{f_o}\right)}\right)\right] \tag{4}$$

where Nu/Nuo, f/f_o are relative Nusselt number and relative Fanning friction factor respectively. This performance parameter consider both the heat transfer augmentation and the friction loss increase, which is frequently used to represent the heat transfer performance of devices.

Entropy generation rate is defined as (Satapathy 2009)

$$S_{gen} = \left[\frac{q^2}{\Pi k_{nf} T^2 Nu}\right] + \left[\frac{32m^3 f}{\Pi^2 \rho_{nf}^2 T D_h^5}\right] \tag{5}$$

where S_{gen} and m are entropy generation rate and mass flow rate respectively.

Equations (1–5) can be used to calculate Reynolds number, Nusselt number, Fanning friction factor, thermal performance and entropy generation rate respectively. By carrying out the simulation on smooth channel the baseline Fanning friction factor, f_0 and Nusselt number, Nu_0 are calculated.

3.3 Solution Method

For the flow of fluid and heat transfer of fluid in microchannel, governing equations are:

Continuity equation

$$\nabla.\vec{u} = 0 \tag{6}$$

Momentum equation

$$\rho_{nf}(\vec{u}.\nabla)\vec{u} = -\nabla p + \mu_{nf}\nabla^2\vec{u} \tag{7}$$

Energy equation

$$\rho_{nf}C_{pnf}(u.\nabla T) = k_{nf}\nabla^2 T \tag{8}$$

FLUENT 15.0 was used to solve the above equations in the steady, incompressible flow by incorporating the boundary conditions. A second-order upwind scheme with a convergence criterion of 1×10^{-5} was used to solve the momentum and energy equations.

4 Results and Discussion

The plots shown below (Figs. 2, 3, 4, 5, 6, 7, 8 and 9) represent the variation of relative friction factor f/f_0, relative Nusselt number Nu/Nu_0, thermal performance and entropy generation rate of different geometries with change in velocity and volume fraction of nanoparticle. For all cases, when the nanoparticle volume fraction (Φ%) increased from 0 to 4 %, there is a slight increment in Nu/Nu_0 and thermal performance values whereas f/f_0 is less, which is mainly because of increase in dynamic viscosity of CuO–water nanofluids when the nanoparticle volume fraction increases. Arrangement of dimple and protrusion reduce friction resistance and enhance form resistance, thus in the laminar region dimple and protrusion reduce the flow resistance with an increase in nanoparticle volume fraction. Thus Nu/Nu_0 of nanofluids is enhanced at the same Reynolds number, and increases more when nanoparticle volume fraction is larger. This means that provision of dimples and protrusions, leads to better flow and heat transfer performance of CuO–water nanofluid compared to that of water.

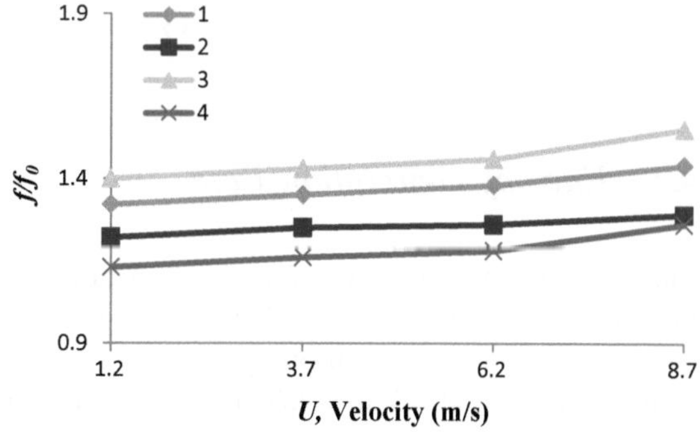

Fig. 2 f/f_0 versus Velocity (1 % CuO). *1* Dimpled + Protrusioned aligned. *2* Dimpled staggered. *3* Dimpled + Protrusioned Staggered. *4* Dimpled aligned

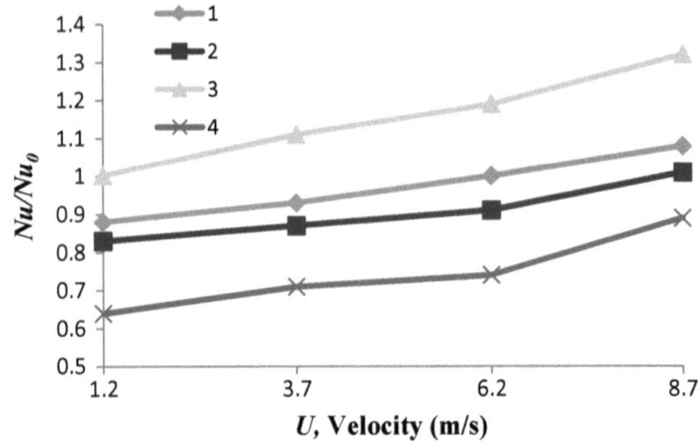

Fig. 3 Nu/Nu_0 versus Velocity (1 % CuO)

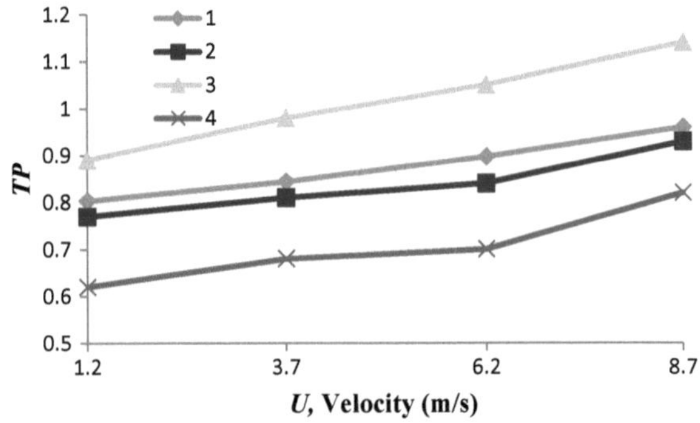

Fig. 4 TP versus Velocity (1 % CuO)

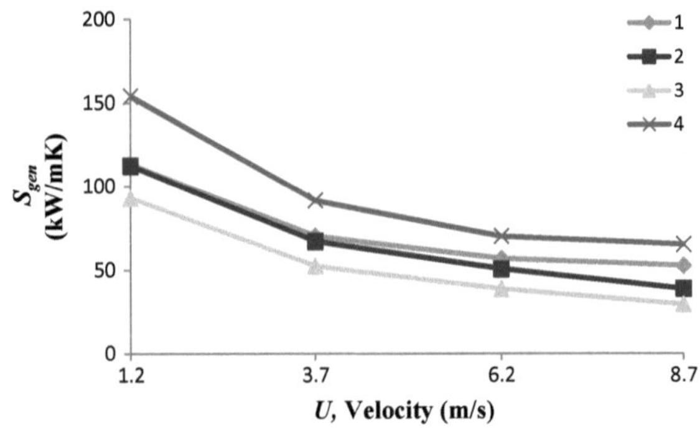

Fig. 5 S_{gen} versus Velocity (1 % CuO)

4.1 Entropy Generation Rate

Entropy generation rate was found out in order to optimize the geometry. It has been found that microchannel having staggered dimple and protrusion has least entropy generation rate. Entropy generation rate is the least at 4 % volume fraction of nanofluid and at 8.7 m/s velocity. This is because, the thermal conductivity of

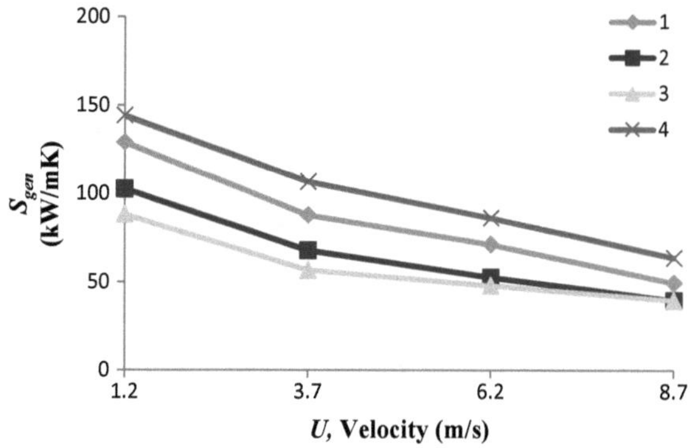

Fig. 6 S_{gen} versus Velocity (4 % CuO)

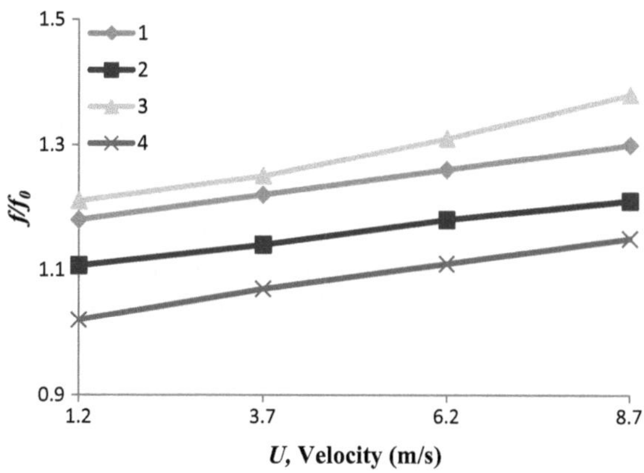

Fig. 7 f/f_0 versus Velocity (4 % CuO)

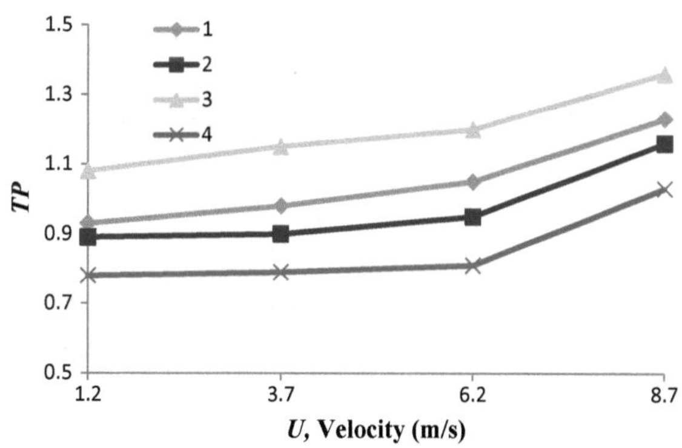

Fig. 8 *TP* versus Velocity (4 % CuO)

nanofluid and the Nusselt number are higher, which reduce the entropy generation rate at higher volume fractions. The arrangement of dimples and protrusion also reduce the entropy generation rate.

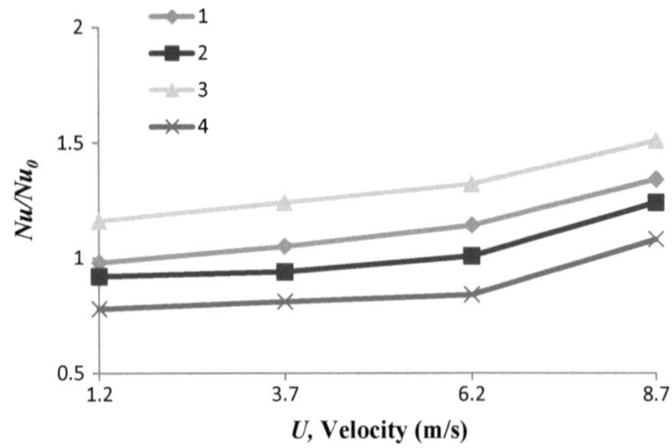

Fig. 9 Nu/Nu_0 versus Velocity (4 % CuO)

4.2 Temperature Profiles

From the temperature profiles of fluid (Fig. 10), we can clearly observe that wall temperature of microchannel is getting reduced for higher values of velocity and volume fraction of nanoparticles, indicating better heat transfer from wall to fluid. Thus more amount of heat is transferred by means of convection, indicating enhancement in heat transfer. The presence of dimples and protrusions leads to wake formation. In Fig. 10, at the end of the dimple/protrusion amount of heat transferred is high (indicated by red or yellow color). Also for case 3, as CuO volume fraction increases from 1 to 4 %, the amount of heat transferred is also increasing, particularly at the end of the dimple in flow direction.

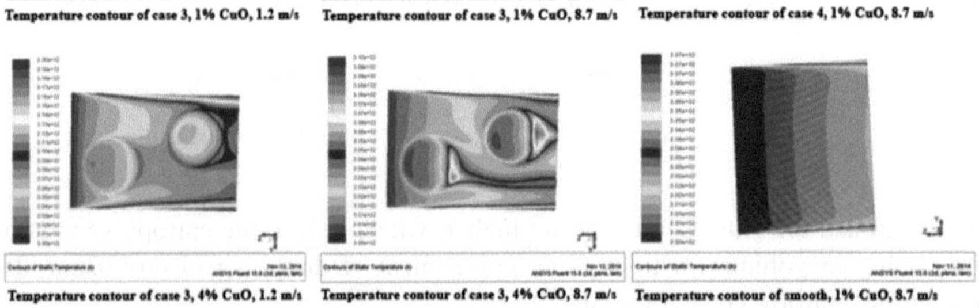

Fig. 10 Temperature profile in fluid in different cases

5 Conclusions

Friction factor and Nusselt number were determined for flow of CuO–water nanofluid through a microchannel having (i) different arrangements of dimples and protrusions and (ii) plane walls. These results were subsequently used to determine thermal performance of the microchannel. Entropy generation rate was determined to optimize the geometry. Temperature contours were reported to clearly visualize the enhancement in heat transfer at higher inlet velocities and higher volume fractions of nanofluid. The findings from the study are as follows:

- Dimples + protrusion geometries provide higher f/f_0, Nu/Nu_0 and thermal performance values compared to dimples only cases.
- f/f_0, Nu/Nu_0 and thermal performance are superior for staggered arrangements than aligned arrangements.
- f/f_0, Nu/Nu_0 and *TP* increase when there is an increase in inlet velocity.
- S_{gen} is found to be least for dimples + protrusion geometries.
- By using nanofluids, heat transfer enhancement was obtained.
- Wall temperature of microchannel is getting reduced for higher values of velocity and volume fraction of nanoparticles, indicating large amount of heat is transferred from wall to fluid. Hence heat transfer enhancement is obtained.

References

Belnap, B., Van Rij, J., Ligrani, P.: A Reynolds analogy for real component surface roughness. Int. J. Heat Mass Transf. **45**(15), 3089–3099 (2002)

Bruggeman, D.A.G.: Calculation of various physical constants of heterogeneous substances. I. dielectric constants and conductivities of the mixed body of isotropic substances. Ann. Phys. **416**(7), 636–664 (1935)

Ho, C.J., Chen, M., Li, Z.: Numerical simulation of natural convection of nanofluid in a square enclosure: effects due to uncertainties of viscosity and thermal conductivity. Int. J. Heat Mass Transf. **51**(17), 4506–4516 (2008)

Lan, J., Xie, Y., Zhang, D.: Flow and heat transfer in microchannels with dimples and protrusions. J. Heat Transf. **134**(2), (2012)

Lee, P.S., Garimella, S.V., Liu, D.: Investigation of heat transfer in rectangular microchannels. Int. J. Heat Mass Transf. **48**, 1688–1704 (2005)

Pak, B.C., Cho, Y.I.: Hydrodynamic and heat-transfer study of dispersed fluid with submicron metallic oxide particles. Exp. Heat Transf. **11**, 151–170 (1998)

Roy, G., Nguyen, C.T., Lajoie, P.-R.: Numerical investigation of laminar flow and heat transfer in a radial flow cooling system with the use of nanofluids. Superlattices Microstruct. **35**(3), 497–511 (2004)

Satapathy, A.K.: Thermodynamic optimization of a coiled tube heat exchanger under constant wallheat flux condition. Int. J. Energy **34**, 1122–1126 (2009)

CFD Simulation of Chemical Looping Combustion of Coal

Kavitha G. Menon and Venkata Suresh Patnaikuni

1 Introduction

One of the most important crises faced by the modern world is the increasing concentration of atmospheric CO_2. CO_2 is produced as a byproduct of the combustion of fossil fuels, which can be considered as the primary source (86 % of total energy produced in world as per IPCC 2005 report (2005)) of energy as there has been a huge depletion in the availability of renewable energy source. Considering the rising energy demand and the economic challenges, focus is shifted towards increasing the efficiency of energy production from fossil fuels, these days. According to the reports (IPCC 2005), fossil fuels accounted for the 75 % of the anthropogenic CO_2 emissions. Chemical Looping Combustion (CLC) is one of the most efficient and promising method among the carbon dioxide capture and sequestration (CCS) technologies for CO_2 separation and thereby preventing its escape to atmosphere.

Chemical Looping Combustion is a process where CO_2 is separated effectively from the flue gases with the aid of an oxygen carrier. It consists of two reactors, namely air reactor and fuel reactor. In air reactor, the metals get oxidized by reacting with air or oxygen. These metal oxides are then carried to the fuel reactor, where the metal oxides react with the fuel (solid of gaseous) to produce CO_2 and H_2O as final products, from which CO_2 can be separated easily by condensing H_2O. The reduced metals are then circulated back to the air reactor and the cycle continues. Using this circulating method, the direct contact of solid fuel and air can be

K.G. Menon · V.S. Patnaikuni (✉)
Department of Chemical Engineering, National Institute of Technology Warangal, Warangal, India
e-mail: pvsuresh@nitw.ac.in

K.G. Menon
e-mail: kavithagmenon@gmail.com

I. Regupathi et al. (eds.), *Recent Advances in Chemical Engineering*,
DOI 10.1007/978-981-10-1633-2_37

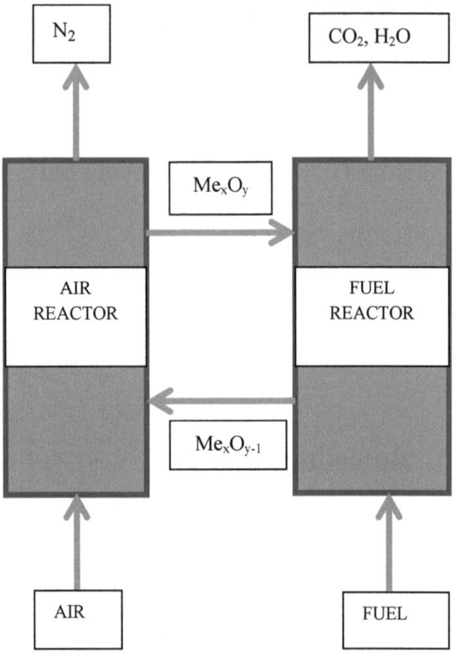

Fig. 1 Schematic representation of CLC process

avoided thus preventing the dilution of CO_2 stream making its separation cumbersome. Schematic diagram representing the chemical looping combustion process is given in the Fig. 1.

2 CFD Model

A CFD model was developed to simulate the fuel reactor of the chemical looping combustion system, where the coal and Fe_2O_3 (Leion et al. 2008; Mattison et al. 2001) metal oxides are put together and allowed to react. An air reactor is more assumed to function as a transport reactor, where the oxides are formed and are transported to fuel reactor in which all major reactions take place. The present simulation was carried out in ANSYS FLUENT 15 software. A 2D geometry with a fine mesh was developed as shown in Fig. 2. The simulation was carried out in an axi-symmetric mesh and for effective visualization; it is mirrored around its axis.

The reactor was initially packed with coal (diameter—10 μm, density—2000 kg/m^3) and oxide particles (diameter—10 μm, density—5240 kg/m^3) to a height of 35 mm (the computational domain is patched with these coal and oxide particles initially) and a 50 % H_2O stream was provided as gasifying agent to aid the char gasification reactions. The gasifying agent was fed at a velocity of 0.0025 m/s, to ensure that no entrainment of solid particles takes place and to

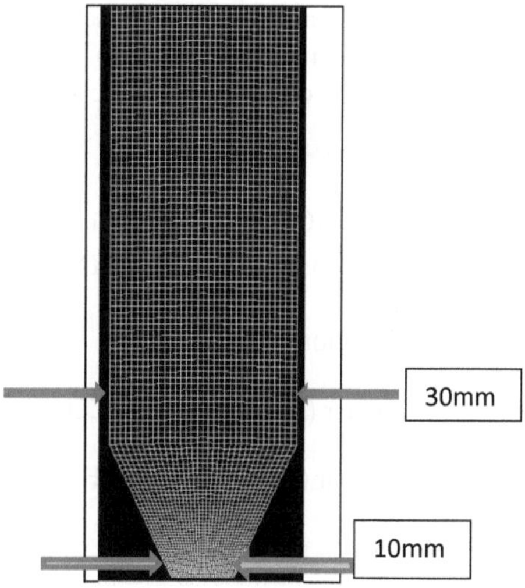

Fig. 2 Lower portion of the fine mesh (enlarged picture)

ensure good mixing of the coal and the metal oxides. The temperature was maintained at 1223 K throughout the reactor.

In the present simulation, multiphase Eulerian approach has been used to model both the solid and gaseous phases. In an Eulerian approach, individual particle properties are not considered and the system is treated mathematically as interpenetrating continua. The complexity of the method depends on the types of phases involved. k-$\mathcal{E}$ model was used to characterize the turbulent flow behavior of components. Heat transfer between the phases is taken into account using Gunn correlation available in fluent database.

3 Reactions in Fuel Reactor

Several reactions are taking place in a fuel reactor, starting with the coal devolatilization, where coal devolatilizes to various species such as char, CH_4, C_2H_6, CO, CO_2, H_2, H_2O and soot. (Ash is not considered in the present simulation). Devolatilization reaction will be followed by char gasification and reduction reactions between metal oxides and the devolatilized components. Subsequently water gas shift reaction takes place, in which CO produced is converted to CO_2. Coal devolatilization is a very fast reaction which occur within a very few seconds, whereas char gasification remains to be the rate limiting step.

In this simulation, components are defined in three different phases, where one phase contains all gaseous components and the other two phases constitutes coal, its residents and metal oxides respectively. Arrhenius rate of reaction mechanism was applied to all reactions. The reaction rates and constants for all reduction reactions and devolatilization reaction were obtained from studies conducted by Mahalatkar et al. (2011). Char gasification reaction kinetics were obtained from studies conducted by Matsumoto et al. (2009).

The reactions considered are the following:

(a) Devolatilization reaction:

$$\begin{aligned} \text{Coal} \rightarrow \quad & 0.46\text{Char} + 0.03\text{CH}_4 + 0.0026\text{C}_2\text{H}_6 + 0.009\text{CO} + 0.003\text{CO}_2 \\ & + 0.05\ \text{H}_2 + 0.08\ \text{H}_2\text{O} + 0.005\ \text{N}_2 \end{aligned} \tag{1}$$

(b) Gasification reactions:

$$\text{Char}\,(\text{CH}_{0.02}\text{O}_{0.005}) + \text{CO}_2 \rightarrow 2\text{CO} + 0.005\text{H}_2\text{O} + 0.005\text{H}_2 \tag{2}$$

$$\text{Char}\,(\text{CH}_{0.02}\text{O}_{0.005}) + \text{HO}_2 \rightarrow \text{CO} + 0.005\text{H}_2\text{O} + 1.005\text{H}_2 \tag{3}$$

(c) Water gas shift reaction:

$$\text{CO} + \text{H}_2\text{O} \rightarrow \text{CO}_2 + \text{H}_2\text{O} \tag{4}$$

(d) Metal oxide reduction reactions:

$$12\text{Fe}_2\text{O}_3 + \text{CH}_4 \rightarrow 8\text{Fe}_3\text{O}_4 + \text{CO}_2 + 2\text{H}_2\text{O} \tag{5}$$

$$21\ \text{Fe}_2\text{O}_3 + \text{C}_2\text{H}_6 \rightarrow 14\ \text{Fe}_3\text{O}_4 + 2\ \text{CO}_2 + 3\ \text{H}_2\text{O} \tag{6}$$

$$3\ \text{Fe}_2\text{O}_3 + \text{CO} \rightarrow 2\ \text{Fe}_3\text{O}_4 + \text{CO}_2 \tag{7}$$

$$3\ \text{Fe}_2\text{O}_3 + \text{H}_2 \rightarrow 2\ \text{Fe}_3\text{O}_4 + \text{H}_2\text{O} \tag{8}$$

4 Results and Discussions

Initially, the reactor is packed with coal and Fe_2O_3 to a height of 35 mm. After one second, as soon as the reaction proceeds, there is a sudden decrease in the concentration of coal in the reactor as can be noticed from Fig. 3, due to the devolatilization reaction. Coal devolatilization is a very fast reaction which happens in matter of seconds, which can be confirmed from the mass fraction of char, one of the main constituent of the volatile species in Fig. 3.

Figure 4 represents the volume fractions of coal and metal oxide at a time of 225 s. From Fig. 4 it can be observed that at lower parts of the reactor, the concentration of the coal is more near the sides of the reactor and it gradually occupies the center part as it goes up. This is because of the fact that the solid particles are pushed away to the sides by the inlet gas as it goes up. But since the velocity of inlet gas is less and it decreases further due to the solid particles, the particle occupies the center region as it goes up. The same trend can be seen in the case of volume

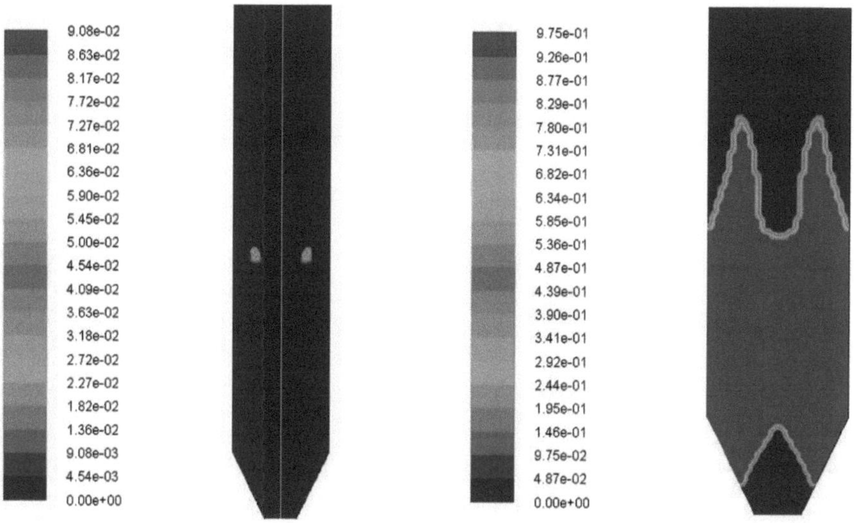

Fig. 3 Mass fraction of coal and char after 1 s and temperature of 1223 K

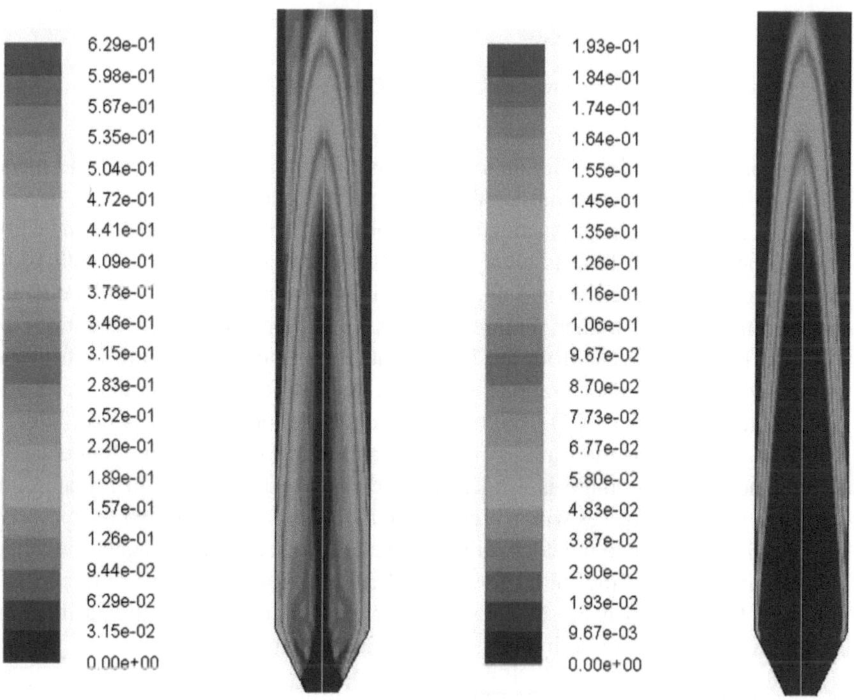

Fig. 4 Volume fractions of coal and metal oxide respectively at 225 s, temperature = 1223 K

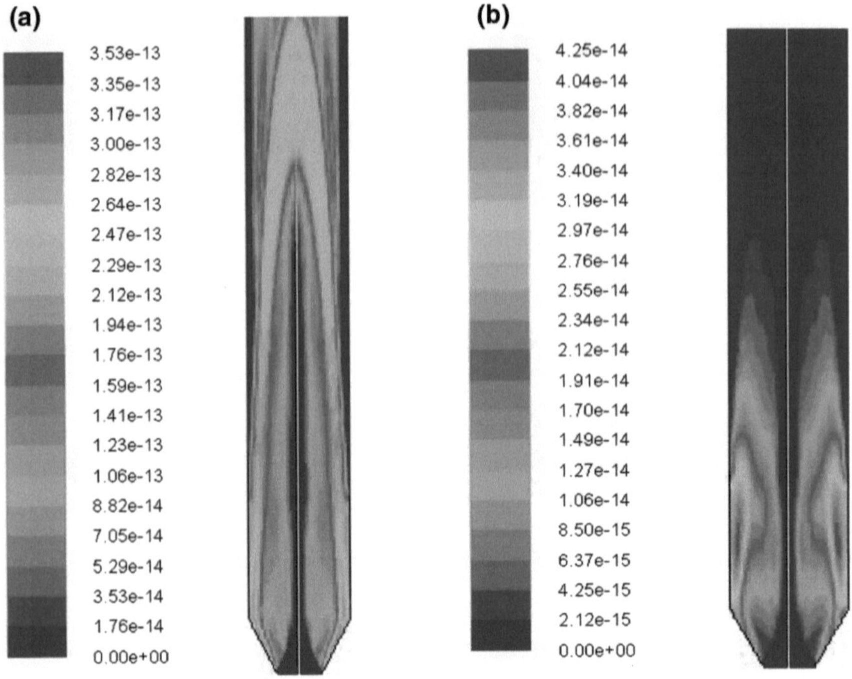

Fig. 5 Rate of char gasification reactions **a** with H_2O **b** with CO_2

fraction of metal oxides. This indicates the perfect mixing of coal and the metal oxides, which is very essential for the better performance of the reactor.

Figure 5 shows the char gasification rates obtained with the aid of H_2O and CO_2. The gasification with H_2O is always greater than the gasification rate with CO_2. This is because the reaction rates of steam are higher than that of CO_2 and the steam concentration is also higher. Figure 6 shows the instantaneous concentrations of CO_2 and H_2O. It is clear from the figure that CO_2 concentrations are high in the regions where H_2O concentrations are lower. As H_2O is utilized in the water gas shift reaction to produce CO_2, the obtained concentration variation is justifiable.

Figure 7 represents the variation in mass fraction of CO_2 with time. Immediately as the reaction starts, there is a hike in the concentration of CO_2 to a maximum value and then it decreases with a higher pace in the initial seconds and later slowly. The initial hike is due to the devolatilization reaction of coal. As coal gets converted almost completely in the initial few seconds, there is an increase in the concentration of CO_2 and as it finishes, a steep decrease occurs. Then in the gasification section, a gradual decrease in concentration can be witnessed.

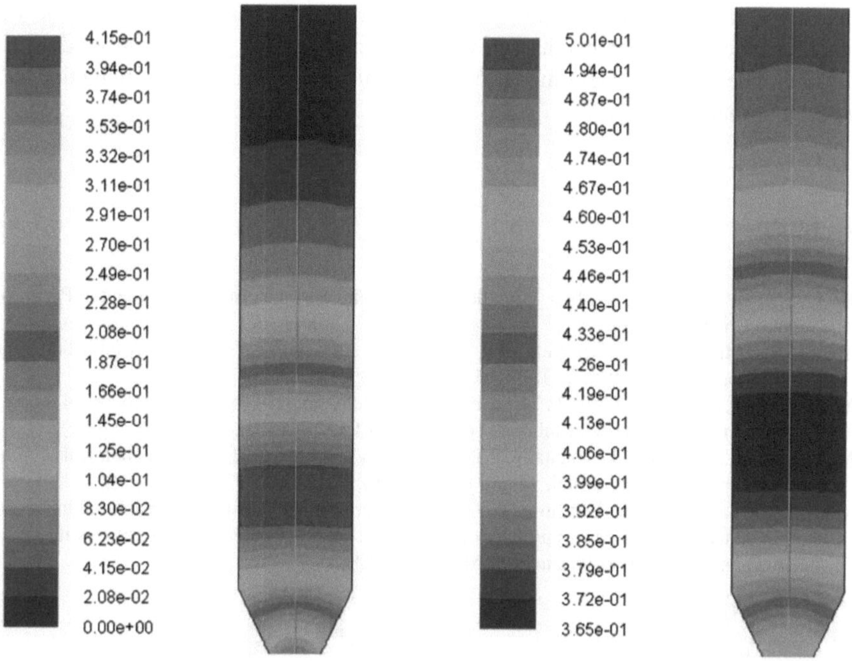

Fig. 6 Instantaneous mass fractions of CO_2 and H_2O respectively at 225 s, temperature = 1223 K

Fig. 7 Variation of mass fraction of CO_2 time

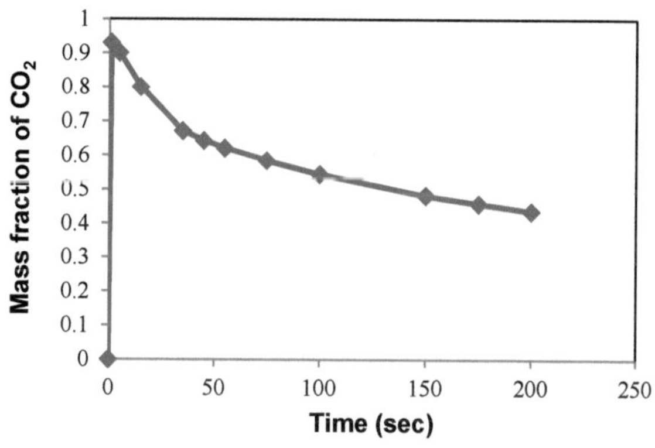

5 Conclusion

A 2D CFD model was developed to simulate the fuel reactor of a chemical looping combustion system by analyzing the overall process including the particle-particle interactions, fluid-particle interactions, reaction rates and the performance of the reactor was predicted. The gasification reactions are taking place at a very low rate than expected; this may be due to the usage of Arrhenius theory to define the reaction kinetics. Better results can be expected if perfect gasification models are defined. The developed 2D model could explain the overall system behavior in a justifiable manner.

References

IPCC Special Report on Carbon Dioxide Capture and Storage, Intergovernmental Panel on Climate Change (2005)

Leion, H., Mattison, T., Lyngfelt, A.: Solid fuels in chemical looping combustion. Int. J. Greenhouse Gas Control **2**, 180–193 (2008)

Mattison, T., Lyngfelt, A., Cho, P.: The use of iron oxide as an oxygen carrier in chemical looping combustion of methane with inherent separation of CO_2. Fuel **80**, 1953–1962 (2001)

Mahalatkar, K., Kuhlman, J., Huckaby, E.D., O'Brien, D.: CFD Simulation of a chemical looping fuel reactor utilizing solid fuel. Chem. Eng. Sci. **66**, 3617–3627 (2011)

Matsumoto, K., Takeno, K., Lehinose, T., Ogi, T., Nakanishi, M.: Gasification reaction kinetics on biomass char obtained as a byproduct of gasification in an entrained flow gasifier with steam and oxygen at 900–1000°C. Fuel **88**, 519–527 (2009)

Analysis of Flow Behaviour in a Jet Ejector Using CFD

D. Arumuga Perumal, M.P. Mydeen, C.S. Sabarish, M. Selvam and B. Duraisamy

1 Introduction

Ejectors are devices used to induce a secondary fluid through momentum and energy from a high speed primary jet (Varga et al. 2009). If the Ejectors are operated with incompressible fluids (liquids)this application is generally referred as jet pumps. On the other hand ejectors are operated with compressible fluids (gases and vapors) the terms ejector and injector are commonly employed. The operating procedure in a liquid jet pump or in a gas ejector are the same. A high-strain fluid with very low speed on the primary inlet is improved to high pace jet through a converging nozzle for the liquid jet pump or a converging-diverging supersonic nozzle for the gas ejector. The supply stress at the inlet is partially converted to be the jet momentum at the nozzle exit in step with the Bernoulli equation. The high velocity, low static strain primary jet induces a secondary flow from the suction port and hurries up it within the course of the using jet. The two streams then combine within the mixing section, and preferably the process is whole with the aid of the give up of this segment. A diffuser is normally set up at mixing chamber exit to raise the static pressure of mixed flow.

Ejectors have easy geometry and no shifting components. Their operation does not require electrical or mechanical shaft energy input. This significantly reduces system mass and increases reliability. The ejector turned into added as an engineering device in the early twentieth century (Sriveerakul et al. 2007). Ejectors are extensively used in energy plant, aerospace, propulsion and refrigeration

D.A. Perumal (✉)
Department of Mechanical Engineering, National Institute
of Technology Karnataka, Surathkal, India
e-mail: perumal@nitk.edu.in

M.P. Mydeen · C.S. Sabarish · M. Selvam · B. Duraisamy
School of Aeronautical Engineering, Noorul Islam University,
Kanyakumari, Tamilnadu, India

I. Regupathi et al. (eds.), *Recent Advances in Chemical Engineering*,
DOI 10.1007/978-981-10-1633-2_38

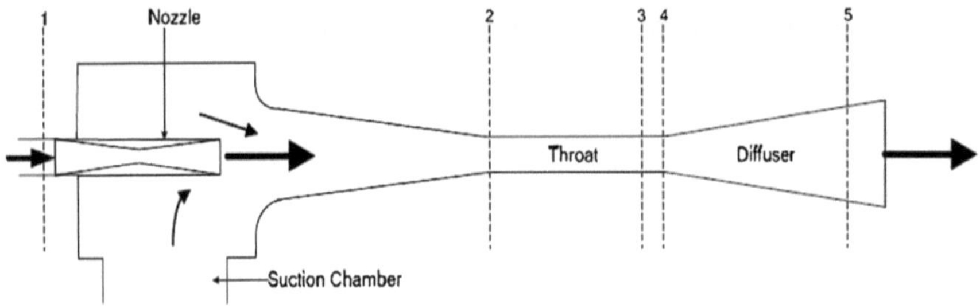

Fig. 1 Schematic view of ejector

applications (Sun and Eames 1995). Moreover, most jet ejectors use steam or compressed air because the motive fluid, that is effortlessly found in chemical plants (Bartosiewicz et al. 2005).

In the past, ejectors had been designed and analyzed based on a classical idea evolved by Keenan and Neumann (1942) which is based on one-dimensional gas dynamic theory. This principle became later modified to account for loss coefficients at the primary nozzle, the mixing chamber and the diffuser. Figure 1 demonstrates the schematic outline of an ejector. Conventional jet ejectors are classified by way of the measurement of the convergence section. There are two types: 1. Constant-area jet ejector, 2. Constant-pressure jet ejector.

Lee et al. (2005) found that the constant-pressure configuration provides a better performance than the constant-area configuration, because turbulent mixing inside the jet-ejector is accomplished extra actively under an adverse pressure gradient, which takes place in the constant-area jet ejector, instead of under constant pressure. Rice and Dandachi (1991) derived equations for steam ejectors to predict the primary flow rate by including the friction and mixing losses which have usually been neglected. From the available literature it is seen that, here is no study of the flow in a whole ejector channel, let alone using CFD to optimize the ejector design. Also, there are few comparisons between CFD predictions and data. In order to validate the CFD simulations, it is necessary to carry out more experimental-CFD and analytical-CFD comparisons. The feasibility of whether CFD methods are capable of dealing adequately with supersonic flow, particularly shock and expansion wave in ejectors, has not been previously investigated.

2 Mathematical Formulation

Computational Fluid Dynamics (CFD) is primarily established as a tool for designing traditional flows which involve empiricism, followed by prototyping-based physical simulation, process evaluation, and component design. Figure 2 shows the computational solution procedure of simulation systems.

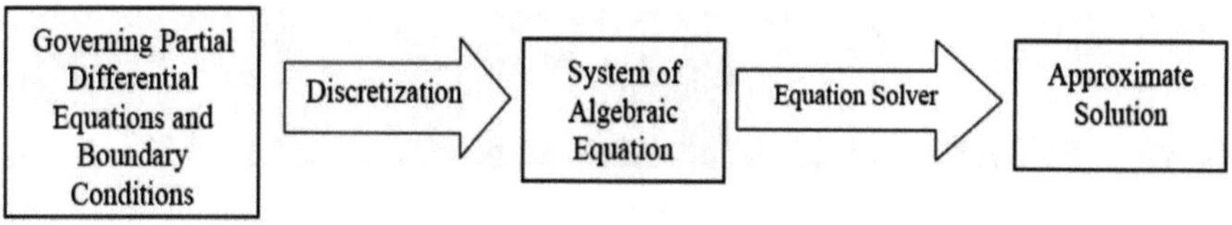

Fig. 2 Overview of computational solution procedure

The first stage is called discretization. The continuous partial differential equations are converted into a discrete system of algebraic equations. The detail of discretization is explained in the following section. In the second stage, a numerical solver is selected to solve a discrete system obtained from the first stage. The solution of algebraic equations are obtained as a consequence.

2.1 Meshing and Analysis

As proposed, GAMBIT is used to create the computational domain and grid elements of the present model. The mesh and model had been created in a two dimensional computational domain. Our simulation contains three stages. They are 1. Pre-processing, a. Geometry, b. blocking, c. Meshing, 2. Solving, 3. Post processing. The ejector design diameters are tabulated in Table 1 Steps involved in meshing are a. Select the edges according to the number of edges required, b. In order to predict the boundary the variable and grid spacing to the given to the interface, c. Quad-map grid is selected for the mesh quality, d. select the face of the throat area and provide the spacing 10. e. After that select the entire face and give a spacing of 20. Figures 3 and 4 shows Meshed geometry model of the constant area and constant pressure ejector. Model is created in CATIA software and imported to GAMBIT software. With a purpose to design and expand a excessive performance ejector, a clear knowledge of the flow and mixing process inside the ejector is first needed. In order to gain a better understanding of the mixing chamber within the ejector, the economic CFD software FLUENT is used to analyze the ejector. FLUENT is a industrial CFD software program package which utilizes the control-volume-based strategies to transform governing equations to algebraic equations that may be solved numerically.

Table 1 Ejector design parameters

Variable	Constant-area ejector (mm)	Constant-pressure ejector (mm)
Throat diameter	15	15
Nozzle exit diameter	70	70
Mixing chamber diameter	100	77.45
Mixing chamber length	2000	2160
Diffuser exit diameter	170	100

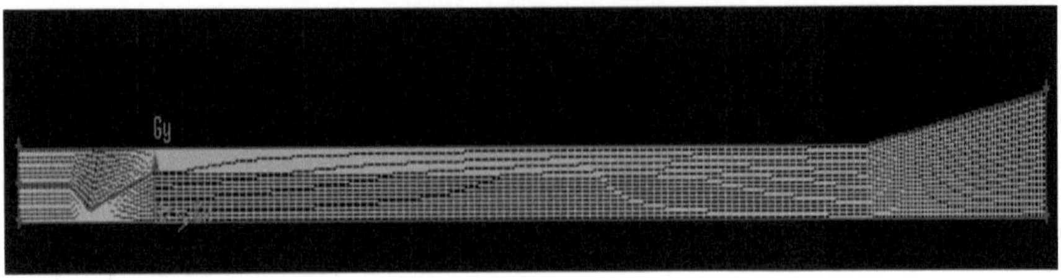

Fig. 3 Constant-area ejector

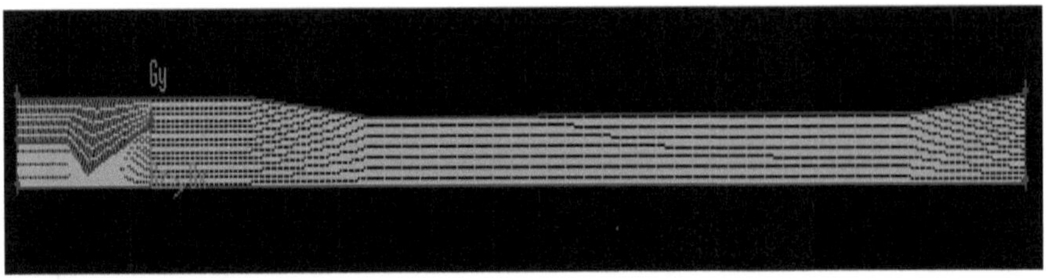

Fig. 4 Constant-pressure ejector

The governing equations in terms of cylindrical coordinates (r, z) and velocity components (Vr, Vz) are given under:

Continuity equation

$$\frac{1}{\mathbf{r}}\frac{\partial}{\partial \mathbf{r}}(\rho\, \mathbf{r} \mathbf{v}_\mathbf{r}) + \frac{\partial}{\partial \mathbf{z}}(\rho\, \mathbf{v}_\mathbf{z}) = 0 \tag{1}$$

r-momentum equation

$$\rho\left[v_r\frac{\partial v_r}{\partial r} + v_z\frac{\partial v_z}{\partial z}\right] = -\frac{\partial P}{\partial r} - \left[\frac{1}{r}\frac{\partial (r\tau_{rr})}{\partial r} + \frac{\partial \tau_{rz}}{\partial z}\right] - \frac{\tau_{\theta\theta}}{r} \tag{2}$$

z-momentum equation

$$\rho\left[\mathbf{v}_\mathbf{r}\frac{\partial \mathbf{v}_\mathbf{z}}{\partial \mathbf{r}} + \mathbf{v}_\mathbf{z}\frac{\partial \mathbf{v}_\mathbf{z}}{\partial \mathbf{z}}\right] = -\frac{\partial \mathbf{P}}{\partial \mathbf{r}} + \left[\frac{1}{\mathbf{r}}\frac{\partial (\mathbf{r}\,\tau_{\mathbf{rz}})}{\partial \mathbf{r}} + \frac{\partial \tau_{\mathbf{zz}}}{\partial \mathbf{z}}\right] \tag{3}$$

Energy equation

$$\rho\left\{v_r\frac{\partial}{\partial r}\left[e + \frac{p}{\rho}\right] + v_z\frac{\partial\left[e + \frac{p}{\rho}\right]}{\partial z}\right\} = -\frac{1}{r}\frac{\partial (rq_r)}{\partial r} - \frac{\partial q_z}{\partial z} + \mu\Phi \tag{4}$$

where $q_r = -k\frac{\partial T}{\partial r}, q_z = -k\frac{\partial T}{\partial z}$, v is the velocity vector. and the shear stresses are given as

$$\tau_{rr} = \mu\left[2\frac{\partial v_r}{\partial r} - \frac{2}{3}(\nabla.v)\right], \tau_{zr} = \tau_{rz} = \mu\left[\frac{\partial v_z}{\partial r} + \frac{\partial v_r}{\partial z}\right],$$

$$\tau_{zz} = \mu\left[2\frac{\partial v_z}{\partial r} - \frac{2}{3}(\nabla.v)\right], \tau_{\theta\theta} = \mu\left[2\frac{v_r}{r} - \frac{2}{3}(\nabla.v)\right].$$

Boundary conditions of two faces entering a primary nozzle and ejector were set as pressure-inlet, even as the one leaving ejector changed into set as pressure-outlet.

3 Results and Discussion

It's far recognized that, two factors are vital in terms of ejector performance. They are Entrainment Ratio (ER), Compression ratio (CR). It is known that, entrainment ratio is the mass flow ratio of secondary to primary, compression ratio is the mixing outlet pressure to secondary fluid inlet pressure.

Table 2 shows the analysis of constant-area ejector. It is seen that, pressure increases entrainment ratio decreases and compression ratio increases.

Table 3 shows the analysis of constant-pressure ejector. It is seen that, pressure increases entrainment ratio decreases and compression ratio initially increases and then deceases. Table 4 shows the analysis of constant-area ejector for pressure inlet

Table 2 Analysis of constant-area ejector

Pressure (Pa) × 10^5	Mixture outlet pressure (Pa)	Mass flow at inlet 1 (kg/s)	Mass flow at inlet 2 (kg/s)	Mass flow at outlet (kg/s)	Entrainment ratio	Compression ratio
22	100688	4.582	4.094	8.676	0.894	0.994
24	100615	4.998	4.076	9.074	0.815	0.993
26	100664	5.415	4.056	9.472	0.749	0.993
28	100815	5.832	4.056	9.868	0.692	0.995
30	101048	6.246	4.015	10.264	0.643	0.997
32	101182	6.665	3.993	10.659	0.599	0.997
34	101167	7.082	3.970	11.053	0.561	0.998
36	101045	7.498	3.497	11.466	0.526	0.999
38	101785	7.915	3.922	11.838	0.496	1.004
40	101923	8.332	3.906	12.337	0.469	1.006

Table 3 Analysis of constant-pressure ejector

Pressure (Pa) × 10^5	Mixture outlet pressure (Pa)	Mass flow at inlet 1 (kg/s)	Mass flow at inlet 2 (kg/s)	Mass flow at outlet (kg/s)	Entrainment ratio	Compression ratio
22	100973	4.557	2.043	6.600	0.448	0.996
24	101132	4.971	2.087	7.059	0.419	0.998
26	102088	5.386	2.133	7.520	0.396	1.007
28	102463	5.800	2.179	7.980	0.375	1.011
30	94232.2	6.215	2.225	8.440	0.358	0.929
32	77388.0	6.629	2.270	8.899	0.342	0.763
34	64319.9	7.044	2.313	9.357	0.328	0.603
36	50859.6	7.459	2.354	9.813	0.315	0.501
38	44266.7	7.873	2.393	10.266	0.304	0.437
40	41713.6	8.287	2.429	10.717	0.293	0.412

Table 4 Analysis of constant-area ejector for pressure inlet 1 is constant (22 bar)

Pressure at inlet 2 (Pa)	Mixture outlet pressure (Pa)	Mass flow at inlet 1 (kg/s)	Mass flow at inlet 2 (kg/s)	Mass flow at outlet (kg/s)	Entrainment ratio	Compression ratio
50000	101214	4.582	1.910	6.492	0.417	2.021
60000	101173	4.582	2.326	6.908	0.507	1.686
70000	101147	4.582	2.758	7.340	0.602	1.445
80000	101134	4.582	3.186	7.765	0.695	1.264
90000	101134	4.582	3.612	8.194	0.788	1.123

1 is constant at 22 bar. It is seen that, entrainment ratio increases and compression ratio decreases. Table 5 shows the analysis of various motive fluids. Figure 5a, b shows the velocity contours for primary inlet pressure = 2,200,000 Pa for constant-area ejector and constant-pressure ejector. Figure 5 shows the velocity contour of constant area ejector for various motive fluids. Figure 6a shows velocity contour of constant area ejector for motive fluid as acetylene. Figure 6b shows velocity contours of constant area ejector for motive fluid a steam.

It is seen that, from Table 5 and Fig. 6a, b, motive fluid containing larger molecular weight will affect the ejector efficiency. To find the performance of a constant-area ejector primary inlet pressure vs entrainment ratio graph was drawn. Figure 7 shows the constant-area ejector performance. It is concluded that, primary

Table 5 Analysis of constant-area ejectors using various motive fluids

Fluid	Pri. inlet pr. bar	Sec. inlet pr. (bar)	Mixture outlet pr. (bar)	Mass flow at inlet 1 (kg/s)	Mass flow at inlet 2 (kg/s)	Mass flow at outlet (kg/s)	Entrainment ratio	Compression ratio
Air	22	1.013	1.006	4.582	4.09	8.67	0.89	0.993
Acetylene	22	1.013	1.006	4.582	4.09	8.67	0.89	0.993
Steam	22	1.013	1.011	3.318	4.07	7.39	1.22	0.997

(a)

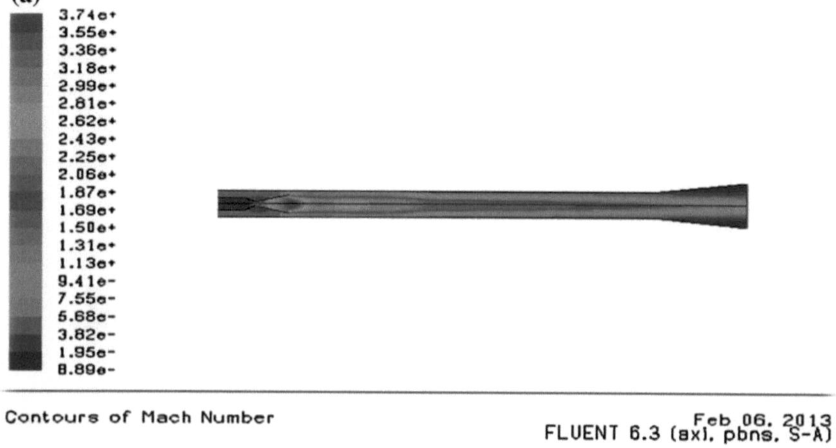

(b)

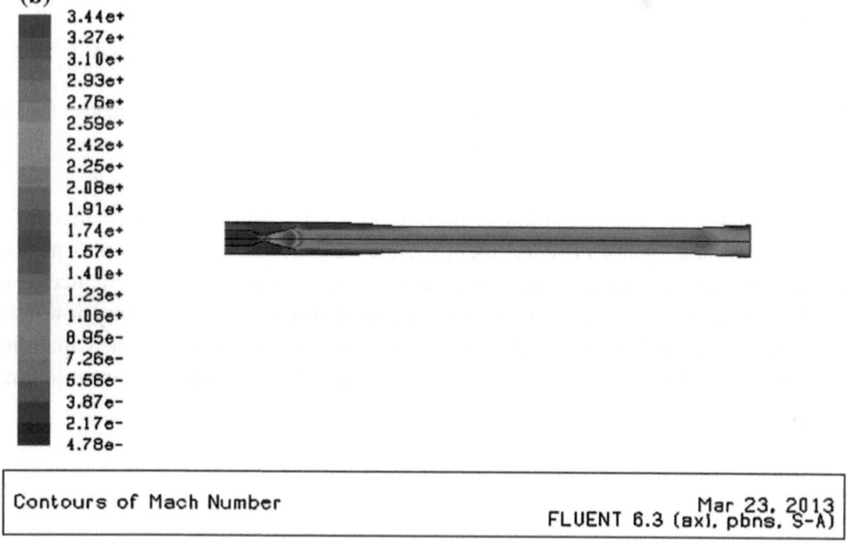

Fig. 5 **a** Velocity contours for constant-area ejector. **b** Velocity contours for constant-pressure ejector

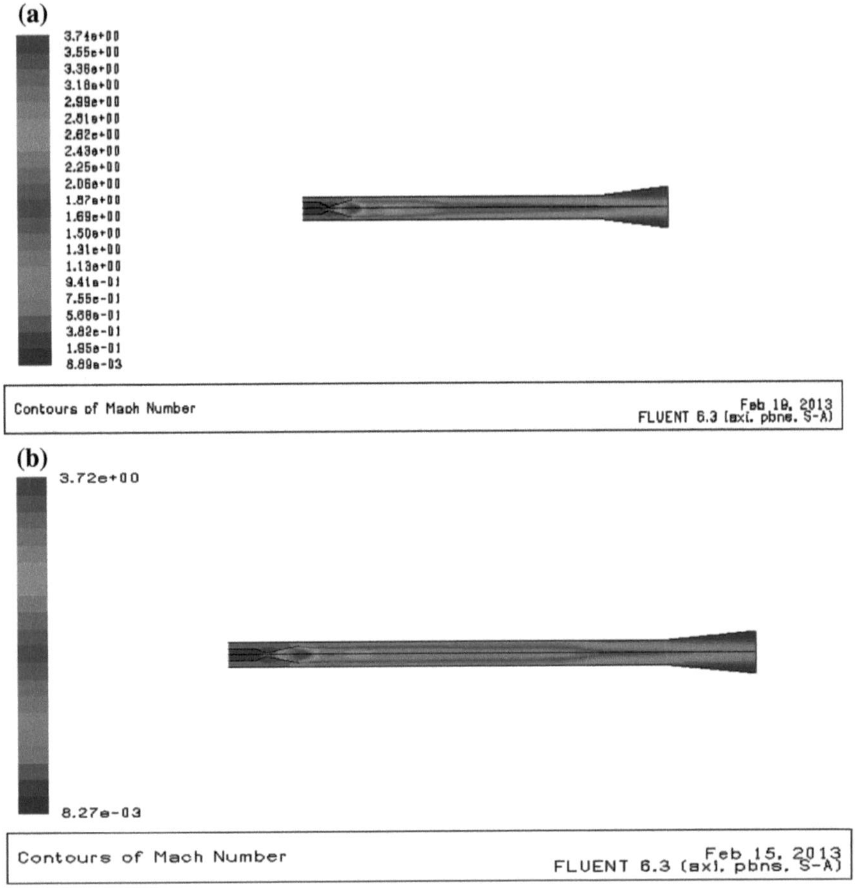

Fig. 6 **a** Constant-area ejector Motive fluid as acetylene. **b** Constant-area ejector motive fluid as steam

pressure increases, the entrainment ratio steeply. To find the performance of a constant-pressure ejector primary inlet pressure vs entrainment ratio graph was also drawn. Figure 8 shows the constant-pressure ejector performance. In constant-pressure ejector the rate of increase in entrainment ratio is very low.

In general, it might be inferred that, the present examination shows the capability of the CFD model in foreseeing an exact execution for both entrainment ratio and compression ratio of an ejector.

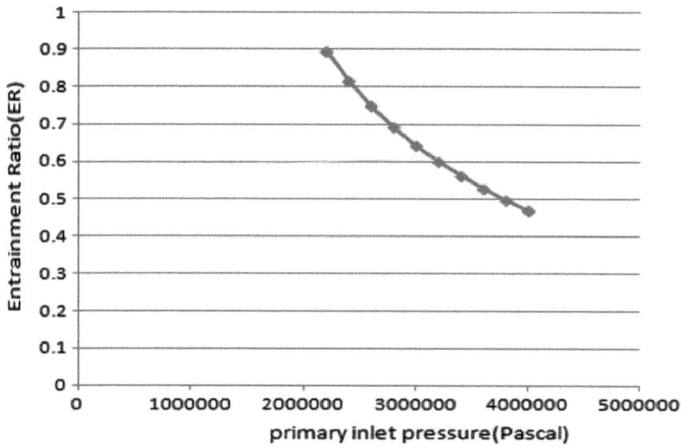

Fig. 7 Primary pressure vs entrainment ratio of constant-area ejector

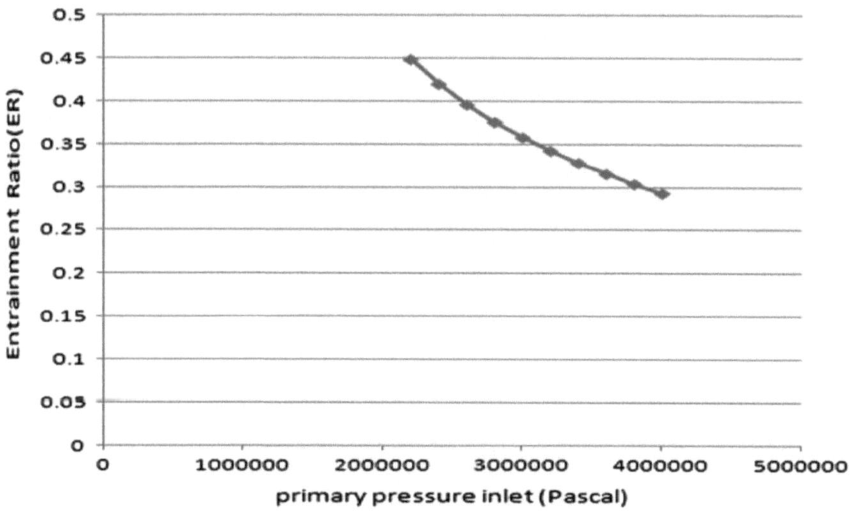

Fig. 8 Primary pressure vs entrainment ratio of constant-pressure ejector

4 Conclusion

The CFD was found to be not solely a tool in predicting ejector performance it additionally offer a far better understanding within the flow and mixing processes in the ejector. After analysis the following conclusions are drawn:

- In constant area ejector, when pressure at inlet section increases accordingly mass flow rate at inlet section also increases, but at the same time mass flow at secondary inlet goes on decreasing. which result in decrease in entrainment ratio, with increase in compression ratio.

- In constant pressure ejector when pressure increases, mass flow at primary inlet increases, which results in decrease in entrainment ratio with increase in compression ratio.
- In this work, two motive fluids are used to analysis the performance of ejector, concluded that motive fluid containing larger molecular weight will affect the ejector efficiency.

References

Bartosiewicz, Y., Aidoun, Z., Desevaux, P., Mercadier, Y.: Numerical and experimental investigations on supersonic ejectors. Int. J. Heat Fluid Flow **26**, 56–70 (2005)

Keenan, J., Neumann, E.: A simple air ejector. ASME J. Appl. Mech. **64**, 75–82 (1942)

Lee, J.H., Sameen, A., Sanal Kumar, V.R., Kim, H.D., Choi, B.G., Kim, K.H.: Studies on ejector systems for hydrogen fuel cell. In: Proceedings of 41st AIAA/ASME/SAE/ASEE Joint Propulsion Conference & Exhibit, Tucson, Arizona (2005)

Rice, P., Dandachi, J.: An equation for the prediction of steam flowrate required in the design of ejectors. Trans I Chem E **69A**, 332–334 (1991)

Sriveerakul, T., Aphornratana, S., Chunnanond, K.: Performance prediction of steam ejector using computational fluid dynamics: Part 2 Flow structure of a steam ejector influenced by operating pressures and geometries. Int. J. Therm. Sci. **46**, 823–833 (2007)

Sun, D.W., Eames, I.W.: Recent developments in the design theories and applications of ejectors —A Review. J. Inst. Energy **475**, 65–79 (1995)

Varga, S., Oliveira, A.C., Diaconu, B.: Numerical assessment of steam ejector efficiencies using CFD. Int. J. Refrig **32**, 1203–1211 (2009)

CFD Studies on Pressure Drop for Low Reynolds Fluid Flows Across Orifice in Similarly Shaped Microchannel

G. Bhuvaneswari, Haritha Madhava Reddy and Venu Vinod Ananthula

1 Introduction

Microfluidic devices have great potential in the fields of automation and miniaturization for handling and analysis of fluids resulting in faster and economical diagnostic procedures. These include enhancement in mass and heat transfer rates, improved energy efficiency, low by-product formation and inherent safety (Zhao et al. 2006; Dessimoz et al. 2008). It has a wide range of applications in the areas of biotechnology, diagnostics, medicine and biosecurity (Jensen 2001; Abgrall and Gue 2007; Haeberle and Zengerle 2007; Melin and Quake 2007). Power requirement of a microfluidic device, which depends on the pressure drop across the device, is an important factor to be considered. This study deals with analysis of pressure drop across an orifice placed in a similarly shaped microchannel.

2 Problem Statement

In the present work, the orifice was positioned mid-way along the microchannel length to give equal upstream and downstream lengths. This arrangement was to ensure that the velocity profiles at the inlet and outlet of the microchannel were fully developed. The studies were carried out to determine pressure drop across an

G. Bhuvaneswari · H.M. Reddy · V.V. Ananthula (✉)
Department of Chemical Engineering, National Institute of Technology, Warangal 506004, Telangana, India
e-mail: avv122@yahoo.co.in

G. Bhuvaneswari
e-mail: gpbhuvaneswari@gmail.com

H.M. Reddy
e-mail: harithareddy157@gmail.com

I. Regupathi et al. (eds.), *Recent Advances in Chemical Engineering*,
DOI 10.1007/978-981-10-1633-2_39

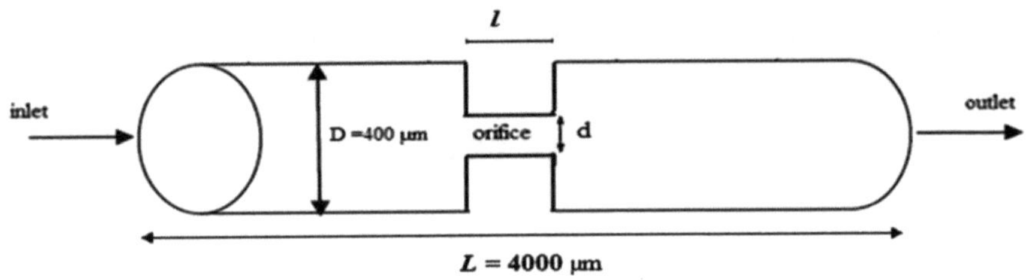

Fig. 1 Circular microchannel with similar shaped orifice

Table 1 Reynolds number at different mean velocities (at 28 °C)

Feed solution	Re(U_m = 50 μm/s)	Re(U_m = 60 μm/s)	Re(U_m = 70 μm/s)	Re(U_m = 80 μm/s)
1.5 % CMC aqueous solution	0.000265	0.000318	0.000371	0.000424
Water	0.024	0.028	0.033	0.038

orifice in a microchannel, varying the fluid velocity (U_m), orifice contraction ratio ($\gamma_{do} = d/D$) and orifice length to diameter ratio ($\lambda_{do} = l/d$). Figure 1 shows the arrangement of micro channel and orifice (Table 1).

3 Numerical Solution

The following assumptions were made in solving the problem of steady laminar flow in microchannel:

(i) The flow is axial (One dimensional), (ii) the flow is laminar, (iii) there is no slip at the wall, (iv) no heat transfer to/from surroundings, (v) the energy dissipation is negligible, (vi) fluid/wall interaction is purely viscous, (vii) straight channel walls, (viii) the smooth walls.

The governing partial differential equations were numerically solved by finite volume method with multi-grid solver with high resolution in the commercial CFD code CFX 14.0. A refined mesh with tetrahedral elements was used in simulation. The RMS convergence criterion used was about 1×10^{-5} for all variables calculated.

4 Results and Discussion

4.1 Mesh Independency Test

Mesh independency test was carried out for circular microchannel of 400 μm diameter and 4000 μm length using water as the fluid. The refined mesh was chosen

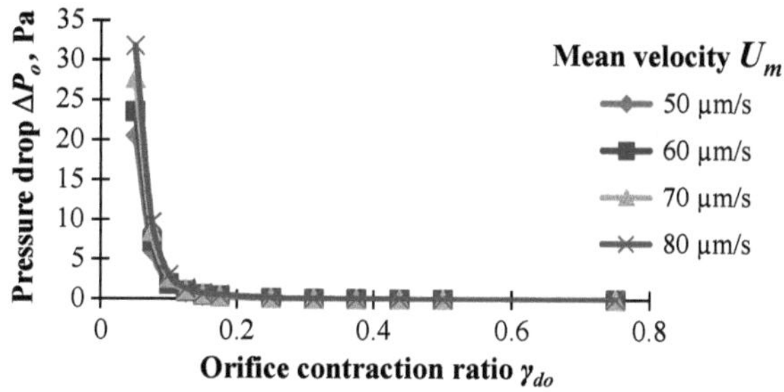

Fig. 2 Orifice pressure drop versus orifice contraction ratio γ_{do} for water (for $\lambda_{do} = 1$)

in such a way that the difference in pressure drop was less than 1 % for two successive meshes.

4.2 *Circular Orifice Pressure Drop Characteristics*

4.2.1 Effect of Orifice Contraction Ratio (γ_{do})

Simulations were carried out for various velocities (U_m) of 50, 60, 70 and 80 μm/s for microchannels corresponding to Reynolds number. Orifice contraction ratio (γ_{do}) was varied in the range of 0.05–0.75 with constant orifice length to diameter ratio (λ_{do}) of 0.05 and the pressure drop values obtained from simulation is in Fig. 2.

Figure 2 shows that as orifice contraction ratio increases, pressure drop across orifice decreases. The decrease in ΔP is steep till ratio reaches 0.175, after which presence of orifice does not affect ΔP significantly.

Figure 3 shows that the pressure drop variation is different from that of Newtonian fluid (water). A sharp decrease in the orifice pressure drop was found to be at orifice contraction ratio of 0.075 and beyond contraction ratio of 0.175 it tends to behave as a microchannel without orifice.

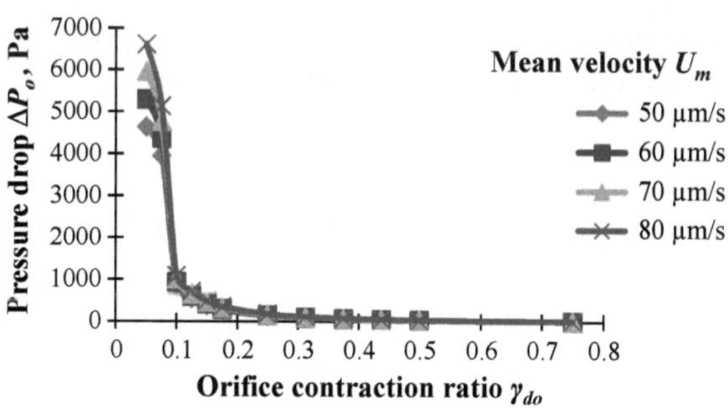

Fig. 3 Orifice pressure drop versus orifice contraction ratio γ_{do} for aqueous CMC solution at ($\lambda_{do} = 1$)

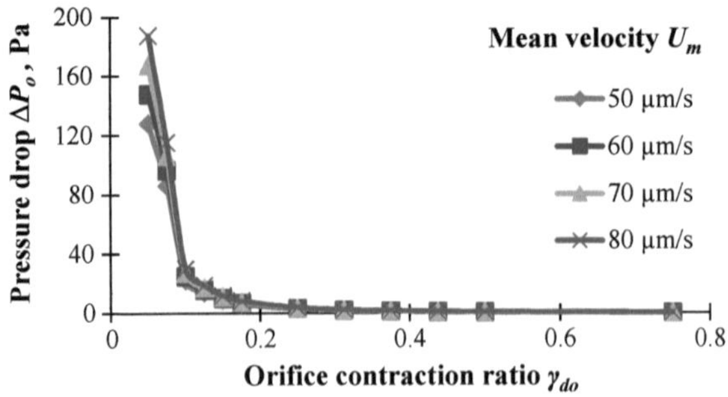

Fig. 4 Orifice pressure drop versus orifice contraction ratio γ_{do} for blood with 45 % haemoglobin ($\lambda_{do} = 1$)

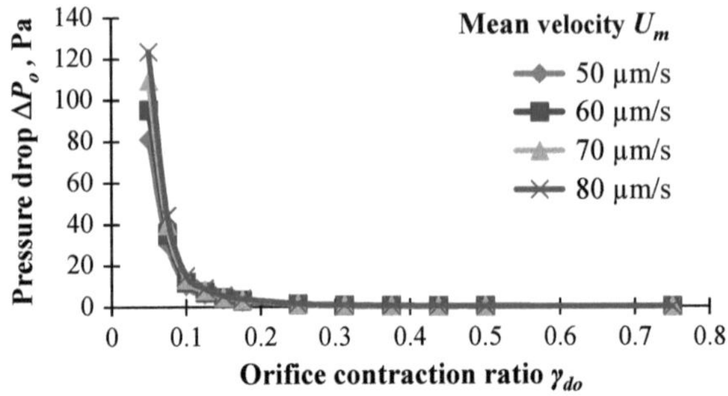

Fig. 5 Orifice pressure drop versus orifice contraction ratio γ_{do} blood with 70 % haemoglobin ($\lambda_{do} = 1$)

From Fig. 4 it is seen that ΔP decreased sharply till orifice contraction ratio of 0.075 and then gradually till orifice contraction ratio of 0.175 beyond which effect of orifice is not felt.

From Fig. 5 it is seen that the critical orifice contraction ratio was found to be 0.175 similar to that of Newtonian fluid (water). Figures 2, 3, 4 and 5 show that, as the orifice contraction ratio increases, pressure drop across orifice has an absolute lower limit, irrespective of the Reynolds number. The orifice pressure drop may become unbounded as the orifice contraction ratio becomes small. Beyond the critical orifice contraction ratio, the orifice in microchannel found to be ineffective and the orifice pressure drop is close to the pressure drop of microchannel. Comparing pressure drop of non-Newtonian fluids to that of Newtonian fluids, it was found that human blood with 70 % haemoglobin have similar pressure drop characteristics as that of Newtonian fluid.

4.3 *Effect of Orifice Length to Diameter Ratio (λ_{do})*

The orifice length to diameter ratio was varied from 0.5 to 15 for Newtonian (water) and non-Newtonian fluids (aqueous CMC solution, human blood with 45 % hemoglobin, human blood with 70 % hemoglobin).

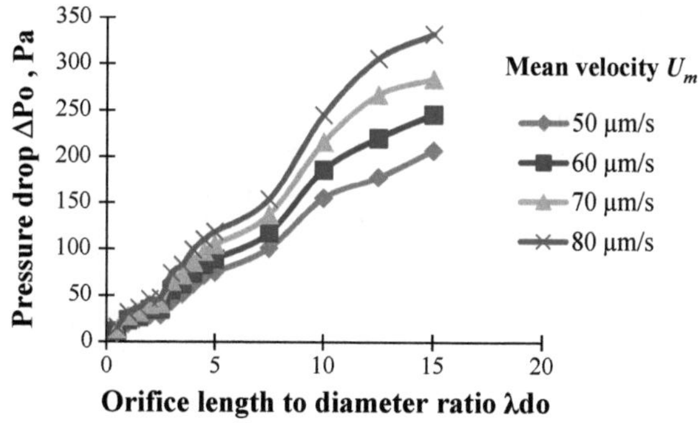

Fig. 6 Orifice pressure drop versus orifice length to diameter ratio λ_{do} for water (γ_{do} = 0.05)

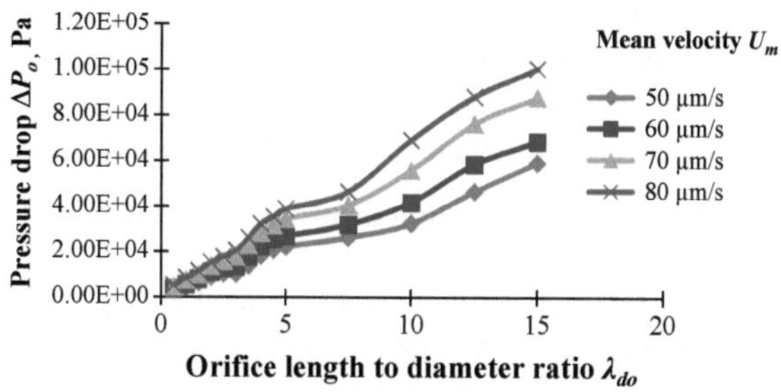

Fig. 7 Orifice pressure drop versus orifice length to diameter ratio λ_{do} for aqueous CMC solution (γ_{do} = 0.05)

Figure 6 shows that the orifice pressure drop increases as orifice length to diameter ratio increases and the pressure drop found to be the maximum at higher orifice length to width ratio.

Figure 7 shows that for higher orifice length to width ratio λ_{do}, the orifice pressure drop becomes unbounded.

It was found that for both Newtonian (water) and for non-Newtonian fluid (aqueous CMC solutions, blood) the orifice pressure drop increases with orifice length to diameter ratio, but for non-Newtonian fluids the pressure drop values found to be high. This has been observed for all non-Newtonian fluids considered in the study. There is no significant change in the pattern of the non-Newtonian flow as compared to the Newtonian flow in all cases (from Figs. 6, 7, 8 and 9).

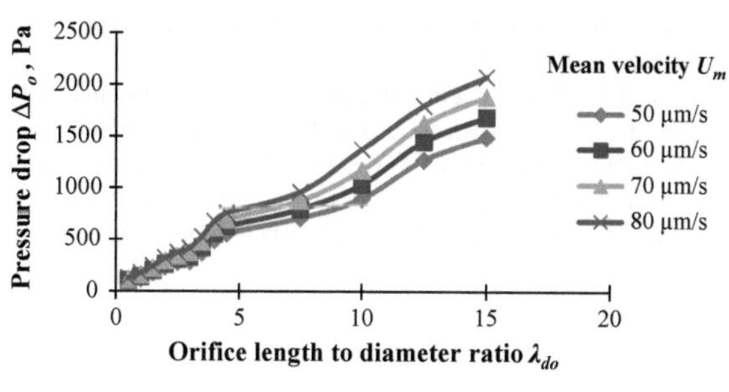

Fig. 8 Orifice pressure drop versus orifice length to diameter ratio γ_{do} for blood with 45 % haemoglobin (γ_{do} = 0.05)

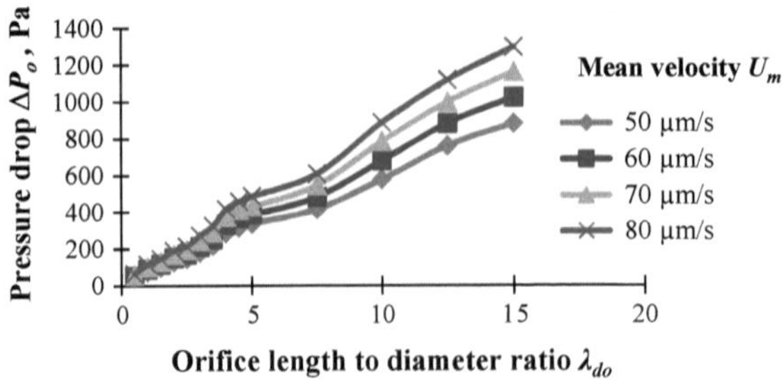

Fig. 9 Orifice pressure drop versus orifice length to width ratio γ_{do} for blood with 70 % haemoglobin (γ_{do} = 0.05)

5 Conclusions

Simulations were carried out for predicting the orifice pressure drop characteristics for various values of orifice contraction ratio and orifice length to diameter ratio at different velocities of the fluid flowing.

The following conclusions can be drawn from this study:

- The simulated pressure drops were found to be higher for non-Newtonian fluid than Newtonian fluid.
- It was found that orifice pressure drop (ΔP_o) decreases as orifice contraction ratio increases. At higher orifice contraction ratios (γ_o), the presence of orifice does not affect orifice pressure drop significantly.
- As orifice length to diameter ratio (λ_o) increases, orifice pressure drop ΔP_o increases and become unbounded for higher ratios.
- It was found that human blood with 70 % haemoglobin has similar pressure drop characteristics as that of Newtonian fluid.

References

Abgrall, P., Gue, A.M.: Lab-on-chip technologies: making a microfluidic network and coupling it into complete microsystems: a review journal. Micromech. Microeng. **17**, 15–49 (2007)

Dessimoz, A., Cavin, L., Renken, A., Kiwi-Minsker, L.: Liquid–liquid two-phase flow patterns and mass transfer characteristics in rectangular glass microreactors. Chem. Eng. Sci. **63**(16), 4035–4044 (2008)

Haeberle, S., Zengerle, R.: Microfluidic platforms for lab-on-a-chip applications. Lab Chip **7**, 1094–1110 (2007)

Jensen, K.F.: Microreaction engineering—is small better? Chem. Eng. Sci. **56**, 293–303 (2001)

Melin, J., Quake, S.R.: Microfluidic large-scale integration: the evolution of design rules for biological automation. Annu. Rev. Biophys. Biomol. Struct. **36**, 213–231 (2007)

Zhao, Y., Chen, G., Yuan, Q.: Liquid-liquid two-phase flow patterns in a rectangular microchannel. AIChE J. **52**(12), 4052–4060 (2006)

MIX
Papier aus verantwortungsvollen Quellen
Paper from responsible sources
FSC® C105338

If you have any concerns about our products,
you can contact us on
ProductSafety@springernature.com

In case Publisher is established outside the EU,
the EU authorized representative is:
Springer Nature Customer Service Center GmbH
Europaplatz 3, 69115 Heidelberg, Germany

Printed by Libri Plureos GmbH
in Hamburg, Germany